Developments in Partial Differential Equations and Applications to Mathematical Physics

Developments in Partial Differential Equations and Applications to Mathematical Physics

Edited by

G. Buttazzo

University of Pisa
Pisa, Italy

G. P. Galdi

University of Ferrara
Ferrara, Italy

and

L. Zanghirati

University of Ferrara
Ferrara, Italy

PLENUM PRESS • NEW YORK AND LONDON

Library of Congress Cataloging-in-Publication Data

Developments in partial differential equations and applications to mathematical physics / edited by G. Buttazzo, G.P. Galdi, and L. Zanghirati.
p. cm.
"Proceedings of an international meeting on New developments in partial differential equations and applications to mathematical physics, held October 14-18, 1991, in Ferrara, Italy"--T.p. verso.
Includes bibliographical references and index.
ISBN 0-306-44311-2
1. Differential equations, Partial--Congresses. 2. Mathematical physics--Congresses. I. Buttazzo, Giuseppe. II. Galdi, Giovanni P. (Giovanni Paolo), 1947- . III. Zanghirati, L.
QA377.D56 1992
515'.353--dc20 92-32748
CIP

Proceedings of an international meeting on New Developments in Partial Differential Equations and Applications to Mathematical Physics, held October 14–18, 1991, in Ferrara, Italy

ISBN 0-306-44311-2

A Division of Plenum Publishing Corporation
233 Spring Street, New York, N.Y. 10013

Printed in the United States of America

PREFACE

During the days 14-18 of October 1991, we had the pleasure of attending a most interesting Conference on New Developments in Partial Differential Equations and Applications to Mathematical Physics in Ferrarra.

The Conference was organized within the Scientific Program celebrating the six hundredth birthday of the University of Ferrarra and, after the many stimulating lectures and fruitful discussions, we may certainly conclude, together with the numerous participants, that it has represented a big success.

The Conference would not have been possible without the financial support of several sources. In this respect, we are particularly grateful to the Comitato Organizzatore del VI Centenario, the University of Ferrarra in the Office of the Rector, Professor Antonio Rossi, the Consiglio Nationale delle Ricerche, and the Department of Mathematics of the University of Ferrarra.

We should like to thank all of the participants and the speakers, and we are especially grateful to those who have contributed to the present volume.

G. Buttazzo, University of Pisa
G.P. Galdi, University of Ferrarra
L. Zanghirati, University of Ferrarra

Ferrarra, May 11th, 1992

CONTENTS

INVITED LECTURES

CONTRIBUTED PAPERS

LIAPUNOV FUNCTIONALS AND QUALITATIVE BEHAVIOUR OF THE SOLUTION TO THE NON LINEAR ENSKOG EQUATION

N. Bellomo*, J. Polewczak** and L. Preziosi*

* Dipartimento di Matematica
Politecnico
Corso Duca degli Abruzzi 24
10129 Torino, Italy

** State University of N. Y.
Stony Brook, U.S.A.

ABSTRACT

This paper deals with the analysis of the initial value problem for the Enskog equation in the whole space $\mathbf{R}^3$. In particular the dissipativity of the system is studied by the analysis of the behaviour of suitable Liapunov functionals. The analysis is related to the qualitative study of the solutions to the Cauchy problem.

1. THE NONLINEAR ENSKOG EQUATION

The Enskog equation is an evolution equation, proposed by Enskog [1, 2], for the one particle distribution function for a gas of hard spheres with diameter a. The derivation of the equation, somewhat at an empirical level, is based upon Boltzmann-like arguments with additional "ad hoc" assumptions able to take into account some effects due to the overall dimensions of the spheres.

Essentially Enskog's derivation is based upon the following two assumptions to be joined to the ones necessary for the derivation of the Boltzmann equation:

i) Only elastic collisions, with conservation of momentum and energy, are taken into account for pairs of particles with their centers at a distance a equal to the diameter of the spheres.

ii) The collision frequency is increased by a factor Y, taken as a function of the gas density at the contact point of two colliding spheres. Y represents the so called pair correlation function corresponding to the system in "uniform" equilibrium.

Formally this equation has the same structure of the Boltzmann equation

$$\frac{\partial f}{\partial t} + \mathbf{v} \cdot \nabla_{\mathbf{x}} f = E(f; f, f) = E^+(f; f, f) - E^-(f; f, f) \,. \tag{1.1}$$

Namely the left-hand-side term, which defines the total derivative of f, is equated to the collision operator expressed in terms of difference between the "gain" and "loss" terms respectively defined as

Developments in Partial Differential Equations and Applications to Mathematical Physics, Edited by G. Buttazzo *et al.*, Plenum Press, New York, 1992

$$E^+(f;f,f)(t,\mathbf{x},\mathbf{v}) = a^2 \int_{\mathbf{R}^3\times S^2_+} Y(f,a)\langle \eta, \mathbf{w}-\mathbf{v}\rangle f(t,\mathbf{x},\mathbf{v}')f(t,\mathbf{x}+a\eta,\mathbf{w}')\,d\eta\,d\mathbf{w}$$

$$E^-(f;f,f)(t,\mathbf{x},\mathbf{v}) = a^2 f(t,\mathbf{x},\mathbf{v}) \int_{\mathbf{R}^3\times S^2_+} Y(f,a)\langle \eta, \mathbf{w}-\mathbf{v}\rangle f(t,\mathbf{x}-a\eta,\mathbf{w})\,d\eta\,d\mathbf{w}\ , \tag{1.2}$$

where Y is a function of the local density at the contact point of two spheres, the post-collisional velocities $\mathbf{v}'$ and $\mathbf{w}'$ are computed as for the Boltzmann equation.

Mathematical aspects, mainly the qualitative analysis of the Cauchy problem, are dealt with in the book [3], which also provides a review of all modifications of the original equations.

This paper deals with the analysis of the qualitative behaviour of the solutions to the initial value problem for the Enskog equation in the whole space $\mathbf{R}^3$. In particular we analyse the time evolution of some Liapunov functional suitable to describe the dissipativity of the system and the stability of equilibrium solutions.

The paper is in four sections. After this introduction, Section 2 deals with the statement of the initial value problem. Section 3 deals with the definition of some Liapunov functionals and with the statement of the problem related to the qualitative analysis of the said functionals. The analysis of the problem follows in Section 4, that provides a review of the existing results, of some new theorems and of the analysis of Enskog-type modelling aspects.

2. THE INITIAL VALUE PROBLEM

Consider the initial value problem for Eq.(1.1) joined to the initial conditions

$$f(t=0,\,\mathbf{x},\,\mathbf{v}) \;=\; f_0(\mathbf{x},\,\mathbf{v})\ . \tag{2.1}$$

Integration along characteristics yields the weaker form

$$f(t) \;=\; \mathcal{U}(t)\,f_0 + \int_0^t \mathcal{U}(t-s)\;E(f(s))\,ds\ , \tag{2.2}$$

where

$$[\mathcal{U}(t)\,h]\;(\mathbf{x},\,\mathbf{v}) \;=\; h(\mathbf{x}-\mathbf{v}t,\,\mathbf{v}) \tag{2.3}$$

for $t \in \mathbf{R}_+$ and $(\mathbf{x},\,\mathbf{v}) \in \mathbf{R}^3 \times \mathbf{R}^3$ is the solution to the problem

$$\frac{\partial g}{\partial t} + \mathbf{v}\cdot\nabla_{\mathbf{x}}\,g = 0\quad ,\quad g(t=0,\,\mathbf{x},\,\mathbf{v}) \;=\; h(\mathbf{x},\,\mathbf{v})\ . \tag{2.4}$$

A solution to (2.4), considered in a function space in which $\mathcal{U}(t)$ acts as an operator and the integral is consistent, is called a "weak solution".

Local existence to the initial value problem in the whole space $\mathbf{R}^3$ has been proved by Lachowicz [4], while Bellomo and Toscani [5] have proved global existence of mild solutions for decaying data small in norm and Polewczak [6] has proved existence of classical solutions under suitable smoothness assumptions on the initial data and smallness assumptions similar to the ones of [5].

Global existence for large initial data in L_1 has been proved by Polewczak [7,8] and Arkeryd [9], using the technique of the celebrated theorem by DiPerna and Lions [10]. Further studies in this direction are due to Esteban and Perthame [11].

All the above mentioned results are well documented in [3]. This paper will give the definitions of the function spaces where the existence results can be proven in view of the analysis of the qualitative behaviour of the Liapunov functionals.

We need mentioning that, in the case of perturbation of vacuum [5,6,7], existence results are linked to uniqueness. Therefore the behaviour of the Liapunov functionals can be followed along the solution. On the other hand, perturbation of vacuum has a limited thermodynamic interest. This is not the case of the analysis of the initial value problem with large L_1 data, in the case of periodic conditions on a torus, however uniqueness is lost.

In order to deal with Problem (2.2) in the case of perturbation of vacuum, we need defining the function spaces which follow.

$\mathcal{C}_b(\mathbf{Z})$, for $\mathbf{Z} \subset \mathbf{R}^n$ and $n \geq 1$ denotes the space of all real bounded and continuous functions defined on $\mathbf{Z}$. For given positive functions h, $m \in \mathcal{C}_b(\mathbf{R}_+)$ let

$$M(h, m) = \Big\{ f \in \mathcal{C}_b(\mathbf{R}_+ \times \mathbf{R}^3 \times \mathbf{R}^3) : \\ |f(t, \mathbf{x}, \mathbf{v})| \leq c h^{-1}(x) m^{-1}(v) \quad \text{for some } c > 0 \Big\} ,$$

which, with norm given by

$$\| f \| = \sup_{(t, \mathbf{x}, \mathbf{v}) \in \mathbf{R}_+ \times \mathbf{R}^3 \times \mathbf{R}^3} | f(t, \mathbf{x}, \mathbf{v}) | \; h(x) m(v) \tag{2.5}$$

becomes a Banach space.

Analogously, $L_\infty(\mathbf{Z})$, for $\mathbf{Z} \subset \mathbf{R}^n$ and $n \geq 1$ denotes the space of all real and essentially bounded functions defined on $\mathbf{Z}$. For given positive functions h, $m \in \mathcal{C}_b(\mathbf{R}_+)$ let

$$N(h, m) = \Big\{ f \in L_\infty(\mathbf{R}_+ \times \mathbf{R}^3 \times \mathbf{R}^3) : \\ |f(t, \mathbf{x}, \mathbf{v})| \leq c h^{-1}(x) m^{-1}(v) \quad \text{a.e. for some } \; c > 0 \Big\} ,$$

which, with norm given by

$$\| f \| = \operatorname*{ess\,sup}_{(t, \mathbf{x}, \mathbf{v}) \in \mathbf{R}_+ \times \mathbf{R}^3 \times \mathbf{R}^3} | f(t, \mathbf{x}, \mathbf{v}) | \; h(x) m(v) , \tag{2.6}$$

becomes a Banach space.

Remark 2.1: Various functions h and m can be used. For instance if one uses

$$h(x) = (1 + x^2)^{\frac{q}{2}} \; ; \quad m(v) = (1 + v^2)^{\frac{p}{2}} , \tag{2.7}$$

$M(h, m)$ will be denoted as $\mathcal{B}_{p,q}$, and $N(h, m)$ will be denoted as $\mathcal{X}_{p,q}$. The norms will be specified by the corresponding indices, i.e. $||| \cdot |||_{p,q}$.

In addition to the functional spaces which have been introduced above, we will need some other definitions in order to deal with classical solutions.

For a non negative integer k, let $C_b^k(\mathbf{R}^3)$ denote the Banach space of k-times continuously differentiable bounded functions on $\mathbf{R}^3$ with bounded derivatives up to order k. The following Banach spaces will be then considered

$$\mathcal{B}_{p,q}^{(k)} = \Big\{ f \in \mathcal{C}_b(\mathbf{R}^3 \times \mathbf{R}^3) : \\ \left[\frac{\partial^{|\gamma|} f}{\partial x^\gamma}(\mathbf{x}, \mathbf{v}) \right] h(x) m(v) \in \mathcal{C}_b(\mathbf{R}^3 \times \mathbf{R}^3) \text{ for each multiindex } \gamma \text{ such that } |\gamma| \leq k \Big\}$$

endowed with the norm

$$\|f\|_{p,q}^{(k)} = \sup_{\substack{|\gamma| \leq k \\ (\mathbf{x}, \mathbf{v}) \in \mathbf{R}^3 \times \mathbf{R}^3}} \left\{ \left[\frac{\partial^{|\gamma|} f}{\partial x^\gamma}(\mathbf{x}, \mathbf{v}) \right] h(x) m(v) \right\} . \tag{2.8}$$

In addition, the space

$$\mathcal{B}_{p,q}^{k} = \Big\{ f \in \mathcal{C}_b(\mathbf{R}_+ \times \mathbf{R}^3 \times \mathbf{R}^3) : \\ \left[\frac{\partial^{|\gamma|} f}{\partial x^\gamma}(t, \mathbf{x}, \mathbf{v}) \right] h(x) m(v) \in \mathcal{C}_b(\mathbf{R}_+ \times \mathbf{R}^3 \times \mathbf{R}^3) \text{ for each multiindex } \gamma \text{ such that } |\gamma| \leq k \Big\}$$

can also be defined and endowed with the norm

$$\||| f |||_{(k)}^{p,q} = \sup_{t \in \mathbf{R}_+} \|f(t)\|_{p,q}^{(k)} . \tag{2.9}$$

Moreover, it is necessary to define the properties on the factor Y which are needed in what follows.

The factor Y is meaningful only for distribution functions corresponding to the moderately condensated densities. In fact, if in some region in $\mathbf{R}_+ \times \mathbf{R}^3$ the local number density $n(t, \mathbf{x}) = \int_{\mathbf{R}^3} f(t, \mathbf{x}, \mathbf{v})\, d\mathbf{v}$ is too large, then $Y(f, a)(t, \mathbf{x}, \eta)$ becomes infinite.

Thus, for all functional spaces $\mathcal{X}$ considered before, we define

$$\mathcal{D}(\mathcal{X}) = \left\{ f \in \mathcal{X} : f \geq 0 \quad \text{and} \quad \sup_{(t,\mathbf{x}) \in \mathbf{R}_+ \times \mathbf{R}^3} n(t, \mathbf{x}) \leq K_c \right\} , \tag{2.10}$$

where the constant K_c depends only on a and defines the physically admissible distribution functions.

Referring to the Banach space $\mathcal{B}_{p,q}$ defined in Remark 2.1, with $p > 3$ and $q > 0$, we can choose $\widetilde{K}_p$ in such a way that $f \in \mathcal{D}(\mathcal{B}_{p,q})$ if $f \geq 0$ and $\||f\||_{p,q} \leq \widetilde{K}_p$. Therefore, we can define the set

$$\mathcal{D}_{p,q} = \left\{ f \in \mathcal{B}_{p,q} : f \geq 0 \quad \text{and} \quad \||f\||_{p,q} \leq \widetilde{K}_p \right\} \tag{2.11}$$

such that $\mathcal{D}_{p,q} \subset \mathcal{D}(\mathcal{B}_{p,q})$.

Moreover we need stating the whole collection of conditions concerning the factor Y.

A1) $\forall a \geq 0 \;,\; Y(0, a) = 1$;

A2) for $a \geq 0$ and $f_1, f_2 \in \mathcal{D}(\mathcal{X})$,

$$0 \leq f_1 \leq f_2 \Rightarrow Y(f_1, a) \leq Y(f_2, a) \;;$$

A3) for $a \geq 0$ and $f \in \mathcal{D}(\mathcal{X})$,

$$\frac{\partial^{|\gamma|} f}{\partial x^\gamma} \in \mathcal{C}_b(\mathbf{R}_+ \times \mathbf{R}^3 \times \mathbf{R}^3) \text{ for all } \gamma \text{ such that } |\gamma| \leq k \Rightarrow$$

$$\Rightarrow \frac{\partial^{|\gamma|} Y}{\partial x^\gamma}(f, a) \in \mathcal{C}_b(\mathbf{R}_+ \times \mathbf{R}^3) \text{ for all } \gamma \text{ such that } |\gamma| \leq k \text{ and } \forall \eta \in \mathbf{S}^2_+ \;;$$

A4) $\forall f_1, f_2 \in \mathcal{D}(\mathcal{X})$

$$\sup_{\substack{|\gamma| \leq k \\ (t,\mathbf{x},\eta) \in \mathbf{R}_+ \times \mathbf{R}^3 \times \mathbf{S}^2_+}} \left| \frac{\partial^{|\gamma|}}{\partial x^\gamma} [Y(f_1, a) - Y(f_2, a)] \right| \leq C_a \sup_{\substack{|\gamma| \leq k \\ (t,\mathbf{x}) \in \mathbf{R}_+ \times \mathbf{R}^3}} \int_{\mathbf{R}^3} \left| \frac{\partial^{|\gamma|}}{\partial x^\gamma} (f_1 - f_2) \right| d\mathbf{v} \;;$$

A5) $Y^{\pm}(f, a)(t, \mathbf{x}, \eta) = Y(\mathbf{x}, \mathbf{x} \pm a\eta, f)$;

A6) $Y(\mathbf{x}_1, \mathbf{x}_2, f) = Y(\mathbf{x}_2, \mathbf{x}_1, f)$.

The conditions **A1** – **A6** are consistent with the original Enskog equation and with its modifications and developments. We will deal, in what follows, with Enskog-type equations which satisfy **A1** – **A6**.

The case of the initial value problem for large L_1–data requires more refined assumptions of the factor $Y(f, a)$ than in the case of perturbation of vacuum. Indeed, the only example of $Y(f, a)$ for which presently one can prove global in time existence theorems for large L_1–data is given by suitable truncations of the exact two–particle correlation function g_2 that is intrinsically related to the revised Enskog equation as proposed by van Beijeren and Ernst in [12]. Before presenting the corresponding $Y(f,a)$, we observe that along with the whole space problem in L_1, one can also treat the case of periodic boundary conditions by considering, for example, the spacial domain Ω to be equal to the three dimensional torus $\mathbf{R}^3/\mathbf{Z}^3$. In what follows Ω is equal either to $\mathbf{R}^3$ or to $\mathbf{R}^3/\mathbf{Z}^3$.

As mentioned above, the revised Enskog equation is obtained when $Y(f,a)$ is given by the following formal Mayer cluster expansion of the two–particle correlation function g_2 for the system of hard–spheres at nonuniform equilibrium

$$g_2 = e^{-\beta\phi(|\mathbf{x}_1-\mathbf{x}_2|)}\left\{1+\sum_{k=3}^{\infty}\frac{1}{(k-2)!}\int_\Omega d\mathbf{x}_3 \cdots \int_\Omega d\mathbf{x}_k\, n(3)\dots n(k)\, V(12\,|\,3\dots k)\right\}. \tag{2.12}$$

Here $n(k)=n(t,\mathbf{x}_k)$, $\beta = 1/k_BT$, $V(12\,|\,3\dots k)$ is the sum of all graphs of k labeled points which are biconnected when the Mayer factor $f_{12}=\exp[-\beta\phi(|\mathbf{x}_1-\mathbf{x}_2|)]-1$ is added. The function ϕ is an interaction potential, which in the case of hard spheres of diameter a is equal to infinity for $|\mathbf{x}_1-\mathbf{x}_2|<a$, and zero for $|\mathbf{x}_1-\mathbf{x}_2|\geq a$. In particular, the Mayer factor $f_{12}=\Theta_{12}-1$, where $\Theta_{12}\equiv\Theta(|\mathbf{x}_1-\mathbf{x}_2|-a)$, and Θ is the Heaviside step function. As an example, we provide below the expressions of $V(12\,|\,3\dots k)$ for $k=3,4$

$$V(12\,|\,3) = (1)f_{13}f_{23}\,,$$

$$V(12\,|\,34) = (2)f_{13}f_{34}f_{24} + (2)f_{13}f_{34}f_{14}f_{24} + (2)f_{13}f_{34}f_{24}f_{23} + $$
$$+ (1)f_{13}f_{24}f_{14}f_{23} + (1)f_{13}f_{34}f_{24}f_{14}f_{23}\,.$$

The numbers in parentheses represent the corresponding symmetry factors. This is precisely the above algebraic structure of g_2 that plays a fundamental role in obtaining various a priori estimations leading to existence and stabilty results for the revised Enskog equation.

For $Y(f,a)$ given by (2.12), the H–function, considered in further details in Section 3, has the form (modulo an additive absolute constant)

$$\Gamma(f)(t) = \int_{\Omega\times\mathbf{R}^3} f(t,\mathbf{x},\mathbf{v})\log f(t,\mathbf{x},\mathbf{v})\,d\mathbf{v}\,d\mathbf{x} +$$

$$\sum_{k=2}^{\infty}\frac{1}{k!}\int_\Omega d\mathbf{x}_1 \cdots \int_\Omega d\mathbf{x}_k\, n(2)\dots n(k)\, V(1\dots k)\,, \tag{2.13}$$

where $V(1\dots k)$ is the sum of all irreducible Mayer graphs which doubly connect k particles. As an illustration, we provide the expressions of $V(1\dots k)$ for $k=2,3,4$

$$V(12) = (1)f_{12}\,,$$

$$V(123) = (1)f_{12}f_{23}f_{13}\,,$$

$$V(1234) = (3)f_{12}f_{23}f_{34}f_{14} + (6)f_{12}f_{23}f_{34}f_{14}f_{13} + (1)f_{12}f_{23}f_{34}f_{14}f_{13}f_{24}\,.$$

As before, the numbers in parentheses represent the corresponding symmetry factors.

Due to difficulties with convergence of the infinite sums in (2.12) and (2.13), at present time, only the truncated problem has been considered, i.e., when the factor $Y(f,a)$ is obtained by truncating the infinite series in (2.12) at some $k<\infty$ (see [3] and [13]). Furthermore, the truncation of the infinite series in (2.13) at the same $k<\infty$ as above provides the corresponding Liapunov functional. We remark that the case of $Y(f,a)$ equal to the first term in the expansion series (2.12) (the corresponding Liapunov functional comes from truncation of (2.13) at $k=2$) has been considered by the authors in [14]. Further analysis can be found in [15] The importance of representation

(2.13) (in the particular case of the above mentioned truncations) is evident when one notice that any finite sum in the second term of (2.13) is bounded, uniformly in t, by a constant that depends only on the conserved quantity $\int_{\Omega\times\mathbf{R}^3} f_0(\mathbf{x},\mathbf{v})\,d\mathbf{v}\,d\mathbf{x}$.

3. ON THE H THEOREM

If one considers the revised Enskog equation proposed by Resibois [13] and the various generalization of the Enskog equation proposed in the work by Polewczak [8], it is possible to derive, in a formal way, an analogue of the classical H–theorem.

Polewczak [8] suggested the following expression for the Liapunov functional

$$\Gamma(f)(t) = H(f)(t) - P^\star(f)(t) \,, \tag{3.1}$$

where $H(f)(t)$ is the classical Boltzmann H–functional

$$H(f)(t) = \int_{\mathbf{R}^3\times\mathbf{R}^3} (f\log f)(t,\,\mathbf{x},\,\mathbf{v})\,d\mathbf{x}\,d\mathbf{v} \tag{3.2}$$

and

$$P^\star(f)(t) = \int_0^t I(f)(s)\,ds \,, \tag{3.3}$$

with

$$I(f)(s) = \frac{1}{2}\int_{\mathbf{R}^3\times\mathbf{R}^3\times\mathbf{R}^3\times\mathbf{S}^2_+} M(f)\,(s,\,\mathbf{x},\,\mathbf{v},\,\mathbf{w},\,\eta)\,d\eta\,d\mathbf{x}\,d\mathbf{v}\,d\mathbf{w} \tag{3.4}$$

and

$$\begin{aligned} M(f)\,(s,\,\mathbf{x},\,\mathbf{v},\,\mathbf{w},\eta) = \{&Y^+(f,\,a)\,(s,\,\mathbf{x},\eta)\,f(s,\,\mathbf{x}+a\eta,\,\mathbf{w}) - \\ &- Y^-(f,\,a)(s,\,\mathbf{x},\eta)\,f(s,\,\mathbf{x}-a\eta,\mathbf{w})\}\,f(s,\,\mathbf{x},\,\mathbf{v})\,\langle\,\mathbf{w}-\mathbf{v},\,\eta\,\rangle \,. \end{aligned} \tag{3.5}$$

Consider at first the case in which, for every $t \geq 0$,

$$\begin{aligned} &M(f) \in L_1(\mathbf{R}^3\times\mathbf{R}^3\times\mathbf{R}^3\times\mathbf{S}^2_+) \,, \\ &(f\log f)\,(t,\,\mathbf{x},\,\mathbf{v}) \in L_1(\mathbf{R}^3\times\mathbf{R}^3) \,. \end{aligned} \tag{3.6}$$

An analogous to the classical H–theorem for the Boltzmann equation can be proven also for the Enskog equation. Such a theorem can be regarded as a rigorous one if both conditions (3.6) hold and the implication

$$t\uparrow \;\Rightarrow\; \Gamma(f)(t)\downarrow \tag{3.7}$$

can be proven, at least for suitable initial conditions.

Another Liapunov functional, proposed by Polewczak and Stell [16], is

$$\mathcal{E}(t) = \int_{\mathbf{R}^3\times\mathbf{R}^3} (\mathbf{x}-t\mathbf{v})^2 f(t,\mathbf{x},\mathbf{v})\,d\mathbf{x}\,d\mathbf{v} \,. \tag{3.8}$$

Also in this case we need

$$(\mathbf{x}-t\mathbf{v})^2 f(t,\mathbf{x},\mathbf{v}) \in L_1(\mathbf{R}^3\times\mathbf{R}^3) \tag{3.9}$$

and

$$t\uparrow \Longrightarrow \mathcal{E}(t)\downarrow \,. \tag{3.10}$$

Implication (3.10) gives the "formal" result. When both conditions (3.9) and (3.10) hold, then one has a "rigorous" theorem. As already mentioned a rigorous theorem can be obtained for perturbation of vacuum.

4. ANALYSIS

The mathematical problem of the qualitative behaviour of the solution to the initial value problem has been stated in Section 3. The analysis of such a problem should bring to evaluate the dissipativity of the system and to analyse the stability of equilibrium solutions.

As a matter of fact the analysis of this equation has rised several questions which still need an answer. In particular the analysis of the qualitative behaviour of Liapunov-type functionals strongly motivates further insights into the mathematical modelling of Enskog-type equations.

Therefore this section attempts to provide some research perspectives into this topic and is organized into three subsections. First the qualitative behaviour of the functionals, defined in Section 3, is dealt with. Then we deal with the analysis of the equilibrium conditions. Finally a new modelling of Enskog-type equations (or a generalized Boltzmann equation) follows.

Qualitative Behaviour of Some Liapunov Functionals

Consider now the problem, stated in Section 3, of proving a rigorous H–type theorem for the functionals $\Gamma(t)$ and $\mathcal{E}(t)$.

The problem referred to the functional (3.1) was studied in [17]. The result consists in the following

Theorem 1: Let $p > 7$, $q > 3$ and $f_0 \in \mathcal{B}^o_{p,q}$ be such that the Enskog equation admits a unique non negative global mild solution in $\mathcal{B}^o_{p,q}$. Then the following conditions hold

$$(f \log f)(t, \mathbf{x}, \mathbf{v}) \in L_1(\mathbf{R}^3 \times \mathbf{R}^3) \qquad \text{for all } t \geq 0 \ , \tag{4.1}$$

$$M(f)(t, \mathbf{x}, \mathbf{v}, \mathbf{w}, \eta) \in L_1(\mathbf{R}^3 \times \mathbf{R}^3 \times \mathbf{R}^3 \times \mathbf{S}^2_+), \ \text{ for all } t \geq 0 \ , \tag{4.2}$$

$$\Gamma(f)(t_2) \leq \Gamma(f)(t_1) \quad \text{for } 0 \leq t_1 \leq t_2 < \infty \ . \ \blacksquare \tag{4.3}$$

Remark 4.1: Existence of solutions in $\mathcal{B}^k_{p,q}$ is assured, under suitable smallness assumption on the norm of the initial conditions

$$||f_o||^{(k)}_{p,q} < c = c(p,q) \ , \tag{4.4}$$

where c is a critical constant. (The proof of this result is reported in Chapter 4 of [18] with reference to papers [5] and [6]).

Remark 4.2: The proof of Theorem 1 is reported in Chapter 2 of [3] with reference to [17].

An analogous theorem can be proven for the functional defined in (3.8). The result is the following

Theorem 2: Let $p > 6$, $q > 3$ and $f_o \in \mathcal{B}^o_{p,q}$. If the norm of the initial conditions is sufficiently small (in the terms stated in Remark 4.1) so that existence and uniqueness of a non-negative mild solution in $\mathcal{B}^o_{p,q}$ is assured, then, conditions (3.9) and (3.10) hold.

PROOF OF THEOREM 2: The proof of the theorem is in two steps: first condition (3.9) is proven, then the second step proves inequality (3.10).

Step 1: Condition (3.9) can be proven exploiting the existence theorems for the solutions to the initial value problem [5,6].

First we note that the statement of the theorem is such that

$$x^2 f_o \in L_1(\mathbf{R}^3 \times \mathbf{R}^3) \ .$$

Moreover, according to [5,6], the solution is such that

$$\||| f \|||_{p,q}^{o} \leq 2\|f_o\|_{p,q}^{o}$$

(4.5)

$$f(t, \mathbf{x}, \mathbf{v}) \leq 2f(t, \mathbf{x} + \mathbf{v}t, \mathbf{v}) .$$

These conditions assure the proof of (3.9).

Step 2: In order to prove the second step of the theorem, it is useful to show some properties of the functional $\mathcal{E}(t)$. Following Polewczak and Stell [16], consider first the equality

$$|\mathbf{x} - t\mathbf{v}'|^2 + |\mathbf{x} - a\eta - t\mathbf{w}'|^2 = |\mathbf{x} - t\mathbf{v}|^2 + |\mathbf{x} - a\eta - t\mathbf{w}|^2 - 2at\langle \eta , \mathbf{w} - \mathbf{v} \rangle \quad (4.6)$$

where the symbols are the ones already used to deal with the collision mechanics.

Multiplying both members of the Enskog equation by $\mathbf{x} - t\mathbf{v}$, using Eq.(4.6) and integrating over $\mathbf{x} \in \mathbf{R}^3$ yields

$$\frac{d\mathcal{E}}{dt} = -at \int_{\mathbf{R}^3 \times \mathbf{R}^3} \langle \eta , \mathbf{w} - \mathbf{v} \rangle \, E^-(t, \mathbf{x}, \mathbf{v}) \, d\mathbf{x} d\mathbf{v} . \quad (4.7)$$

Therefore, we have that the functional $\mathcal{E}$ is always monotone decreasing, as already observed in [16]. We need proving that $(\mathbf{x} - \mathbf{v}t)f(t, \mathbf{x}, \mathbf{v})$ is defined over $L_1(\mathbf{R}^3 \times \mathbf{R}^3)$ and is continuous in time.

Following the proof given in [19] for the Boltzmann equation, we can now choose a nonincreasing sequence of functions

$$f_{o,j}(\mathbf{x}, \mathbf{v}) = \max \left\{ f_o(\mathbf{x}, \mathbf{v}), \frac{\alpha}{j+1}(1 + x^2)^{-q/2}(1 + v^2)^{-p/2} \right\} , \quad (4.8)$$

where α is chosen so small that the initial value problem for the Enskog equation, with $f_o(x, v) = \alpha(1 + x^2)^{-\frac{q}{2}}(1 + v^2)^{-\frac{p}{2}}$ has a unique global solution. Obviously,

$$f_{o,j}(\mathbf{x}, \mathbf{v}) \in \mathcal{C}_b\left(\mathbf{R}^3 \times \mathbf{R}^3\right) ,$$

(4.9)

$$\lim_{j \to \infty} \|f_{o,j} - f_o\|_{p,q}^{(o)} = 0 .$$

By the existence theorem [5,6], it follows that there exists a sequence $\{f_j\} \in \mathcal{B}_{p,q}^{o}$ of unique non negative solutions f_j of the Cauchy problem with initial data $f_{o,j}$ instead of f_o.

Looking at the proof of the existence theorem, and in particular to the Kaniel and Shinbrot iteration scheme, it follows that f_j is bounded from below and above.

Then, the proof follows the same steps of [17], (cf. [3], Chapter 2). In particular one proves

$$\mathcal{E}(f_j) = \mathcal{E}(f_j^\sharp) \in C^1([0, T]) \quad \text{for all } T \in [0, \infty) ,$$

where $f_j^\sharp = f_j(\mathbf{x} + \mathbf{v}t)$, and

$$\frac{d\mathcal{E}}{dt}(f_j) \leq 0 \qquad \text{for all } j .$$

Thus, for all j and $0 \leq t_1 \leq t_2 < \infty$, we have

$$\mathcal{E}(f_j)(t_2) \leq \mathcal{E}(f_j)(t_1) \quad (4.10)$$

and, since $\|f_j - f\|_{p,q}^{(o)} \to 0$, the Fatou's lemma gives the result. ∎

Analysis similar to the one developed in Theorems 1 and 2 can be developed having in the background the existence result for large L_1 data by Polewczak [8], which starts from the existence result by DiPerna and Lions [10] for the Boltzmann equation.

Indeed, we observe first that in the case of $Y(f,a)$ obtained from the truncations of (2.12), the corresponding Liapunov functional $\Gamma(f)(t)$ (in fact an H–function for the problem) is a convex function of f. Thus the kinetic part of the Liapunov functional, $H(f)(t)$, (simply equal to Boltzmann's H–function) is bounded from above, uniformly in t. Next one shows monotonicity of $\Gamma(f)(t)$ in t. Monotonicity implies that the function $t \longrightarrow \Gamma(t)$ is absolutely continuous as a function from $[0,T]$ into $\mathbf{R}$, thus differentiable a.e. in t. Hence, at least a.e. in $t \in [0,T]$ one has

$$\frac{d\Gamma(t)}{dt} \leq 0 \, .$$

Further details of the proof are in progress.

Remark 4.3: The result of Theorem 1 holds for a gas in the whole space. The presence of boundaries can, however, modify the result. A correct formulation of the boundary conditions is necessary in order to deal with the qualitative analysis of the functionals and with the initial-boundary value problem. This topic has not been dealt with yet in the literature of the Enskog equation.

On the Stability of Equilibrium Solutions

As already mentioned, one can analyse the qualitative behaviour of the functionals in order to obtain suitable information on the stability properties of solutions in the whole space problem.

An analysis of this type is dealt with in [13], where, referring to the functional $\mathcal{E}$, it has been shown that

$$t > 0 \, , \quad f > 0 \quad \Longrightarrow \quad \frac{d\mathcal{E}}{dt} \leq 0 \, . \tag{4.11}$$

Therefore, considering that $\mathcal{E}$ is decreasing for all times, namely $\frac{d\mathcal{E}}{dt}$ is never equal to zero for positive f, trend to equilibrium is described only for $f = 0$.

This aspect cannot be applied to the Boltzmann equation. In fact, as already observed in [13], for the Boltzmann model one has $\frac{d\mathcal{E}}{dt} = 0$ for all times.

On the other hand, some relevant advantages of this model are known: better accuracy, with respect to the Boltzmann equation, for predicting transport coefficients [2], capability of describing some moderately dense gas effects. Nevertheless, several problems remain open, like hydrodynamic description, local Maxwellian, among others.

A Generalized Model of the Enskog Equation

The crucial difference between the Boltzmann and the Enskog equations is, besides the factor Y, that the distribution function of the two colliding particles is computed, in the Boltzmann equation, in the same point $\mathbf{x}$, while the Enskog equation is such that the function of the field particle is estimated at a distance $2a$ from the center $\mathbf{x}$ of the test particle.

Both models introduce at a microscopic level an approximation with respect to the true collision mechanics. Therefore a conceivable alternative modelling consists in evaluating the distribution function of the field particle $f(\,\cdot\,,\mathbf{w})$ distributed in the domain of action of the test particle. The proposed model is

$$\frac{\partial f}{\partial t} + \mathbf{v} \cdot \nabla_{\mathbf{x}} f = B(f,f) = B^{+}(f,f) - B^{-}(f,f) \, , \tag{4.12}$$

where

$$B^{+}(f,f)(t,\mathbf{x},\mathbf{v}) = \int_{[0,R]\times\mathbf{R}^3\times\mathbf{S}^2_+} Y(f,r)Q(\eta,\mathbf{w}-\mathbf{v})f(t,\mathbf{x},\mathbf{v}')f(t,\mathbf{x}+r\eta,\mathbf{w}')P(r)\,d\eta\,d\mathbf{w}\,dr \tag{4.13}$$

and

$$B^-(f,f)(t,\mathbf{x},\mathbf{v}) = \int\limits_{[0,R]\times\mathbf{R}^3\times\mathbf{S}^2_+} Y(f,r)Q(\eta,\mathbf{w}-\mathbf{v})f(t,\mathbf{x},\mathbf{v})f(t,\mathbf{x}-r\eta,\mathbf{w})P(r)\,d\eta\,d\mathbf{w}\,dr \tag{4.14}$$

where Q is the pair interaction potential (with a cut-off), r is the radius of a spherical coordinate system with center in $\mathbf{x}$, R is the action radius of the test particle and $P(r)$ is the probability distribution over r.

We observe that when $Y(f,r) \equiv 1$ and $P(r) = \delta(r-0)$, where δ is the Dirac delta function, Eq.(4.12) reduces to the Boltzmann equation with the potential interaction corresponding to Q. Furthermore, by assuming that $Y(f,r) \equiv 1$ in (4.13–4.14), one obtains the collision operator similar to the one considered considered by Povzner [20]. There is, however, one important difference of the above model (with $Y(f,r) \equiv 1$) as compared to Povzner's collision operator. The alternation of signs in $\mathbf{x} \pm r\eta$ that appears in $B^{\pm}$ makes the equation itself and the Liapunov functional for it different from the corresponding equation and the Liapunov functional considered by Povzner [20]. Although Povzner had considered it, the Liapunov functional for his equation is equal to the Boltzmann's H–function.

The model equation (4.12) with the right hand side given by (4.13–4.14) describes the collision process in which collisions are being "smeared" in the domain (determined by $P(r)$) around the points of impact. As it is well known, this modification introduces an additional integration in the collision operator B. This new integration together with the two–dimensional integration with respect to η amounts to three–dimensional integration in the spatial variable $\mathbf{x}$ of the distribution function f. Recently, the authors in [21] have used the above model in the proof of convergence and factorization of solutions of the mollified BBGKY–hierarchy to the solutions of Equation (4.12) (with $Y(f,r) \equiv 1$ and $Q = \langle\eta,\mathbf{w}-\mathbf{v}\rangle)$ when Grad's limit is considered.

We propose to use this model with particular application to the Enskog equation, i.e., when $Q(\eta,\mathbf{w}-\mathbf{v}) = \langle\eta,\mathbf{w}-\mathbf{v}\rangle$ and $Y(f,r) \not\equiv 1$. Equation (4.12) with B given by (4.13–4.14) reduces formally to the corresponding Enskog equation when $a \in (0,R)$ and $P(r) = \delta(r-a)$, with a equal to the diameter of hard spheres. It seems to us that the above model can provide better understanding of solutions, with particular emphasis on asymptotic behaviour (as $t \to \infty$) and hydrodynamical limits. The fact that solutions are unique, also for large L_1–data, can be considered as an additional advantage of the model.

Our first observation regarding equation (4.12–4.14) points out an existence of Liapunov functionals for the problem. Indeed, similar arguments as in the case of the Enskog equation imply that the Liapunov functional of (4.12–4.14) is

$$\Gamma_{\mathrm{mod}}(f)(t) = H(f)(t) - P^*_{\mathrm{mod}}(f)(t), \tag{4.15}$$

where $H(f)(t)$ is given by (3.2),

$$P^*_{\mathrm{mod}}(f)(t) = \int_0^t I_{\mathrm{mod}}(f)(s)\,ds\ ,$$

and

$$I_{\mathrm{mod}}(f)(s) = \frac{1}{2} \int\limits_{[0,R]\times\mathbf{\Omega}\times\mathbf{R}^3\times\mathbf{R}^3\times\mathbf{S}^2_+} M_{\mathrm{mod}}(f)(s,r,\mathbf{x},\mathbf{v},\mathbf{w},\eta)\,d\eta\,d\mathbf{w}\,d\mathbf{v}\,d\mathbf{x}\,dr, \tag{4.16}$$

with $M_{\mathrm{mod}}(f)$ equal to the product of $P(r)$ and $M(f)$, given by (3.5), where the parameter a is replaced by r. In the same fashion as in the case of the Enskog equation (see [13], and also [16] for further generalizations) one shows that stationary points of $\Gamma_{\mathrm{mod}}(f)(t)$ coincide with the stationary points of $\Gamma(f)(t)$, the Liapunov functional for the Enskog equation. Hence, Equation (4.12), formally at least, should provide the same equilibrium solutions as in the case of the Enskog equation. Furthermore, in the case of the whole space problem ($\mathbf{\Omega} = \mathbf{R}^3/\mathbf{Z}^3$) one can also show that $\mathcal{E}(t)$, defined in (3.8), can be considered as another Liapunov functional when problems related to escape of the gas from the space are important.

The existence and uniqueness theorems for Equation (4.12) can be proven along the following lines. First, one considers an approximation of Q by a sequence of bounded functions $Q_n = \min\{n, Q\}$, for $n \geq 1$. For the just defined approximation of Q, one considers a sequence of equations defined by (4.12), with Q replaced by Q_n. We call them $(4.12)_n$. Using, for example, Theorem 5.1 of Chapter 3 in [3] one shows existence of unique solutions, f_n, for $(4.12)_n$ in the set

$$D_M = \{f \in L_1 : f \geq 0, \quad \int_{\Omega\times\mathbf{R}^3} (1+\mathbf{v}^2) f \, d\mathbf{v} \, d\mathbf{x} \leq M\} ,$$

with $0 < M < \infty$. For above mentioned theorem to work one needs some Lipschitz conditions of $Y(f,r)$. An important class of $Y(f,r)$ with such a property is provided by any truncations of (2.12), as explained in details in Section 2. In addition, bounded Y's satisfying condition A4) of Section 2, with $\gamma = 0$, also have the Lipschitz condition required in Theorem 5.1 of Chapter 3 in [3].

Next, due to the presence of the additional integration with respect to r in (4.16), it is easy to show that for any bounded $Y(f,r)$, $\sup_n |P^*_{\mathrm{mod}}(f_n)(t)| < \infty$, uniformly in $t \in [0,T]$. Now, using an analog of (3.7) for $(4.12)_n$, we obtain that a finite upper bound of the positive part of $H(f_n)(t)$, uniform in n and $t \in [0,T]$, exists as long it does initially, i.e., for $f_0(\mathbf{x},\mathbf{v})$. This shows that condition (i) of Theorem 4.1 of Chapter 3 in [3] (the main convergence and stability result for the Enskog equation) is fulfilled. Finally, Theorem 4.1 of Chapter 3 in [3] implies that some subsequence of $\{f_n\}$ converges weakly to a solution of (4.12). We observe that in the case of the whole space problem, one needs an additional bound of $\int_{\mathbf{R}^3\times\mathbf{R}^3} \mathbf{x}^2 f_n \, d\mathbf{v}\, d\mathbf{x}$, uniformly in n and $t \in [0,T]$. This bound follows from property (3.10) of the functional $\mathcal{E}(t)$.

Uniqueness of solutions can be proven by using similar arguments to those used in the cases of the homogenouos Boltzmann equation [22], the mollified Povzner's equation [20], or the case of the equation considered in [21]. The uniqueness requires additional boundedness of higher moments of an initial value f_0. Summarizing, we have (for $|Q| \leq \mathrm{const}(|\mathbf{v}|^\lambda + |\mathbf{w}|^\lambda)$ and $0 \leq \lambda < 2$)

<u>Theorem 3:</u> Assume that the factor Y is bounded on the set D_M and $f_0 \geq 0$ satisfies

$$\int_{\Omega\times\mathbf{R}^3} (1+\mathbf{v}^2+\mathbf{x}^2+|\log f_0|) f_0 \, d\mathbf{v}\, d\mathbf{x} = C_0 < \infty. \tag{4.17}$$

Then there exists a mild, nonnegative solution $f(t,\mathbf{x},\mathbf{v})$ of (4.12).

<u>Theorem 4:</u> In addition to the conditions of Theorem 3, assume that the factor Y is given by truncating the infinite series in (2.12) at some $k < \infty$, and

$$\int_{\Omega\times\mathbf{R}^3} |\mathbf{v}|^4 f_0 \, d\mathbf{v}\, d\mathbf{x} = C_1 < \infty. \tag{4.18}$$

Then a solution obtained in Theorem 3 is unique.

We remark that when $\Omega = \mathbf{R}^3/\mathbf{Z}^3$ the $\mathbf{x}^2$ term in (4.17) is superfluous.

This mathematical result certainly encourages further studies of this model, which allows to introduce general pair interaction potentials in an Enskog-type equation.

<u>Acknowledgements:</u> Partially supported by the Italian Ministry for University and Scientific Research MURST and by the National Council for Research C.N.R.–G.N.F.M.; J. P. acknowledges a partial support of the Division of Chemical Sciences Office, Office of Basic Energy Sciences, Office of Energy Research, U.S. Department of Energy.

REFERENCES

1. D. Enskog, Kinetiske Theorie, Svenska Akad. 63 (1921).

2. P. Resibois and M. De Leener, "Classical Kinetic Theory of Fluids", Wiley, London (1977).

3. N. Bellomo, M. Lachowicz, J. Polewczak and G. Toscani, "Mathematical Topics in Nonlinear Kinetic Theory II: The Enskog Equation", World Scientific, London, Singapore (1991).

4. M. Lachowicz, On the local existence and uniqueness of solution of initial-value problem for the Enskog equation, Bull. Polish Acad. Sci. Math. 31:89 (1983).

5. G. Toscani and N. Bellomo, The Enskog-Boltzmann equation in the whole space $\mathbf{R}^3$: Some global existence, uniqueness and stability results, Comp. Math. Appl. 13:851 (1987).

6. J. Polewczak, Global existence and asymptotic behaviour for the nonlinear Enskog equation, SIAM J. Appl. Math. 49:952 (1989).

7. J. Polewczak, Global existence in L_1 for the modified nonlinear Enskog equation in $\mathbf{R}^3$, J. Statist. Phys. 56:159 (1989).

8. J. Polewczak, Global existence in L_1 for the generalized Enskog equation, J. Statist. Phys. 59:461 (1990).

9. L. Arkeryd, On the Enskog equation with large initial data, SIAM J. Math. Anal. 21:631 (1990).

10. R.L. DiPerna and P.L. Lions, On the Cauchy problem for the Boltzmann equation: Global existence and weak stability, Ann. of Math. 130:321 (1989).

11. M.J. Esteban and B. Perthame, Global solutions for the modified Enskog equation with elastic and inelastic collisions, Comp. Rend. Acad. Sci. Paris I, 309:897 (1988).

12. H. van Beijeren and M. H. Ernst, The modified Enskog equation, Physica 68:437 (1973)

13. P. Resibois, H-theorem for the (modified) nonlinear Enskog equation, J. Statist. Phys. 19:593 (1978).

14. L. Arkeryd and C. Cercignani, Global existence in L^1 for the Enskog equation and convergence of solutions to solutions of the Boltzmann equation, J. Statist. Phys. 59:845 (1990).

15. M. Cannone and C. Cercignani, The inverse conjecture for the revised Enskog equation, J. Statist. Phys. 63:363 (1991).

16. J. Polewczak and G. Stell, New properties of a class of generalized kinetic equations, J. Statist. Phys. 64:437 (1991).

17. N. Bellomo and M. Lachowicz, On the asymptotic theory of the Boltzmann and Enskog equations. A rigorous H-theorem for the Enskog equation, in "Mathematical Aspects of Fluid and Plasma Dynamics", G. Toscani, V. Boffi and S. Rionero, eds., Springer Lecture Notes in Mathematics n. 1460 (1991).

18. N. Bellomo, A. Palczewski and G. Toscani, "Mathematical Topics in Nonlinear Kinetic Theory", World Scientific, London, Singapore (1988).

19. G. Toscani, H-Theorem and asymptotic trend of the solution for a rarefied gas in the vacuum, Arch. Rational Mech. Anal. 100:1 (1987).

20. A. Povzner, The Boltzmann equation in the kinetic theory of gases, Amer. Math. Soc. Transl. 47:533 (1955).

21. M. Lachowicz and M. Pulvirenti, A stochastic system of particles modelling the Euler equations, Arch. Rational Mech. Anal. 109:81 (1990).

22. L. Arkeryd, On the Boltzmann equation. Part *I*: Existence. Part *II*: The full initial value problem, Arch. Rational Mech. Anal. 45:1 (1972).

DISCONTINUOUS MEDIA AND DIRICHLET FORMS OF DIFFUSION TYPE

Marco Biroli

Department of Mathematics
Politecnico di Milano
P. Leonardo de Vinci, 32 I-20133 Milano

Umberto Mosco

Department of Mathematics
University of Rome "La Sapienza"
I-00185 Roma

Abstract. We suggest that Dirichlet forms may be used in describing the variational behaviour of possibly highly nonhomogeneous and nonisotropic bodies and prove structural Harnack inequalities and Saint-Venant type decays for local solutions.

We consider a body X with a very irregular internal structure, highly nonhomogeneous and possibly nonisotropic. We suppose that $u : X \to \mathbb{R}$ describes a physical state of X, whose equilibrium is subjected to a *variational principle* of a suitable nature.

In order to formulate such a principle with very few requirements about the internal structure of X, we shall assume that the *energy functional* minimized by u can be written as a *Dirichlet form* $a(u,v)$ in the Hilbert space

$$H = L^2(X,m),$$

for a suitable choice of a locally compact separable Hausdorff topology on X and of a positive Radon measure m on X, with $\operatorname{supp} m = X$.

We recall that a **Dirichlet form** on H is a bilinear form $a(u,v)$ defined on a dense linear subspace $D[a]$ of H, which can be written as

$$a(u,v) = (\sqrt{L}u, \sqrt{L}v) \text{ for every } u,v \in D[a],$$

where $(.,.)$ denotes the inner product of H and L is a positive semi-definite selfadjoint operator in H with $D_{\sqrt{L}} = D[a]$, and which has in addition the following fundamental *Markovianity property*: $T \circ u \in D[a]$ and

$a(T \circ u, T \circ u) \leq a(u,u)$, whenever $u \in D[a]$ and $T : \mathbb{R} \to \mathbb{R}$, $T(0) = 0$, $|T(x) - T(y)| \leq |x - y|$ for every $x, y \in \mathbb{R}$.

We shall restrict our study to Dirichlet forms of **diffusion type**, that is to forms a that have the following *strong local property*:

$a(u,v) = 0$ for every $u, v \in D[a]$ with v constant on a neighborhood of $\operatorname{supp} u$,

where $\operatorname{supp} u$ is taken to be the support of the measure $u \cdot m$ in X.

Furthermore, we shall assume that the form a is **regular** in H, that is, there exists a subset C of $D[E] \cap C_0(X)$ which is both dense in $C_0(X)$ with the uniform norm and dense in $D[E]$ with the *intrinsic norm* $(a(u,u) + (u,u))^{1/2}$. Such a set C, that without restriction can be assumed to be a subalgebra of $D[E] \cap C_0(X)$, is called a *core* of a in H. The functions that belong to C play the role of *test functions* in our variational theory. By $C_0(X)$ we are denoting the space of continuous functions with compact support in X.

From the physical point of view taken at the beginning of this paper, the choice of the class of regular Dirichlet forms of diffusion type in order to state our variational principle for X is motivated by the fact that any such form can be given the following integral expression:

$$a(u,v) = \int_X \mu(u,v)(dx), \tag{1}$$

for every $u, v \in D[a]$, where μ is a Radon-measure-valued positive-semidefinite symmetric bilinear form on $D[a]$, uniquely associated with a, called the *energy measure* of a

One of the most important property of μ is its *local* character: the restriction of the measure $\mu(u,v)$ to any open subset A of X only depends on the restrictions of u and v to A. This property entitles us to interpret μ as a measure-valued description of the *physical characteristics* of the body X on which our variational principle relies. Moreover, it enables us to define in a natural way the space of functions u that *belong locally* to the domain of the form on a given open subset A of X. We shall denote this space by $D[a, A]_{\mathrm{loc}}$ and simply by $D[a]_{\mathrm{loc}}$ if $A = X$. We may suppress the explicit reference to the form a in our notation, if we are dealing with forms that have a common domain.

Furthermore, the local character of μ allows us to define a function u to be a *local minimizer* of the *energy functional*

$$E[u] = \frac{1}{2} a(u,u)$$

in a given arbitrary open subset X_0 of X, if u is a function on X that satisfies the minimality condition:

$$u \in D[a, X_0]_{\mathrm{loc}} : \frac{1}{2} \int_{X_0} \mu(u,u)(dx) \leq \frac{1}{2} \int_{X_0} \mu(u + \phi, u + \phi)(dx) \tag{2}$$

for every $\phi \in C$ with $\operatorname{supp} \phi \subset X_0$.

Clearly, u is a solution of (2) if and only if u is a solution of

$$(3) \qquad u \in D[a, X_0]_{\text{loc}} : \quad a(u, v) = 0 \text{ for every } v \in D_0[a, X_0],$$

where for any open subset A of X we denote by $D_0[a, A]$ the closure of the $D[E] \cap C_0(A)$ in $D[a]$ with its intrinsic norm. We will refer to any solution of (3) as to a *local solution* in X_0 of the equation formally written as

$$(4) \qquad Lu = 0.$$

Before going on, let us point out that our present interpretation of μ as the measure-valued characteristics of the body is further supported by the special coordinate-invariant expression that is taken by the form a, whenever we are ready to introduce in X, or in some open portion of it, the additional structure of a *differentiable manifold*. In this case, in fact, if there exist coordinate functions $x_1, \ldots, x_n$, that belong locally to $D[a]$ on their domain of definition, then any differentiable function u on X also belongs locally to $D[a]$ and the energy measure $\mu(u, v)$, for every $u, v \in C^1(X)$, can be written in local coordinates as

$$\mu(u, v) = \sum_{ij=1}^{n} (\partial u/\partial x_i)(x_1, x_2, \ldots x_n)(\partial v/\partial x_j)(x_1, x_2, \ldots, x_n)\nu^{ij},$$

where

$$\nu^{ij} = \mu(x_i, x_j), \quad i, j = 1, \ldots, n,$$

defines a positive-semidefinite symmetric tensor v on X. The form a, for every $u, v \in C_0^1(X)$, takes now the following more familiar invariant integral expression:

$$a(u, v) = \int_X \sum_{ij=1}^{n} (\partial u/\partial x_i)(x_1, x_2, \ldots x_n)(\partial v/\partial x_j)(x_1, x_2, \ldots, x_n)\nu^{ij}(dx),$$

where the integral at the right hand side has to be intended as reduced to the coordinate domains by means of a partition of unit in C and the following (degenerate) *ellipticity condition*

$$\sum_{ij=1}^{n} \xi_i \xi_j \nu^{ij} \geq 0 \text{ for every } \xi \in R^n$$

is satisfied in the sense of measures on X.

We refer for instance to M. Fukushima [F] for a general treatment of the theory of Dirichlet forms and their probabilistic interpretation.

Let us now come back to our general setting and to equation (4).

Our aim is to describe the behaviour of an arbitrary local solution u of (4) in X_0, in a neighborhood of an arbitrary given point x_0 of X_0. Moreover, we want our theory and estimates have a *structural* character. By this we mean that the properties we will establish have to hold *uniformly* for a whole family of equivalent Dirichlet forms of diffusion type, in a sense that will be made precise below.

We will suppose then that we are given a whole family of regular Dirichlet forms of diffusion type, a, which have a common domain $D[a] = D$ in $L^2(X, m)$ and are mutually *equivalent* in the following sense:

However we choose a form in the family, say b, there exist two constants $0 < \lambda \leq \Lambda$, depending on b but whose ratio Λ/λ is independent of b, such that any other form a of the family is related to b by the condition

(5) $$\lambda b(u,u) \leq a(u,u) \leq \Lambda b(u,u) \quad \text{for every } u \in D[a] = D[b] = D.$$

We remark that, by a well known *domination principle* for energy measures, condition (5) is equivalent to the condition

(6) $$\lambda \mu_b(u,u) \leq \mu_a(u,u) \leq \Lambda \mu_b(u,u) \text{ in } X, \text{ for every } u \in D[a] = D[b] = D,$$

where μ_a, μ_b are the energy measures of a, b, respectively.

We will develop our theory under the assumption that the set of all test function $\phi \in C$ whose energy measures have a *bounded density* with respect to the measure m, is rich enough to separate the points of X. More precisely, we suppose that there is a form of the family (5), say b, that admits a m-**separating** core, that is, a core C that has the following separating property:

(7) $$\text{For every } x, y \in X, x \neq y, \exists \phi \in C \text{ with } \mu(\phi,\phi) \leq m \text{ on } X, \text{ such that } \phi(x) \neq \phi(y),$$

where $\mu = \mu_b$. Clearly, if a set C is a core of b, then C is also a core for any other form a of the family (5); moreover, in view of (6), if in addition C has the separating property (7) with respect to the given form b, then C has the same property with respect to any other form a of the family.

We are now in a position to introduce the basic notion that is at the heart of our theory, that is, a family of (equivalent) *metrics* induced on the space X by the forms of our family. For related notions we refer to the fundamental papers by A. Nagel, E.M. Stein and S. Wainger [NSW] and by L. Rothschild and E.M. Stein [RS].

Given a form a, we define the *distance function* $d = d_a : X \to [0, +\infty]$ by setting for every $x, y, \in X$:

(8) $$d(x,y) = \sup\{\phi(x) - \phi(y) : \phi \in C, \mu(\phi,\phi) \leq m \text{ on } X\},$$

and we denote by $B = B_a$ the metric balls associated with d:

(9) $$B(x,r) = \{y \in X : d(x,y) < r\},$$

for every $x \in X$ and $r > 0$. Clearly, under our assumptions (6) and (7), for every form a of the family (5) the function d just defined has indeed the properties of a distance. Moreover, the distance functions associated with two arbitrary forms of the family induce equivalent metric structures on X.

We point out that, while the energy measures in (1) are intrinsically defined in terms of the form itself, their densities occurring in (7), hence the metric induced by the form on X, can be highly affected by the initial choice of the measure m. In any case, the metric balls $B_a(x,r)$ of a given form a of the family single out special regions of the space X, on which the form a, as well as its local solutions, should be expected to enjoy special properties that might not hold on other regions of X.

Moreover, when taken up to the metric equivalence pointed out before, the intrinsic balls $B(x,r)$ play a basic role in two main regards. From one side, they allow us to formulate a *compatibility condition*, relating the whole family (5) to the initial topology of X and to the measure m initially chosen on X. This compatibility condition is expressed by **Assumption I** below. From the other side, the system of balls $B(x,r)$ allows us to formulate special *scaling* and *imbedding* properties of the forms and their domains, that also have a structural character for the family (5). These properties, in the form of suitably scaled Poincaré and Sobolev-Poincaré inequalities, are expressed by **Assumption II** below.

Assumption I. The forms (5) admit a m-separating core C in the space $L^2(X,m)$ and the following two properties hold:

(i) The metric topology induced by the distance (8) on X is *equivalent* to the initial topology of X;

(ii) The measure m is *doubling* with respect to the balls (9), that is, there exists a constant c_0 such that $0 < m(B(x,2r)) \leq c_0 m(B(x,r)) < +\infty$ for every $x \in X$ and $r > 0$.

We remark that under this assumption the space X with the distance d acquires the structure of a *homogeneous space*, according to R. Coifman and G. Weiss [CW], Ch. III, sec. 1.

Assumption II. Given a relatively compact open subset X_0 of X, there exists constants $c_1 > 0$, $c_2 > 0, s > 2$ and an integer $k \geq 1$, such that for every $x \in X_0$ and every $r > 0$ with $B(x,r) \subset X_0$, the following inequalities hold:

(j)
$$\int_{B(x,r/k)} |u - \bar{u}|^2 m(dx) \leq c_1 r^2 \int_{B(x,r)} \mu(u,u)(dx),$$

for every $u \in D[X_0]_{\mathrm{loc}}$, where $\bar{u} = \frac{1}{B(x,r/k)} \int_{B(x,r/k)} u m(dx)$;

(jj)
$$\left(\frac{1}{B(x,r)} \int_{B(x,r)} |u|^s m(dx) \right)^{1/s} \leq c_2 r \left(\int_{B(x,r)} \mu(u,u)(dx) \right)^{1/2},$$

for every $u \in D[X_0]_{\mathrm{loc}}$ with $\operatorname{supp} u \subset B(x,r)$.

It is easily checked that if Assumptions I and II are satisfied by a given form of the family (5), with some constants c_0, c_1, c_2, k and s, then they are also satisfied by any other form of the family, with the same constant s and with possibly new constants c'_0, c'_1, c'_2, k' depending on the initial c_0, c_1, c_2, k and on the ratio Λ/λ.

We can now state our main results. We denote below by X_0 a connected relatively compact open subset of X and by u an arbitrary solution of (3), where a is any form of a given family (5) for which Assumptions I and II hold. The *structural constants* c and α in the estimates below, possibly different ones, only depend on the given constants c_0, c_1, c_2, k and s occurring in the Assumptions I and II and on the ratio Λ/λ.

THEOREM 1 (Harnack inequality). *If u is positive, then*

$$\sup_{B(x,r)} \leq c \inf_{B(x,r)} u$$

for every $B(x,r) \subset X_0$.

A standard consequence of Theorem 1 is the following

COROLLARY. *There exists $\alpha > 0$ such that*

$$\operatorname*{osc}_{B(x,r)} u \leq c \left(\frac{r}{R}\right)^{\alpha} \operatorname*{osc}_{B(x,R)} u,$$

for every $0 < r \leq \frac{R}{4} \leq \frac{R_0}{10}$, $B(x,R_0) \subset X_0$.

Therefore, u is Hölder continuous with respect to the intrinsic distance of X, hence u is also continuous with respect to the initial topology of X.

By taking also into account the following L^∞-estimate:

$$\sup_{B(x,r)} |u| \leq cR^2 m(B(x,R))^{-1/p} \|f\|_{L^p(B(x,R),m)},$$

that holds in $B(x,R) \subset X_0$ with $p > \max\left\{\frac{s}{s-2}, 2\right\}$ for every solution u of the equation

$$(10) \quad u \in D_0[a, B(x,R)] : a(u,v) = \int_{B(x,R)} fv m(dx) \text{ for every } v \in D_0[a, B(x,R)],$$

the Corollary enables us to define the *Green function* $G^{x_0}_{B_R}$, for $x_0 \in B(x,R) \subset X_0$, as the unique function $G \in L^{p'}(B(x,R),m) \cap C(B(x,R) \setminus \{x_0\})$ such that

$$u(x_0) = \int_{B(x,R)} G f m(dx),$$

for every $f \in L^p(B(x,R),m)$ and $u = u_f$ solution of (10).

THEOREM 2 (Size of Green's functions). *For every* $B(x_0, R) \subset X_0$ *and every* $0 < r \leq \frac{R}{2}$, *the following estimate holds on* $\partial B(x_0, r)$

$$\frac{1}{c\Lambda} \int_r^R \frac{s^2}{m(B(x_0, s))} \frac{ds}{s} \leq G^{x_0}_{B(x_0,R)} \leq \frac{c}{\lambda} \int_r^R \frac{s^2}{m(B(x_0, s))} \frac{ds}{s}.$$

The structural estimates in Theorem 3 and in the Corollary below are the analogue in our present setting of the *Saint-Venant principle* for the energy decay in linear elasticity.

THEOREM 3 (Saint-Venant principle). *There exists* $\alpha > 0$ *such that*

$$\frac{\int_{B(x_0,r)} G^{x_0}_{B(x_0,q^{-1}r)} \mu(u,u)(dx)}{\int_{B(x_0,R)} G^{x_0}_{B(x_0,q^{-1}R)} \mu(u,u)(dx)} \leq c \left(\frac{r}{R}\right)^{\alpha}$$

for every $0 < r \leq 2q^2 R \leq 2q^3 R_0, B(x, R_0) \subset X_0,$ *and some fixed* $q \in (0, 1/6)$ *depending on* k.

Furthermore,

$$\int_{B(x_0,R)} G^{x_0}_{B(x_0,q^{-1}R)} \mu(u,u)(dx) \leq c \frac{1}{m(B(x_0, R_0))} \|u\|^2_{L^2(B(x_0,R_0),m)}.$$

From Theorem 2 and Theorem 3, we obtain the more explicit special estimate:

COROLLARY. *For* r, R *and* R_0 *as in Theorem 3, we have*

$$\frac{\frac{r^2}{m(B(x_0,r))} \int_{B(x_0,r)} \mu(u,u)(dx)}{\frac{R^2}{m(B(x_0,R))} \int_{B(x_0,R)} \mu(u,u)(dx)} \leq c \left(\frac{r}{R}\right)^{\alpha}.$$

The proof of Theorem 1 is obtained by adapting De Giorgi's truncation method to the metric structure of the homogeneous space X. The proof of Saint-Venant principle is obtained by a modified "hole-filling" argument that takes into account the relative size of Green's functions on homothetic balls, as given by Theorem 2. Detailed proofs will be given in [BM].

We illustrate the previous results from the point of view of PDE by referring to two important classes of operators in $\mathbb{R}^n$: (a) *Weighted* uniformly elliptic operators in divergence form with measurable coefficients, (b) *Subelliptic* (selfadjoint) second order operators with *bounded measurable* coefficients. Moreover, (c), we shall give an example of an operator L in $\mathbb{R}^2$ whose local solutions are Hölder continuous with respect to the intrinsic distance of L, and only continuous with respect to the euclidean metric of $\mathbb{R}^2$.

(a) We consider the form

$$a(u,v) = \int_{\mathbb{R}^n} \sum_{ij=1}^{n} \partial u/\partial x_i \partial v/\partial x_j a^{ij}(x)dx,$$

on $C_0^1(\mathbb{R}^n)$, where the measurable coefficients $a^{ij}(x) = a^{ji}(x)$, $i,j = 1,\dots,n$, satisfy the condition

$$\lambda|\xi|^2 w(x) \le \sum_{ij=1}^{n} \xi_i\xi_j a^{ij}(x) \le \Lambda|\xi|^2 w(x) \qquad \text{a.e. in } \mathbb{R}^n,$$

for every $\xi \in \mathbb{R}^n$. Here w is a weight in the Muckenhoupt class A_2, or the weight $w(x) = |\det F'|^{1-2/n}$ associated with a quasi-conformal transformation F in $\mathbb{R}^n$. The domain $D[a]$ of this form is obtained by completion of $C_0^1(\mathbb{R}^n)$ with respect to the norm $(a(u,u) + (u,u))^{1/2}$, where $(.,.)$ is taken to be the inner product of the Hilbert space $H = L^2(\mathbb{R}^n, w(x)dx)$, and $D[a]$ is found to be injected in H. This class of operators has been considered by E. Fabes, C. Kenig and R. Serapioni [FKS] and E. Fabes, D. Jerison and C. Kenig [FJK], who proved Theorems 1 and 2, where now $X = \mathbb{R}^n$, the balls $B(x,r)$ are the usual euclidean balls and $m(dx) = w(x)dx$. The doubling property of our Assumption I is a well known property of the weights considered above and the scaled Poincaré and Sobolev-Poincaré inequalities of Assumption II are given in [FKS].

(b) In the space $H = L^2(\mathbb{R}^n, dx)$, we consider a family of forms (5), where b is a given selfadjoint subelliptic form with smooth coefficients in $\mathbb{R}^n$, that is b is given on $C_0^1(\mathbb{R}^n)$ by

$$b(u,v) = \int_{\mathbb{R}^n} \sum_{ij=1}^{n} \partial u/\partial x_i \partial v/\partial x_j b^{ij}(x)dx,$$

where $b^{ij}(x) = b^{ji}(x)$, $ij = 1,\dots,n$, are smooth functions that satisfy the (degenerate) ellipticity condition

$$0 \le \sum_{ij=1}^{n} \xi_i\xi_j b^{ij}(x) \qquad \text{in } \mathbb{R}^n \text{ for every } \xi \in \mathbb{R}^n,$$

and b satisfies in addition the following subellipticity estimate for some $\varepsilon \in (0,1)$

$$c\|u\|_{H_\varepsilon}^2 \le b(u,u) + \|u\|_{L^2}^2 \qquad \text{for every } u \in C_0^1(\mathbb{R}^n),$$

where H_ε denotes the usual fractional Sobolev space of order ε. The form a is any form of the following type

$$a(u,v) = \int_{\mathbb{R}^n} \sum_{ij=1}^{n} \partial u/\partial x_i \partial v/\partial x_j a^{ij}(x)dx,$$

with measurable coefficients $a^{ij} = a^{ji}$ that satisfy the *uniform subellipticity condition*

$$\lambda \sum_{ij=1}^{n} \xi_i\xi_j b^{ij}(x) \le \sum_{ij=1}^{n} \xi_i\xi_j a^{ij}(x) \le \Lambda \sum_{ij=1}^{n} \xi_i\xi_j b^{ij}(x) \qquad \text{a.e. in } \mathbb{R}^n,$$

for some given constants $0 < \lambda \leq \Lambda$.

The distance d on $\mathbb{R}^n$ induced by the form b according to our definition (8) turns out to be equal to the distance d^* associated to the form b according to C.L. Fefferman and D.H. Phong [FP], which is shown by these authors to satisfy the condition

$$\frac{1}{c}|x-y| \leq d^*(x,y) \leq c|x-y|^{\varepsilon}.$$

By noting that $m(dx) = dx$, this shows that Assumption I above is satisfied on any bounded open domain X of $\mathbb{R}^n$.

The scaled Poincaré inequalities on the intrinsic balls for the form b have been proved by D. Jerison [J] and D. Jerison and A. Sanchez Calle, [JSC] Th. 5.1. These authors also provide examples that show that Poincaré inequality may not hold on regions that do not coincide with an intrinsic ball, what may serve as an illustration of a remark we made before. The Poincaré-Sobolev inequality on X is an immediate consequence of the subellipticity estimate and of Sobolev imbedding theorems. The scaled Poincaré-Sobolev inequalities, as occurring in Assumption II, are then obtained by using the diffeomorphism Φ associated with the form b, given by Theorem 3.1 of [JSC]. For related results see also S. Kusuoka and D. Stroock [KS].

Theorem 1 to 3 above seem to be new in this subelliptic setting. The main example to which they apply is that of an operator of *Hörmander type*

$$L = -\sum_{hk=1}^{m} X_k^* a^{hk}(x) X_h, \quad x \in \mathbb{R}^n,$$

where $X_h, h = 1, \ldots, m$ are m smooth vector fields in $\mathbb{R}^n$ that satisfy the Hörmander condition and where $a = (a^{hk})$ is any symmetric $m \times m$ matrix of measurable functions on $\mathbb{R}^n$, such that $\lambda|\eta|^2 \leq a\eta \cdot \eta \leq \Lambda|\eta|^2$ for every $\eta \in \mathbb{R}^m$, a.e. on $\mathbb{R}^n$.

(c) In the space $L^2(B, m(dxdy))$, where $B = \left\{(x,y) \in \mathbb{R}^2 : \frac{1}{(\log|x|)^2} + |y|^2 \leq \frac{1}{4}\right\}$ and $m(dxdy) = \frac{1}{|x|(\log|x|)^2} dxdy$, we consider the form

$$a(u,v) = \int_B \{x^2(\log|x|)^4 \partial u/\partial x \partial v/\partial x + \partial u/\partial y \partial v/\partial y\} \frac{1}{|x|(\log|x|)^2} dxdy.$$

Assumptions I and II can be easily checked by suitably rescaling the form on a euclidean ball of $\mathbb{R}^2$. The intrinsic balls $B(o,r)$ shrink to $\{o\}$ as $\exp(-1/r)$ in the x-direction, as $r \to 0$.

We finally mention that any family of forms satisfying a condition like (5) enjoys special *variational compactness properties*, that can be expressed in terms of Γ*-convergence* of the functionals $a(u,u)$ now defined on the whole of H by extending them to $+\infty$ outside their domain. In case the domains are uniformly compactly imbedded in H, as for instance in the uniform subelliptic case (b), these convergence properties can be expressed in terms of the *resolvent operators* associated with the forms and are thus related to convergence of spectra and semigroups. The compactness properties play an important role in *homogenization theory* and, more generally, in the asymptotic variational approach to *composite media*, see [M]. The structural estimates presented before describe some properties of local solutions that are kept in the variational limit and this is of particular relevance in the asymptotic theory, due to the fact that the explicit expression of the limit energy form in terms of the *effective characteristics* of the composite body may not always be easily determined.

ACKNOWLEDGEMENTS

The authors wish to thank the Institute for Mathematics and its Applications at the University of Minnesota, where this note was written, for hospitality and support.

REFERENCES

BM. M. Biroli and U. Mosco, to appear.

CW. R.R. Coifman and G. Weiss, *Analyse harmonique non-commutative sur certains éspaces homogènes*, Lecture Notes in Math. 242, Springer Verlag, 1971.

FJK. E. Fabes, D. Jerison, C. Kenig, *The Wiener test for degenerate elliptic equations*, Ann. Inst. Fourier, 3 (1982), 151–183.

FKS. E. Fabes, C. Kenig, R. Serapioni, *The local regularity of solutions of degenerate elliptic equations*, Comm. in Part. Diff. Eq., 7 (1) (1982), 77–116.

FP. C.L. Fefferman and D.H. Phong, *Subelliptic eigenvalue problems*, Proceedings of the Conference on Harmonic Analysis in honor of A. Zygmund, Wadsworth Math. Series, 1981, pp. 590–606.

F. M. Fukushima, *Dirichlet forms and Markov Processes*, North-Holland Math Library, Vol. 23, 1980.

H. L. Hörmander, *Subelleptic operators*, Annals of Math. Studies 91, Princeton Univ. Press (1978), 127–208.

J. D. Jerison, *The Poincaré inequality for vector fields satisfying Hörmander's condition*, Duke Math. J. 53 N. 2 (1986), 503–523.

J. D. Jerison and A. Sanchez-Calle, *Subelliptic, second order differential operators*, Springer Lecture Notes in Math. 1277 (1987), 46–77.

KS. S. Kusuoka and D. Stroock, *Application of the Malliavin Calculus, Part III*, J. Fac. Sci. Univ. Tokyo Sect. IA Math. 34 (1987), 391–442.

KS. S. Kusuoka and D. Stroock, *Long time estimates for the heat kernel associated with a uniformly subelliptic symmetric second order operator*, pre-print.

M. U. Mosco, *Composite Media and Dirichlet forms*, Proc. Workshop on Composite Media and Homogenization, ICTP Triest 1990, G. Dal Maso and G. Dell'Antonio edts., Birkhaüser 1991.

M. U. Mosco, *Composite media and asymptotic Dirichlet forms*, to appear.

NWS. A. Nagel, E.M. Stein and S. Wainger, *Balls and metrics defined by vector fields I: Basic Properties*, Acta Math. 155 (1985), 103–147.

RS. L. Rothschild and E.M. Stein, *Hypoelliptic differential operators and nilpotent Lie groups*, Acta Math 137 (1977), 247–320.

ON AN ABSTRACT WEAKLY HYPERBOLIC EQUATION MODELLING THE NONLINEAR VIBRATING STRING

Piero D'Ancona

Dip. di Matematica
Università di Roma II
Via F.di Carcaricola
I-00133 Roma

Sergio Spagnolo

Dip. di Matematica
Università di Pisa
Via Buonarroti
I-56127 Pisa

§1. Introduction

Let V be a complex Banach space continuously embedded into its dual V', and denote by $\langle \cdot, \cdot \rangle$ the (sesquilinear) pairing on $V' \times V$. Then we can consider the Hilbert space H, completion of V with respect to the product $(u, v)_H = \langle u, v \rangle$, and the increasing triple of spaces (V, H, V') thus obtained is usually called a *Hilbert triple.*

In this framework, we consider the following abstract Cauchy problem:

$$u'' + m(\langle Au, u \rangle)Au = 0 \tag{1}$$

$$u(0) = u_0, \quad u'(0) = u_1 \tag{2}$$

where A is a bounded operator from V to V' such that, for all v, w in V,

$$\langle Av, w \rangle = \overline{\langle Aw, v \rangle} \tag{3}$$

$$\langle Av, v \rangle \geq \nu \|v\|_V^2, \quad \nu > 0 \tag{4}$$

while $m : [0, +\infty[\to \mathbf{R}$ satisfies

$$m(s) \text{ is continuous, } m(s) \geq 0 \text{ on } [0, +\infty[. \tag{5}$$

Since the operator A is hermitian and coercive and the function $m(s)$ is nonnegative, eq.(1) is of *weakly hyperbolic* type.

A concrete version of (1),(2), namely the problem

$$u_{tt} - m\left(\int_\Omega |\nabla_x u|^2 dx\right) \Delta u = 0 \qquad (t, x) \in \mathbf{R}^+ \times \mathbf{R}^n \tag{6}$$

$$u(0, x) = u_0(x), \quad u_t(0, x) = u_1(x) \tag{7}$$

where $\Omega = [0, 2\pi]^n$ (and the unknown $u(t, x)$ is 2π-periodic in the space variables), or alternatively Ω is an open subset of $\mathbf{R}^n$ (and $u(t, \cdot) \in H_0^1(\Omega)$), initially proposed

Developments in Partial Differential Equations and Applications to Mathematical Physics, Edited by G. Buttazzo *et al.*, Plenum Press, New York, 1992

by Kirchhoff [K] in a special case, has been studied by many authors, usually under the stronger assumption

$$m(s) \text{ is Lipschitz continuous, } m(s) \geq \mu > 0 \text{ on } [0, +\infty[. \tag{8}$$

We mention, among the others, the papers [B], [D], [M], [P], [R] where the *local existence* in suitable Sobolev spaces is proved under condition (8).

A different kind of result is the *global existence* with C_0^∞ *small data* (with $\Omega = \mathbf{R}^n$); this was proved in the case $n = 1$ and $m(s) = 1 + s$ by J.M.Greenberg and S.C.Hu ([GH]), and in the general case in [DS1].

A third kind of result is the *global existence with real analytic initial data* of unrestricted size. This was originally proved in one space dimension by S.Bernstein ([B]), under the stronger assumption (8), and later extended to several space dimensions by S.I.Pohožaev. Since the analytic class is a suitable framework for the *weak* hyperbolicity, at least in the linear case (see [CDS]), Bernstein's result raised the natural question whether it is possible to relax assumption (8) to the weaker (5) (cf. J.L.Lions, [L]).

This was done in two steps. In [AS1] the global existence in the analytic category was proved, assuming that $m(s)$ satisfies (5) and in addition

$$\textit{either} \int_0^{+\infty} m(s)ds = +\infty \quad \textit{or} \quad \sup_{s\geq 0} m(s) < +\infty; \tag{9}$$

finally, in [DS2] condition (9) was dropped, thus giving a positive answer to the above question. We remark that when $m(s)$ is merely continuous one can lose the uniqueness of the solution; only the existence can be proved.

Coming back to the abstract setting, first of all we have to mention the paper of S. Pohožaev [Po], who extended Bernstein's global existence result to any problem of the form (1),(2), assuming that the function $m(s)$ satisfies (8) and that the initial data are *A-analytic* vectors of V. This means that, for $v = u_0, u_1$, for some constants K, Λ one has

$$A^j v \in V \quad \text{and} \quad |\langle A^j v, v\rangle|^{1/2} \leq K\Lambda^j j! \qquad \forall j \geq 0. \tag{10}$$

This is the counterpart, in the abstract case, of the analyticity condition in the concrete case; see the more precise considerations after the statement of the theorem.

Our aim here is to extend the result of [DS2] to any abstract problem (1), (2), by proving that there is a global solution under the weak hyperbolicity assumption (5), provided the initial data are A-analytic. More precisely, we prove

Theorem 1.

Assume the injection $V \subseteq V'$ is compact, and that (3), (4), (5) hold. Then for any A-analytic initial data $u_0, u_1 \in V$, there exists a solution $u \in C^2([0, +\infty[, V)$ to problem (1),(2), such that $u(t), u'(t)$ are A-analytic for all $t > 0$.

Moreover, if $m(s)$ is Lipschitz continuous the solution is unique.

We remark that the proof given in [DS2] for the concrete case uses a complex extension argument due to Garabedian, which does not apply to the abstract problem (1),(2) (however, the method of [DS2] applies with small changes to more general integro-differential equations, which cannot apparently be reduced to the form (1)).

On the other hand, from Theorem 1 one can derive the global existence of a solution to the concrete problem (6),(7) with $\Omega = [0, 2\pi]^n$ under assumption (5), provided

$$u_0(x), u_1(x) \textit{ are real analytic, } 2\pi\textit{-periodic in } x_1, \ldots, x_n. \tag{11}$$

To this end, it is sufficient to choose $A = -\Delta$, $V = H^1(\mathbf{T}^n)$, $V' = H^{-1}(\mathbf{T}^n)$, where $\mathbf{T}^n$ is the n-dimensional torus, so that the A-analytic vectors are simply the real analytic periodic functions.

Another application of Theorem 1 is the following: choose $A = -\Delta$, $V = H_0^1(\Omega)$, $V' = H^{-1}(\Omega)$, where Ω is an open subset of $\mathbf{R}^n$; then we get the global existence for problem (6),(7) under assumption (5), provided the initial data belong to the class of the A-analytic vectors in the sense of (10). Now, a subclass of this class is given by the functions $v(x)$, real analytic in a neighbourhood of $\overline{\Omega}$, such that

$$\Delta^j v = 0 \quad \text{on} \quad \partial\Omega, \quad j \geq 0;$$

this subclass coincides with the whole class of A-analytic vectors when $\partial\Omega$ is a real analytic manifold of dimension $n-1$, leaving Ω on one side ([LM]).

§2. Proof of Th.1 (sketch).

A) THE LOCAL EXISTENCE

Since the embedding $V \mapsto V'$ is compact, recalling (3), (4), we can find an orthonormal basis $\{v_k\}$ of H such that $v_k \in V$ and

$$Av_k = \lambda_k^2 v_k, \quad \lambda_k > 0, \quad \lambda_k \to +\infty.$$

Moreover (see [AS1]), a vector $v = \sum_k y_k v_k \in V$ is A-analytic if and only if, for some $\delta > 0$,

$$\sum_{k=1}^{\infty} e^{2\delta\lambda_k} |y_k|^2 < \infty. \tag{12}$$

Thus equation (1) can be transformed, writing $u(t) = \sum_k y_k(t) v_k$, in the system

$$y_k'' + \lambda_k^2 a(t) y_k = 0 \qquad (k = 1, 2, \ldots) \tag{13}$$

where

$$a(t) = m(\langle Au, u \rangle).$$

The initial data $u_0 = \sum y_k^{(0)} v_k$ and $u_1 = \sum y_k^{(1)} v_k$ satisfy an assumption like (12).

Then, using finite dimensional approximations and a suitable continuation argument, one can prove the following lemma (see Theorem 2 of [AS1], p.8):

Lemma 1.

Under the assumptions of Theorem 1, problem (1),(2) has a local solution $u \in C^2([0, T], V)$ for some $T = T(u_0, u_1) > 0$, such that $Au \in C^0([0, T], V)$; moreover $u(t), u'(t)$ are A-analytic for all $t \in [0, T]$.

We emphasize that in the proof of Lemma 1 it is not necessary to assume that $m(s)$ be nonnegative; indeed Lemma 1 is nothing but an abstract version of the Cauchy-Kowalewsky theorem, which holds without any assumption of hyperbolicity.

B) THE LINEARIZED EQUATION

Consider the *linear* Cauchy problem

$$u'' + a(t)Au = 0 \tag{14}$$
$$u(0) = u_0, \quad u'(0) = u_1 \tag{15}$$

where the coefficient $a(t)$ satisfies

$$a(t) \geq 0, \quad a \in L^1_{\rm loc}(0, +\infty) \tag{16}$$

while the operator A is as above. The following lemma can be proved by the method of the *perturbed energy of infinite order*, firstly introduced in [CDS] (where the equations like $u_{tt} - a(t)\Delta u$ were considered, with $a(t)$ satisfying (16)):

Lemma 2.
Under assumptions (3),(4),(16), for any A-analytic initial data u_0, u_1 (see (10)), problem (14),(15) has a unique global solution $u \in C^1([0, +\infty[, V)$. Moreover, $u(t), u'(t)$ are A-analytic for all $t \geq 0$.

For a proof of Lemma 2, we refer to [AS1], pp.6-9. To be precise, in that paper the coefficient $a(t)$ is assumed to be continuous, but in fact, all that is used in the proof is that $a(t) \in L^1(0, T)$ for all $T > 0$. A more general result of global existence, including in particular Lemma 2, holding for a wider class of equations of the form $u'' + A(t)u = 0$ (with $A(t) \geq 0$) can be found in [AS2]. We refer to this paper also for a detailed exposition of the infinite order energy method.

C) THE GLOBAL EXISTENCE

Let $u(t)$ be the local solution to (1),(2) whose existence is ensured by Lemma 1. By an inductive argument, we can prolonge u to a *maximal* solution on $[0, T^*[$. We want to show that $T^* = +\infty$.

Arguing by contradiction, assume that $T^* < \infty$. Consider then the following *energy function*

$$E(t) = \frac{1}{2}|u' + u|_H^2 + \frac{1}{2}|u|_H^2 + \frac{1}{2}M(\langle Au, u\rangle) \tag{17}$$

where $M(s)$ is a nonnegative primitive of $m(s)$, namely

$$M(s) = \int_0^s m(\sigma)d\sigma. \tag{18}$$

By differentiating $E(t)$ we get, recalling (1),

$$\begin{aligned} E'(t) &= \mathrm{Re}\langle u'' + u', u' + u\rangle + \mathrm{Re}\langle u', u\rangle + m(\langle Au, u\rangle) \cdot \mathrm{Re}\langle Au, u'\rangle \\ &= 2\,\mathrm{Re}\langle u', u\rangle + |u'|_H^2 + \mathrm{Re}\langle u'', u\rangle \end{aligned}$$

and hence, as $2E(t) \geq |u'|_H^2 + 2\,\mathrm{Re}\langle u', u\rangle$,

$$E'(t) \leq 2E(t) + \mathrm{Re}\langle u'', u\rangle; \tag{19}$$

using again (1), this gives

$$E'(t) \leq 2E(t) - m(\langle Au, u\rangle)\langle Au, u\rangle. \tag{20}$$

Writing for brevity

$$\phi(t) = \langle Au(t), u(t) \rangle, \tag{21}$$

we can rewrite (20) as follows:

$$e^{-2t} m(\phi(t)) \cdot \phi(t) \leq -(e^{-2t} E(t))'.$$

Integration on $[0, T^*]$ gives

$$e^{-2T^*} \int_0^{T^*} m(\phi(t)) \cdot \phi(t) dt \leq E(0) < \infty. \tag{22}$$

It is easy to see that (22) implies

$$m(\phi(t)) \in L^1(0, T^*); \tag{23}$$

indeed, we have

$$\begin{aligned} \int_0^{T^*} m(\phi(t)) dt &= \int_{[0,T^*] \cap \{\phi \geq 1\}} m(\phi(t)) \cdot \phi(t) dt + \int_{[0,T^*] \cap \{\phi < 1\}} m(\phi(t)) \cdot \phi(t) dt \\ &\leq \int_0^{T^*} m(\phi(t)) \cdot \phi(t) dt + T^* \cdot \max_{[0,1]} m(s) \\ &\leq e^{2T^*} E(0) + T^* \cdot \max_{[0,1]} m(s). \end{aligned}$$

Now, consider the function

$$a(t) = \begin{cases} m(\phi(t)), & t < T^*, \\ 0, & t \geq T^*; \end{cases}$$

by (23) and (15) we see that $a(t)$ is integrable and nonnegative on $\mathbf{R}^+$, hence the linear Cauchy problem

$$v'' + a(t) A v = 0 \tag{24}$$
$$v(0) = u_0, \quad v'(0) = u_1 \tag{25}$$

satisfies all the assumptions of Lemma 2. Thus (24),(25) has a unique global solution for $t \geq 0$, in particular there exist (in V) the limits $v(T^*-)$, $v'(T^*-)$ and they are A-analytic vectors. Recalling that, by the definition of $a(t)$, $u(t)$ is also a solution of (24),(25) on $[0, T^*[$, by the uniqueness (see Lemma 2) we have $u \equiv v$. Now, choosing $v(T^*-)$, $v'(T^*-)$ as initial data, we can apply again the local existence result (Lemma 1) for $t \geq T^*$ and thus prolonge the solution u beyond T^*, which is the desired contradiction. We conclude that $T^* = +\infty$.

The uniqueness in the Lipschitz case is proved by a standard linearization argument, whereas we refer to [AS1] for a discussion of the non-uniqueness in the case of a merely continuous $m(s)$.

References

[AS1] Arosio,A. and Spagnolo,S., Global solution to the Cauchy problem for a nonlinear equation, in Nonlinear PDE's and their applications, Collège de France Seminar, H.Brezis and J.L.Lions (eds.), Vol.VI, *Research Notes Math.* **109**, Pitman, Boston (1984), 1-26

[AS2] Arosio,A. and Spagnolo,S., Global existence for abstract evolution equations of weakly hyperbolic type, *J. Math. pures et appl.* **65** (1986), 263-305

[B] Bernstein,S., Sur une classe d'équations fonctionnelles aux dérivées partielles, *Izv. Akad. Nauk SSSR*, Sér. Math. **4** (1940), 17-26

[CDS] Colombini,F., De Giorgi,E. and Spagnolo,S., Sur les équations hyperboliques avec des coefficients qui ne dépendent que du temps, *Ann. Sc. Norm. Sup. Pisa* **6** (1979), 511-559

[D] Dickey,R.W., Infinite systems of nonlinear oscillation equations related to the string, *Proc. Am. Math. Soc.* **23** (1969), 459-468; Infinite systems of nonlinear oscillation equations, *J. Diff. Eq.* **8** (1970), 16-26

[DS1] D'Ancona,P. and Spagnolo,S., A class of nonlinear hyperbolic problems with global solutions, *Pubbl. Dip. Mat. Univ. Pisa* **631** (1992), 1-22

[DS2] D'Ancona,P. and Spagnolo,S., Global solvability for the degenerate Kirchhoff equation with real analytic data, *Invent. Math.* **108** (1992), 247-262

[GH] Greenberg,J.M and Hu,S.C., The initial value problem for a stretched string, *Quart. Appl. Math.* (1980), 289-311

[K] Kirchhoff,G., Vorlesungen über Mechanik, Teubner, Leipzig (1883)

[L] Lions,J.L., On some questions in boundary value problems of mathematical physics, in Contemporary developments in continuum mechanics and partial differential equations, G.M.De La Penha and L.A.Medeiros (eds.), *North-Holland Math. St.* **30** (1978), 284-346

[LM] Lions,J.L. and Magenes,E., Espaces de fonctions et distributions du type de Gevrey et problèmes aux limites paraboliques, *Ann. Mat. pura e appl.* **68** (1965), 341-417

[M] Medeiros,L.A., On a new class of nonlinear wave equations, *J. Math. Anal. Appl.* **69** (1979), 252-262

[P] Perla Menzala,G., On classical solutions of a quasilinear hyperbolic equation, *Nonlinear Anal.* **3** (1979), 613-627

[Po] Pohožaev,S.I., On a class of quasilinear hyperbolic equations, *Mat. Sbornik.* **96** (138) (1975) N.1, 152-166 [Transl.: *Math. USSR Sbornik* **25** (1975) N.1]

[R] Rivera Rodriguez,P.H., On local strong solutions of a non-linear partial differential equation, *Appl. Anal.* **8** (1980), 93-104

ON THE RELAXATION OF FUNCTIONALS DEFINED ON CARTESIAN MANIFOLDS

Ennio De Giorgi

Scuola Normale Superiore
Pisa, Italy

In this conference I shall present some conjectures concerning the measure of manifolds of cartesian type. I think that solving these conjectures could have a remarkable interest both in the development of geometric measure theory and in the study of variational problems where integrals of functions of the gradients of vector–valued functions are involved.

I have tried to formulate all the statements in such a way to make the conference completely self–contained and fully understandable even to mathematicians whose interests are far from geometric measure theory. Nevertheless, results of such theory will likely be useful in proving or disproving the conjectures.

Before stating any conjecture, I shall recall the classical definition of Hausdorff measure.

Definition 1. Let be $E \subset \mathbf{R}^n$ and $h > 0$ a real number; the h–dimensional Hausdorff measure of E is defined by

$$\mathcal{H}^h(E) = \omega(h) \lim_{\epsilon \to 0+} \inf \left\{ \sum_{i=1}^{\infty} (\mathrm{diam} E_i)^h \, ; \; E \subset \bigcup_{i=1}^{\infty} E_i, \; \mathrm{diam} E_i < \epsilon \right\},$$

where $\omega(h) = \dfrac{2^{1-h}\pi^{h/2}}{h\Gamma(h/2)}$ and $\Gamma(s) = \displaystyle\int_0^{\infty} e^{-t} t^{s-1} dt$ for $s > 0$; this choice of $\omega(\cdot)$ implies that for every positive integer n it holds $\mathcal{H}^n([0,1]^n) = 1$. Notice also that, more generally, in $\mathbf{R}^n$ the Hausdorff measure $\mathcal{H}^n$ coincides with the exterior Lebesgue measure.

We also set $\mathcal{H}^0(E) = \mathrm{card}(E)$ if E is finite, $\mathcal{H}^0(E) = +\infty$ if E is infinite.

Remark that in the definition of Hausdorff measure only the metric space structure of $\mathbf{R}^n$ is involved and that many definitions and results concerning the Hausdorff measure may be extended to every metric space. For instance, for any $h > 0$ the h-dimensional Hausdorff measure enjoys the following properties (see *e.g.* [3]):

1. it is countably subadditive;
2. all the Borel sets are measurable;
3. if F is a contractive map between two metric spaces, then the measure of the image of a set E is not greater than the measure of E;

Developments in Partial Differential Equations and Applications to Mathematical Physics, Edited by G. Buttazzo *et al.*, Plenum Press, New York, 1992

4. it is h-homogeneous, *i.e.* multiplying the distance by a factor ρ changes the measure by a factor ρ^h.

Recall also that, if A is an open subset of $\mathbf{R}^n$, and $f \in C^1(A)$, then

$$\mathcal{H}^n\left(\{(x,y) : x \in A,\ y = f(x)\}\right) = \int_A \sqrt{1+|\nabla f|^2}dx \tag{1}$$

and that an analogous formula holds for a vector valued function $u \in [C^1(A)]^k$:

$$\mathcal{H}^n\left(\{(x,y) : x \in A,\ y = u(x)\}\right) = \int_A \varphi(\nabla u)dx, \tag{2}$$

where $\varphi(\nabla u)$ indicates the well-known square root of 1 plus the sum of the squares of the minors of the jacobian matrix of u.

A first problem, which is still open to the best of my knowledge, is presented in the following conjecture.

Conjecture 1. There exists a function $\sigma : \mathbf{N}^2 \to \mathbf{R}$ such that

1. $n \geq 1,\ k \geq 1 \quad \Longrightarrow \quad \sigma(n,k) > 0$,
2. if $B \subset \mathbf{R}^n$ is an open ball, $f \in [L^\infty(B)]^k$, $|f(x)| \leq \sigma(n,k)$ for any x, then the following minimum is achieved:

$$\min\left\{\int_B (\varphi(\nabla u) + |u-f|)\,dx;\ u \in [C^1(B)]^k\right\}.$$

In order to study cases where the minimum in conjecture 1 does not exists (or it is difficult to prove its existence) it seems to be useful to consider a suitable relaxed functional. Following [2], I shall then recall a definition.

Definition 2. Let A be an open subset of $\mathbf{R}^n$, and let be $u \in [C^1(A)]^k$; we set

$$F(u,A) = \int_A \varphi(\nabla u)dx;$$

if $w \in [L^1(A)]^k$ we set

$$\overline{F}(w,A) = \min\left\{\liminf_{h\to+\infty} F(u_h,A); u_h \in [C^1(A)]^k,\ \lim_{h\to+\infty}\int_A |u_h - w|dx = 0\right\}.$$

Remark 1. In connexion with conjecture 1, notice that it is not difficult to see that the following minimum is achieved for every $A \subset \mathbf{R}^n$ with $\mathcal{H}^n(A) < +\infty$ and for every $f \in [L^1(A)]^k$:

$$\min\left\{\overline{F}(w,A) + \int_A |w-f|dx;\ w \in [L^1(B)]^k\right\}.$$

Let us present an example showing that in general the function giving the minimum in the previous remark is not continuous.

Example. Let be $n = k = 1$, $A =]-2,2[$, and $f(x) = 2x/|x|$. It is not difficult to see that the function u minimizing the functional

$$w \mapsto \overline{F}(w,A) + \int_A |w - f|dx$$

is an increasing function such that $\lim_{x\to 0+} u(x) = 1$, $\lim_{x\to 0-} u(x) = -1$ (or it is Lebesgue equivalent to such a function).

Moreover, for a.a. $x \in]0,2[$ one has $u(x) = -u(-x)$, and precisely

$$u(x) = \begin{cases} 1 + \sqrt{2x - x^2} & \text{if } 0 < x \leq 1, \\ 2 & \text{if } 1 \leq x < 2. \end{cases}$$

Remark 2. One could also consider functionals like

$$\overline{F}(w,A) + \int_A |f - w|^p dx, \quad \text{or} \quad \overline{F}(w,A) + \int_A g(x,w)dx,$$

and, even more generally, relaxed of functionals where integrands with linear growth with respect to the partial derivatives and the minors of the jacobian matrix of w are considered instead of $\varphi(\nabla w)$.

Problems like those considered in the previous remark are more difficult if $\min\{n,k\} > 1$; in fact, if $\min\{n,k\} = 1$ the functional $\overline{F}(w,A)$ is convex with respect to w and it is (the trace of) a Borel regular measure with respect to the open set A, whereas if $\min\{n,k\} > 1$ then it is not convex and probably it is not a measure. To check this latter statement, it could be useful to consider the following conjecture.

Conjecture 2. Let a, b, c be three points in $\mathbf{R}^2$ such that $|a-b| = |b-c| = |a-c| = \tau > 0$. Suppose that the origin is the center of the equilateral triangle with vertices a, b, c, and put

$$A = \{x \in \mathbf{R}^2 : x = \lambda b + \mu c,\ \lambda, \mu > 0\},$$
$$B = \{x \in \mathbf{R}^2 : x = \lambda a + \mu c,\ \lambda, \mu > 0\},$$
$$C = \{x \in \mathbf{R}^2 : x = \lambda a + \mu b,\ \lambda, \mu > 0\}.$$

It is easy to see that the three sets above are disjoint and that their union is the whole plane with three halflines removed. Set, for $\rho > 0$, $B_\rho = \{x \in \mathbf{R}^2 : x_1^2 + x_2^2 < \rho^2\}$, and consider the function u defined by

$$u(x) = \begin{cases} a & \text{if } x \in A \\ b & \text{if } x \in B \\ c & \text{if } x \in C. \end{cases}$$

Then, for any $0 < \sigma < \rho < +\infty$, the following properties are true:

$$\overline{F}(u, B_\rho \setminus \overline{B}_\sigma) = \pi(\rho^2 - \sigma^2) + 3\tau(\rho - \sigma),$$
$$\overline{F}(u, B_\rho) > \pi\rho^2 + 3\tau\rho$$
$$\lim_{\rho\to 0+} \overline{F}(u, B_\rho) = 0.$$

It then immediately follows that for every $\rho > 0$ there exists $\sigma > 0$ such that

$$\overline{F}(u, B_\rho \setminus \overline{B}_\sigma) + \overline{F}(u, B_{2\sigma}) < \overline{F}(u, B_\rho),$$

hence $\overline{F}(u,\cdot)$ is not a Borel measure.

Remark 3. In order to discuss the previous conjecture and other similar problems, it would be interesting to study in a sistematic way the functional $\overline{F}(w, A)$ when w takes only a finite numbers of values.

Since it is likely true that when $\min\{n, k\} > 1$ the functional $\overline{F}$ is not always a measure, it seems to be useful to put the following definition.

Definition 3. Let A be an open subset of $\mathbf{R}^n$, and let be $w \in [L^1(A)]^k$. We set

$$\overline{\overline{F}}(w, A) = \inf\left\{\sum_{h=1}^{\infty} \overline{F}(w, A_h);\ A_h \text{ open},\ A = \bigcup_{h=1}^{\infty} A_h\right\}.$$

Remark 4. For any $w \in [L^1(A)]^k$ the function $\overline{\overline{F}}(w, \cdot)$ defined on the open sets is the trace of a Borel regular measure.

In order to compare the conjectures presented here with those in [2], it would be interesting to solve the following conjectures.

Conjecture 3. If $A \subset \mathbf{R}^n$ is open, $w \in [L^1(A)]^k$ and $\overline{F}(w, A) < +\infty$ then

$$\overline{\overline{F}}(w, A) = \inf\left\{\overline{F}(w, A \setminus C);\ C \text{ closed},\ \mathcal{H}^{n-1}(A \cap C) = 0\right\}.$$

Related to the previous conjecture are the following ones.

Conjecture 4. If $A, A' \subset \mathbf{R}^n$ are open, $A \subset A'$, $w \in [L^1(A')]^k$ and $\overline{F}(w, A') < +\infty$, then

$$\overline{F}(w, A) - \overline{\overline{F}}(w, A) \leq \overline{F}(w, A') - \overline{\overline{F}}(w, A').$$

It is natural to ask whether the infimum in conjecture 3 is a minimum; I think that a sufficient condition could be that one in the following conjecture.

Conjecture 5. If $A \subset \mathbf{R}^n$ is an open set with $\mathcal{H}^n(A) < +\infty$, $f \in [L^1(A)]^k$ and u minimizes the functional

$$w \mapsto \overline{F}(w, A) + \int_A |f - w|\,dx,$$

then there exists a closed set C such that

$$\overline{F}(u, A \setminus C) = \overline{\overline{F}}(u, A) \quad \text{and} \quad \mathcal{H}^{n-1}(C \cap A) = 0.$$

Notice that under the same conditions as in conjecture 5, it is foreseen to be possible to choose an *optimal* closed set C.

Conjecture 6. If $A \subset \mathbf{R}^n$ is an open set with $\mathcal{H}^n(A) < +\infty$, $f \in [L^1(A)]^k$ and u minimizes the functional

$$w \mapsto \overline{F}(w, A) + \int_A |f - w|\,dx,$$

set

$$K = \bigcap_C \left\{C;\ C \text{ closed},\ \overline{F}(u, A \setminus C) = \overline{\overline{F}}(u, A) \text{ and } \mathcal{H}^{n-1}(C \cap A) = 0\right\},$$

it holds $\overline{F}(u, A \setminus K) = \overline{\overline{F}}(u, A)$.

Remark 5. The set K in conjecture 6 may be considered as the set of "defects" of u; it would be interesting to make a comparison with analogous "defects" which have been studied in different contexts (see *e.g.* [1]).

Besides these conjectures, whose statements are all independent of the notions of differential form and current, I shall point out a conjecture connected to the theory of rectifiable currents presented in [3]. To this end, I shall recall some definitions from such theory.

For any $p \in \{1, \ldots, m\}$, the p-dimensional currents in $\mathbf{R}^m$ ($\mathcal{D}_p(\mathbf{R}^m)$) are the linear and continuous functionals on the space $\mathcal{D}^p(\mathbf{R}^m)$ of differential p-forms on $\mathbf{R}^m$ with $C_0^\infty(\mathbf{R}^m)$ coefficients; the boundary ∂T of a p-dimensional current T is the $(p-1)$-dimensional current defined by $\langle \partial T, \omega \rangle = \langle T, d\omega \rangle$ for any $\omega \in \mathcal{D}^{p-1}(\mathbf{R}^m)$, where $d\omega$ denotes the exterior differential of the differential form ω. The mass of the p-dimensional current T in the open set $\Omega \subset \mathbf{R}^m$ is defined by

$$\|T\|(\Omega) = \sup\{\langle T, \omega \rangle,\ \omega \in \mathcal{D}^p(\mathbf{R}^m),\ \sup_z |\omega(z)| \leq 1,\ \mathrm{supp}\omega \subset \Omega\}.$$

We are now ready to present the following conjecture.

Conjecture 7. Let $A \subset \mathbf{R}^n$ be open, and let $w \in [L^1(A)]^k$, $E \subset A$; $W_1, W_2 \subset A \times \mathbf{R}^k$, with:

$$\mathcal{H}^n(A \setminus E) = 0, \quad \text{and} \quad W_1 = \{(x, w(x)),\ x \in E\},$$

$$\mathcal{H}^n(\{x \in A, \exists y \in \mathbf{R}^k : (x,y) \in W_2\}) = 0;$$

let further be $\nu : W_2 \to \mathbf{N}$, and $T \in \mathcal{D}_n(\mathbf{R}^{n+k})$ such that

$$\|T\|(\Omega) = \mathcal{H}^n(W_1 \cap \Omega) + \int_{W_2 \cap \Omega} \nu d\mathcal{H}^n \qquad \text{for every } \Omega \subset A \times \mathbf{R}^k,$$

$$\|\partial T\|(A \times \mathbf{R}^k) = 0.$$

Then

$$\overline{F}(w, A) \leq \|T\|(A \times \mathbf{R}^k) = \mathcal{H}^n(W_1) + \int_{W_2} \nu d\mathcal{H}^n.$$

Remark 6. During the conference, E. Acerbi and G. Dal Maso have pointed out to me that the following result can be deduced from some results of M.Giaquinta–G.Modica–J.Souček.

Let $A \subset \mathbf{R}^n$ be open, and $w \in [L^1(A)]^k$ such that $\overline{F}(w, A) < +\infty$. Then, there exist E, W_1, W_2, ν, T as in conjecture 7 such that

$$\overline{F}(w, A) \geq \|T\|(A \times \mathbf{R}^k) = \mathcal{H}^n(W_1) + \int_{W_2} \nu d\mathcal{H}^n.$$

Remark 6. From the previous statement it easily follows conjecture 1 of [2].

References

[1] **F.Bethuel–H.Brezis–J.M.Coron:** *Relaxed energies for harmonic maps,* Preprint 1989.

[2] **E. De Giorgi:** *Riflessioni su alcuni problemi variazionali,* Proc. of the Conference "Equazioni differenziali e calcolo delle variazioni", Pisa, september 9–12, 1991, to appear.

[3] **H.Federer:** *Geometric Measure Theory.* Springer Verlag, Berlin, 1969.

[4] **H.Federer–W.Fleming:** *Normal and integral currents,* Annals of Math. **72** (1960), 458–520.

[5] **M.Giaquinta–G.Modica–J.Souček:** *Weak diffeomorphisms and nonlinear elasticity,* Arch. Rational Mech Anal., **106** (1989), 97–159, and *Erratum and Addendum,* ibid., **109** (1990), 385–392.

[6] **M.Giaquinta–G.Modica–J.Souček:** *Cartesian currents and variational problems for mappings into spheres,* Annali Scuola Norm. Sup., **16** (1989), 393–485.

[7] **M.Giaquinta–G.Modica–J.Souček:** *The Dirichlet energy of mappings with values into the sphere,* Manuscripta Math., **65** (1989), 489–507.

[8] **M.Giaquinta–G.Modica–J.Souček:** *Partial regularity of cartesian currents which minimize certain variational integrals,* PDE and the Calculus of Variations, Essays in Honor of Ennio De Giorgi, Birkhäuser, Boston, 1989.

[9] **M.Giaquinta–G.Modica–J.Souček:** *Liquid crystals: relaxed energies, dipoles, singular lines and singular points,* preprint 1990.

GREEN'S IDENTITIES AND PENDENT LIQUID DROPS, I

Robert Finn

Mathematics Department
Stanford University
Stanford, CA 94305–2125 USA

Abstract: It is proved that the "bubble solutions" of the capillary equation, whose existence was shown in an earlier paper with Concus, converge, as the vertex height tends to negative infinity, to a solution of the equation which is a graph over the entire base plane with an isolated singularity. This partly settles our previous conjecture, and also provides a new and conceptually preferable proof for existence of such a singular solution. The discussion is based on a form of Green's Identity that has an independent interest.

1. MISE EN SCÈNE; THE IDENTITIES

In this paper we derive a particular form of Green's Identity and use it to prove, in part, an earlier conjecture. The identity applies specifically to configurations governed by an equation

$$\Delta \boldsymbol{x} = \kappa z \boldsymbol{N} \tag{1.1}$$

where Δ is the Laplace-Beltrami operator (intrinsic Laplacian) on a two dimensional orientable surface S. $\boldsymbol{x}(\alpha, \beta)$ in Euclidean space $\mathbb{R}^3$, κ a prescribed positive constant and $\boldsymbol{N}$ a unit normal to S. It will be assumed that S has a finite minimum $(0, 0, u_0)$ and that in an interval $u_0 < z < u_1$ every plane Π: $z = h$ cuts S in a finite number of simple closed arcs Σ, enclosing simple domains Ω on Π, and cutting off topological disks on S below Π. These requirements are more restrictive than they need be; they are designed for the particular application we have in mind. We choose $\boldsymbol{N}$ to be the continuation of the vector that points vertically downward at the minimum point. With this normalization, there follows immediately $u_0 \leq 0$. See Figure 1.

Every such configuration will be said to represent, in the interval $u_0 < z < u_1$, a *simple pendent liquid drop*, with the equation in normalized form. For background discussion, see §2.

For a particular h in the range considered, we restrict attention to a single component on S below Π (which we denote again by S) and we refer to Π as

Developments in Partial Differential Equations and Applications to Mathematical Physics, Edited by G. Buttazzo *et al.*, Plenum Press, New York, 1992

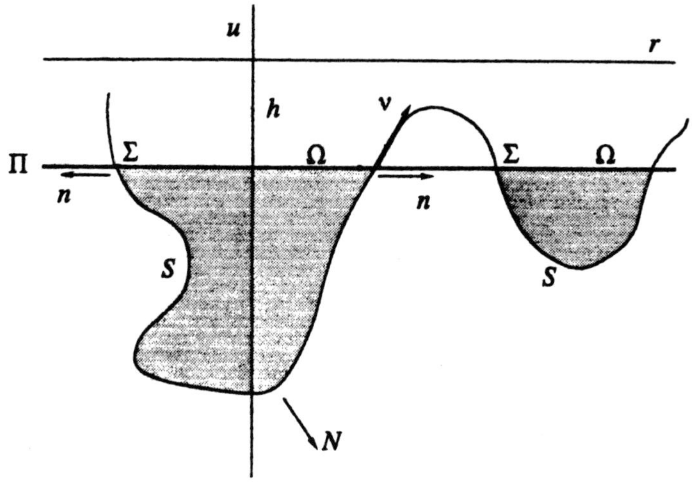

Figure 1. Configuration for Green's Identity

the *support plane*. We multiply (1.1) by $\boldsymbol{x}$ and integrate over S, obtaining, since $|\nabla \boldsymbol{x}|^2 \equiv 2$ on S,

$$
\begin{aligned}
\int_S \boldsymbol{x} \cdot \Delta \boldsymbol{x} dS = \kappa \int_S z\boldsymbol{x} \cdot \boldsymbol{N} dS &= -\int_S |\nabla \boldsymbol{x}|^2 dS + \oint_\Sigma \boldsymbol{x} \cdot \frac{\partial \boldsymbol{x}}{\partial \nu} ds \\
&= -2|S| + \oint_\Sigma \boldsymbol{x} \cdot \boldsymbol{\nu} ds
\end{aligned} \tag{1.2}
$$

where $\boldsymbol{\nu}$ is the conormal vector on Σ. We write

$$
\int_S z\boldsymbol{x} \cdot \boldsymbol{N} dS = \int_{S \cup \Omega} z\boldsymbol{x} \cdot \boldsymbol{N} dS - h \int_\Omega \boldsymbol{x} \cdot \boldsymbol{k} d\Omega = 4 \int_V z dV - h^2 |\Omega| \tag{1.3}
$$

by the divergence theorem and definition of Ω. Here $\boldsymbol{k}$ is the vertical unit vector $\langle 0, 0, 1 \rangle$. Also,

$$
\oint_\Sigma \boldsymbol{x} \cdot \boldsymbol{\nu} ds = \oint_\Sigma (\boldsymbol{x} - \boldsymbol{x} \cdot \boldsymbol{k}\boldsymbol{k}) \cdot \boldsymbol{\nu} ds + h \oint_\Sigma \boldsymbol{k} \cdot \boldsymbol{\nu} ds \tag{1.4}
$$

since $\boldsymbol{x} \cdot \boldsymbol{k} = h$ on Π. Letting $\boldsymbol{n}$ denote the unit exterior normal to Σ in Π, we find

$$
(\boldsymbol{x} - (\boldsymbol{x} \cdot \boldsymbol{k})\boldsymbol{k}) \cdot \boldsymbol{\nu} = \boldsymbol{x} \cdot \boldsymbol{n} \cos \gamma , \quad \boldsymbol{k} \cdot \boldsymbol{\nu} = \sin \gamma , \tag{1.5}
$$

where γ is the angle between S and Π on Σ; thus, we arrive at the identity

$$
\oint_\Sigma \boldsymbol{x} \cdot \boldsymbol{n} \cos \gamma ds + h \oint_\Sigma \sin \gamma ds - z|S| = 4\kappa \int_V z dV - \kappa h^2 |\Omega| . \tag{1.6}
$$

Observe now that the z-coordinate on S satisfies an equation

$$
\Delta z = \kappa z \boldsymbol{k} \cdot \boldsymbol{N}. \tag{1.7}
$$

Proceeding analogously as above, we obtain first

$$\int_S z\Delta z dS = -\int_S |\nabla z|^2 dS + \oint_\Sigma z\frac{\partial z}{\partial \nu}ds = \kappa\int_S z^2\boldsymbol{k}\cdot\boldsymbol{N}dS\,. \tag{1.8}$$

We have

$$\int_S z^2\boldsymbol{k}\cdot\boldsymbol{N}dS = \int_{S\cup\Omega} z^2\boldsymbol{k}\cdot\boldsymbol{N}dS - h^2|\Omega| = 2\int_V zdV - h^2|\Omega| \tag{1.9}$$

by the divergence theorem. Also

$$\oint_\Sigma z\frac{\partial z}{\partial \nu}ds = h\oint_\Sigma \sin\gamma ds \tag{1.10}$$

and we observe that

$$|\nabla z|^2 = 1 - \frac{1}{W^2} \tag{1.11}$$

on S, with $W^2 = 1 + |\nabla u|^2$, $u(x,y)$ being a local defining function for S; if S is vertical at the point considered, then $|\nabla z|^2 = 1$ at that point. We obtain the identity

$$\int_S \frac{1}{W^2}dS + h\oint_\Sigma \sin\gamma ds - |S| = 2\kappa\int_V zdV - \kappa h^2|\Omega|\,. \tag{1.12}$$

It is remarkable that the two terms adjacent to the equality sign in (1.12) are exactly half the corresponding terms in (1.6). Thus, if the double of (1.12) is subtracted from (1.6), both these terms (which in practice are difficult to estimate) disappear. *We are left with the identity*

$$\frac{1}{2}\oint_\Sigma \boldsymbol{x}\cdot\boldsymbol{n}\cos\gamma ds = \int_S \frac{1}{W^2}dS + \frac{1}{2}\left(\kappa h^2|\Omega| + h\oint_\Sigma \sin\gamma ds\right), \tag{1.13}$$

a relationship that will be central to the further discussion.

The surface integral appearing in (1.13) has a geometric connotation, as $1/W$ is the magnitude of the cosine of the angle ψ formed between the normal $\boldsymbol{N}$ to S and the vertical direction, see Figure 1.

The case $h = 0$ has a special interest (see §2), and (1.13) then yields a simple explicit expression for the surface integral. Additionally, if $\gamma \equiv$ const, then according to a theorem of Wente [1], S is rotationally symmetric about the z-axis. We see immediately from (1.13) that *the meridian curve C crosses the radial axis with positive slope* (for an estimate of that slope, see [2]). *If in addition S projects simply onto Ω, then (1.13) yields*

$$\cos\gamma = \frac{1}{|\Omega|}\int_\Omega \cos\psi d\Omega\,, \tag{1.14}$$

that is, $\cos\gamma$ *becomes the mean value of* $\cos\psi$ *over* Ω. Since γ is the value of ψ on Π, we conclude further *that C must have an inflection below* Π.

Beyond the identities just considered, a formal integration of Δz over S yields the relation

$$\oint_\Sigma \sin\gamma ds = \kappa V - \kappa h|\Omega| \tag{1.15}$$

2. PHYSICAL CONSIDERATIONS

We consider a connected drop of liquid, of prescribed volume V, suspended under a horizontal plate and in mechanical equilibrium in a vertically downward uniform gravity field (Figure 1). In a given coordinate frame, the principle of virtual work leads to an equation

$$\Delta \boldsymbol{x} = (\kappa z + \lambda)\boldsymbol{N} \tag{2.1}$$

for the free surface interface, and the condition that γ is prescribed, depending only on the materials at the points of contact. Here λ is a Lagrange parameter to be determined by the volume constraint, and $\kappa = \frac{\rho g}{\sigma}$, $\rho =$ density change across the free surface, $g =$ gravitational acceleration, $\sigma =$ surface tension. For details, see [3], Chapters 1 and 4. The transformation $z \to z - \lambda/\kappa$ puts (2.1) into the standard form (1.1), without changing γ. It follows that *every physical drop can be found among the solutions of (1.1).* This circumstance motivates the definitions in §1. If the liquid is homogeneous and the supporting plane of uniform material, then γ will be constant, and by [1] the configuration will be axially symmetric. Thus, in order to obtain all pendent drops with uniform boundary condition, it suffices to characterize all solutions of (1.1) with axial symmetry and with finite minimum u_0.

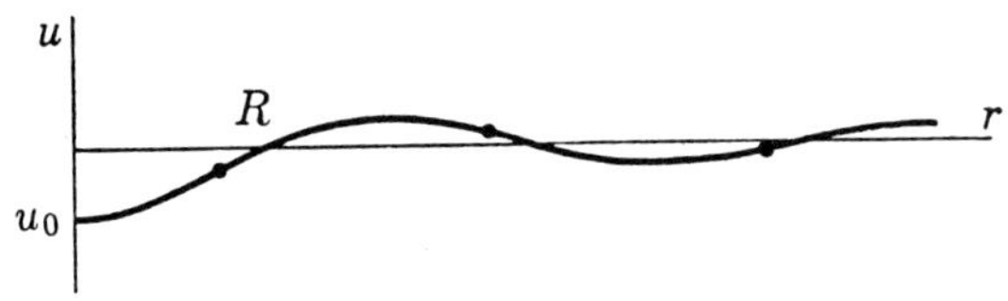

Figure 2. The case $u_0 > u_{0c}$; inflections

To a large extent, this program was completed in [4]. It was shown that for any initial point u_0 on the negative z-axis, there exists a unique globally embedded meridian curve $\mathcal{C}$, corresponding to a solution of (1.1). If $|u_0|$ is small enough, then $\mathcal{C}$ projects simply onto the r-axis. But as u_0 decreases past a critical value u_{0c}, a vertical appears on the profile, and eventually solutions appear that exhibit large numbers of "near bubbles" joined together by narrow necks. All solution profiles lie above the hyperbola $\kappa r u = -2$ until the r-axis is crossed, after which they oscillate out to infinity along that axis. Figures 2 and 3 illustrate the general character of the behavior, in the two cases, for which the projection on the r-axis is simple or multiple.

In an independent (and earlier) work [5], the existence was proved of a symmetric solution of (1.1) with simple projection onto the entire deleted base plane $r > 0$ and with an isolated singularity at $r = 0$. See Figure 4. Figure 5 shows computer plots of the cases $u_0 = -4, -8, -16$, compared with the singular solution. These plots led us to the conjecture that in any compact region, the "bubble" solutions converge uniformly to a singular solution of the type considered, as $u_0 \to -\infty$. In this paper we prove the conjecture.

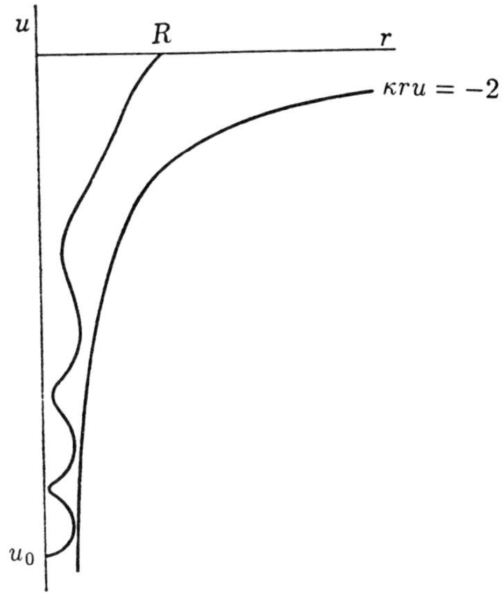

Figure 3. The case $u_0 < u_{0c}$; initial segment

3. THE BUBBLE PROFILES: GENERAL PROPERTIES

At non vertical points on a rotationally symmetric bubble surface, (1.1) takes the simple form, in terms of the local representation $u(r)$,

$$(r \sin \psi)_r = -\kappa r u\,; \tag{3.1}$$

here ψ can be interpreted as the inclination angle relative to the r-axis. Equivalently we may write

$$\frac{\sin \psi}{r} + (\sin \psi)_r = -\kappa u \tag{3.2}$$

which splits the mean curvature $-\kappa u/2$ into its latitudinal and meridional components.

At vertical points we replace (3.2) by

$$\frac{\sin \psi}{r} - (\cos \psi)_u = -\kappa u\,. \tag{3.3}$$

LEMMA 3.1. *The initial portion of a bubble profile, from the vertex $(0, u_0)$ to the initial crossing $(R, 0)$ with the r-axis, lies strictly between the u-axis and the hyperbola $\kappa r u = -2$, see Figure 3.*

This lemma is proved in [4], p. 318. □

LEMMA 3.2. *The vertical points on a bubble profile are isolated.*

Proof. At a vertical point, (3.3) takes the form

$$(\cos \psi)_u = \frac{1}{r}(1 + \kappa r u) \tag{3.4}$$

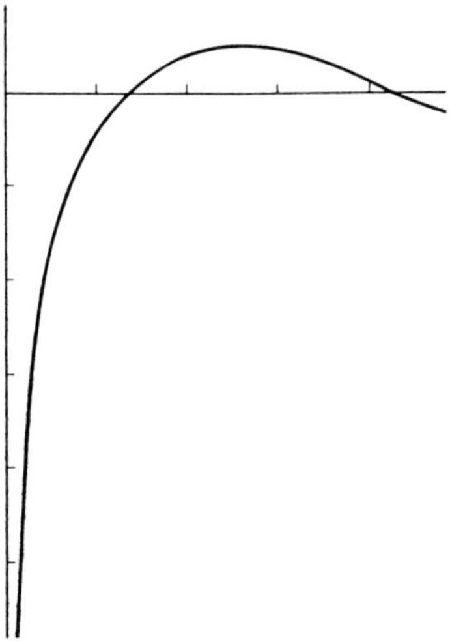

Figure 4. Singular solution

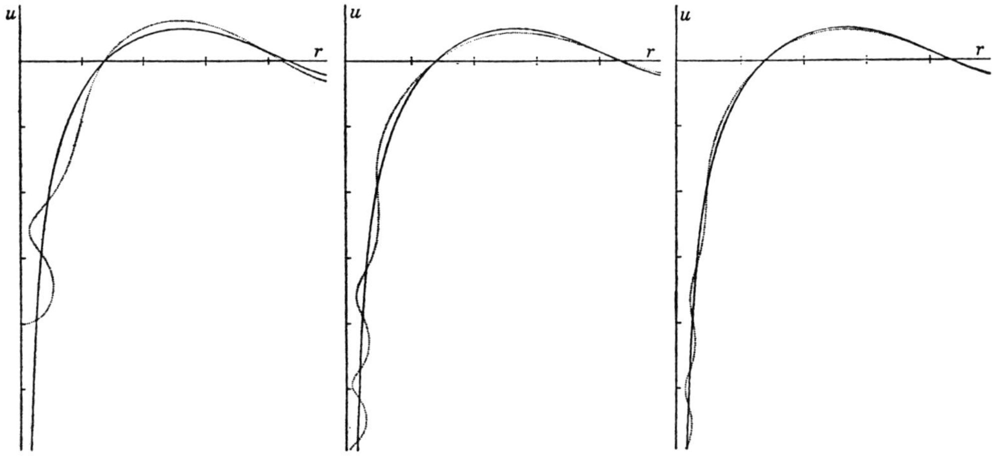

Figure 5. The cases $u_0 = -4, -8, -16$; singular solution

so that the result is immediate if $\kappa r u \neq -1$. If $\kappa r u = -1$ at the vertical, we find from (3.3)

$$(\cos\psi)_{uu} = \kappa > 0 \tag{3.5}$$

by hypothesis. □

LEMMA 3.3. *let (r_α, u_α) be a vertical point on a bubble profile. If S_α denotes the part of the surface S below u_α, there holds*

$$\kappa \int_{S_\alpha} \frac{1}{W^2} dS = -\frac{\pi}{2} \kappa r_\alpha u_\alpha (2 + \kappa r_\alpha u_\alpha). \tag{3.6}$$

The proof is immediate from (1.13). □

From Lemma 3.3 we find easily:

LEMMA 3.4. *Let* (r_α, u_α), (r_β, u_β) *be verticals on a bubble profile,* $u_\beta > u_\alpha$. *If* $S_{\alpha\beta}$ *denotes the part of* S *between the heights* u_α *and* u_β, *then*

$$(\kappa r_\beta u_\beta + 1)^2 = (\kappa r_\alpha u_\alpha + 1)^2 - \frac{2\kappa}{\pi}\int_{S_{\alpha\beta}} \frac{1}{W^2}dS\,. \tag{3.7}$$

That is, the successive verticals lie in successively narrower "hyperbolic regions" surrounding the hyperbola $\kappa r u = -1$.

COROLLARY 3.4. *Suppose a vertical point* (r_α, u_α) *of a bubble profile lies on the hyperbola* $\kappa r u = -1$. *Then there can be no verticals above* (r_α, u_α).

Proof. Immediate from (3.7).

4. THE COMPARISON SURFACES

We intend to estimate the behavior of solutions of (3.1) for large $|u|$ by comparing locally with profiles of rotation surfaces of constant mean curvature H. These surfaces are conveniently classified as nodoids, catenoids, and unduloids, of which our particular interest is in the unduloids. These are periodic (in z) surfaces without double points, and are completely determined by the vertical points (r_α, v_α), (r_β, v_β) on a period loop, see Figure 6.

Unduloid profiles were originally characterized by Delaunay [6] as the curves traced by the focal points of ellipses, as the ellipse is rolled along the z-axis. We obtain the curves analytically as solutions of

$$(r \sin\psi)_r = 2rH\,, \quad H > 0\,. \tag{4.1}$$

that are vertical at the given points. We may assume $r_\alpha < r_\beta$, $v_\alpha < v_\beta$. An integration from r_α to r_β yields

$$H = \frac{1}{r_\alpha + r_\beta} \tag{4.2}$$

while from the form

$$\frac{\sin\psi}{r} - (\cos\psi)_u = 2H \tag{4.3}$$

of (4.1), we see that an inflection appears at

$$r_i = \sqrt{r_\alpha r_\beta}\,. \tag{4.4}$$

Under our normalizations, we may suppose $0 < \psi \le \pi/2$. Solving for $r(\psi)$, we find

$$r = \frac{\sin\psi \pm \sqrt{k^2 - \cos^2\psi}}{2H}, \qquad k = \frac{r_\beta - r_\alpha}{r_\beta + r_\alpha}, \tag{4.5}$$

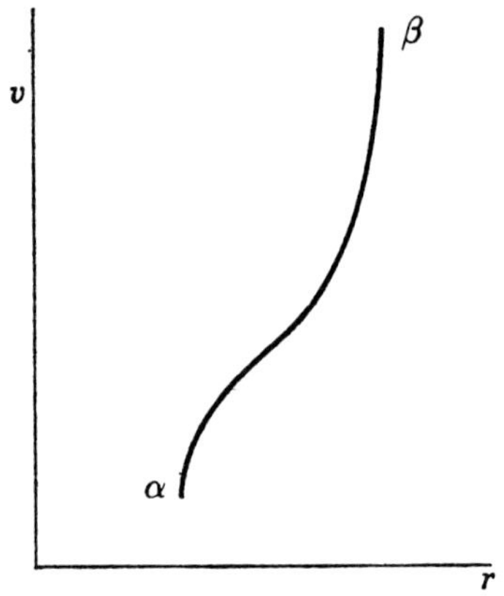

Figure 6. Unduloid

where the upper (lower) sign is to be chosen, according as $r > (<)\ r_i$. We see that

$$k = \cos \psi_i\,. \tag{4.6}$$

Setting $\cos \psi = k \sin \varphi$, $0 \leq \varphi \leq \pi/2$, and using $u_r = \tan \psi$, we can integrate (4.5) to obtain

$$v - v_\alpha = \frac{1}{2H}[-k \sin \varphi + \mathsf{E}(\varphi, k)] \tag{4.7}$$

when $v_\alpha \leq v \leq v_i$; here $\mathsf{E}(\varphi, k)$ is the incomplete elliptic integral of the second kind, of modulus k. If $v_i \leq v \leq v_\beta$, then

$$v - v_\alpha = \frac{1}{2H}\,[2\mathsf{E}(k) - k \sin \varphi - \mathsf{E}(\varphi, k)]\ , \tag{4.8}$$

$\mathsf{E}(k)$ being the complete elliptic integral of the second kind.

At the vertical points we find $\varphi = 0$, at the inflection $\varphi = \pi/2$. Setting $\varphi = 0$ in (4.8) yields the half period

$$v_\beta - v_\alpha = \frac{1}{H}\mathsf{E}(k)\,. \tag{4.9}$$

Since $\mathsf{E}(k)$ decreases monotonically from $\pi/2$ to 1 as k increases from 0 to 1, we obtain from (4.9) a property of the unduloids that will be essential for the estimates to follow:

LEMMA 4.1. *The period τ of any unduloid satisfies*

$$\frac{2}{H} \leq \tau \leq \frac{\pi}{H}\,. \tag{4.10}$$

Since the unduloids include as limiting cases the cylinders and spheres, (4.9) allows us to define a "period" for such surfaces, satisfying the bounds (4.10).

To complete the profile of Figure 6 to a full period, we need only reflect the figure in the line $v = v_\alpha$ (or $v = v_\beta$). The identical discussion applies to the reflected profile, with minor notational changes.

We specialize now the curvature H so that it corresponds to appropriate heights on a solution curve of (3.1), and we derive estimates for four configurations that will be used as comparison surfaces. The first two of these configurations are obtained directly from the above description:

LEMMA 4.2. *Suppose* $1 + \kappa r_\alpha u_\alpha = \varepsilon$, $0 < \varepsilon < 1$, $u_\alpha < 0$. *Then an unduloid* $v(r)$ *of mean curvature* $-\frac{1}{2}\kappa u_\alpha$ *can be found, which is vertical at* (r_α, u_α), (ρ_β, v_β), *with*

$$\rho_\beta = -\frac{1+\varepsilon}{\kappa u_\alpha}, \quad v_\beta = u_\alpha - \frac{2}{\kappa u_\alpha}\mathsf{E}(\varepsilon). \tag{4.11}$$

If $1 + \kappa r_\alpha u_\alpha = -\varepsilon$, $u_\alpha < 0$, *then an unduloid* $v(r)$ *of mean curvature* $-\frac{1}{2}\kappa u_\alpha$ *can be found, which is vertical at* (r_α, u_α), (ρ_β, v_β), *with*

$$\rho_\beta = -\frac{1-\varepsilon}{\kappa u_\alpha}, \quad v_\beta = u_\alpha - \frac{2}{\kappa u_\alpha}\mathsf{E}(\varepsilon), \tag{4.12}$$

In both cases, $|\cos\psi| < \varepsilon$ *throughout the trajectory.*

The interest in the result stems from the fact that $-\frac{1}{2}\kappa u$ is exactly the mean curvature of the "bubble solution" at the height u. We are interested in behavior for large $|u|$, where the terms involving $\mathsf{E}(\varepsilon)$ in (4.11) and in (4.12) are (relatively) small. For large $|u|$ we obtasin comparison surfaces in the opposite direction as follows:

LEMMA 4.3. *Given* $\varepsilon_0 > 0$. *Suppose* $1 + \kappa r_\alpha u_\alpha = \varepsilon$, *with* $\varepsilon_0 \le \varepsilon < 1$. *If* $u_\alpha < 0$ *such that* $\kappa u_\alpha^2 > 8\mathsf{E}(0) = 4\pi$ *then there is an unduloid* $w(r)$ *which is vertical at* (r_α, u_α), (σ_β, w_β), *with mean curvature* $-\frac{\kappa}{2}w_\beta$, *such that*

$$\sigma_\beta = -\frac{2}{\kappa w_\beta} - r_\alpha \tag{4.13}$$

$$w_\beta = u_\alpha - \frac{2\mathsf{E}(k)}{\kappa w_\beta} \tag{4.14}$$

$$k = \frac{\sigma_\beta - r_\alpha}{\sigma_\beta + r_\alpha} = 1 + \kappa r_\alpha w_\beta = \varepsilon + \frac{2(1-\varepsilon)\mathsf{E}(k)}{\kappa u_\alpha w_\beta}, \tag{4.15}$$

and such that

$$w_\beta < u_\alpha - \frac{\pi}{\kappa u_\alpha} - \frac{4\pi^2}{\kappa^2 u_\alpha^3} \tag{4.16}$$

and such that

$$0 < \cos\psi < \varepsilon + \frac{\pi(1-\varepsilon)}{\kappa u_\alpha^2 - \pi - \dfrac{4\pi^2}{\kappa u_\alpha^2}} \tag{4.17}$$

throughout the trajectory.

We note that the conditions imply that $u_\alpha < w_\beta < 0$, $\sigma_\beta > r_\alpha$.

Proof. We seek a solution pair (k, w_β) of (4.14) and (4.15) from among the lower intersections, in the (z, w_β) plane, of the line $z = w_\beta - u_\alpha$ with the hyperbolae $z = -\frac{2\mathsf{E}(k)}{\kappa w_\beta}$, $k \in [0,1]$, see Figure 7. If $\kappa u_\alpha^2 > 8\mathsf{E}(0) = 4\pi$, then the line will meet all hyperbolae in the family, thus providing a continuous mapping $\mathcal{F}(k) : k \to w_\beta$ of $[0,1]$ to the values of w at the intersections of the line with the k-hyperbola.

We have

$$\mathcal{F}(k) = u_\alpha - \frac{2\mathsf{E}(k)}{\kappa w_\beta} = w_\beta \tag{4.18}$$

and thus

$$w_\beta = \frac{1}{2} u_\alpha \left[1 + \sqrt{1 - \frac{8\mathsf{E}(k)}{\kappa u_\alpha^2}} \right] \tag{4.19}$$

From (4.19) follows, since $1 + \kappa r_\alpha u_\alpha = \varepsilon$,

$$0 < 1 + \kappa r_\alpha w_\beta < 1; \tag{4.20}$$

thus the mapping

$$\Phi(k) \equiv 1 + \kappa r_\alpha \mathcal{F}(k)$$

takes the unit interval continuously into itself, and necessarily has a fixed point. Corresponding to that value of k and $w_\beta = \mathcal{F}(k)$, we now construct an unduloid through (r_α, u_α) with mean curvature $-\frac{\kappa}{2} w_\beta$, and verify immediately that (4.13) holds. The relation (4.16) follows from (4.19) and the general inequality $\sqrt{1-x} \geq 1 - \frac{1}{2}x - \frac{1}{2}x^2$ in the range $0 \leq x \leq 1$; the final relation (4.17) is obtained since the maximum of $\cos\psi$ occurs at the inflection ψ_i, and $\cos\psi_i = k$.

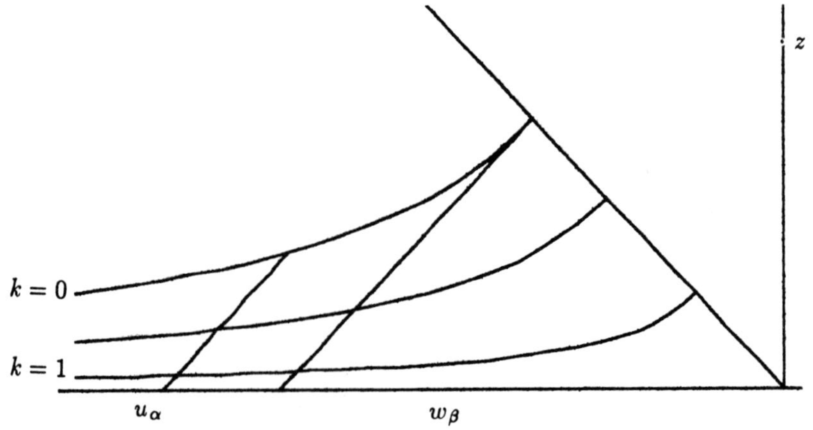

Figure 7. Proof of Lemma 4.3

LEMMA 4.4. *Given $\varepsilon_0 > 0$. Suppose $1 + \kappa r_\alpha u_\alpha = -\varepsilon$, with $\varepsilon_0 \le \varepsilon < 1$, $u_\alpha < 0$. Then if $\kappa u_\alpha^2 > \frac{1+\varepsilon}{\varepsilon}\frac{\pi}{2}\left[1+\sqrt{1+\frac{16\varepsilon}{1+\varepsilon}}\right]$ there is an unduloid $v(r)$ which is vertical at (r_α, u_α), (ρ_β, v_β), with mean curvature $-\frac{\kappa}{2}v_\beta$, such that*

$$\rho_\beta = \frac{-2}{\kappa v_\beta} - r_\alpha \tag{4.21}$$

$$v_\beta = u_\alpha - \frac{2\mathsf{E}(k)}{\kappa v_\beta} \tag{4.22}$$

$$k = \frac{r_\alpha - \rho_\beta}{r_\alpha + \rho_\beta} = -\kappa r_\alpha v_\beta\,, \quad 0 < k < 1\,, \tag{4.23}$$

and such that

$$u_\alpha < v_\beta < u_\alpha - \frac{\pi}{\kappa u_\alpha} - \frac{4\pi^2}{\kappa^2 u_\alpha^3}\,. \tag{4.24}$$

As before, we find $u_\alpha < w_\beta < 0$.

Proof. We begin as in the proof of Lemma 4.3, seeking a solution pair (k, v_k) of (4.22), (4.23). We obtain the mapping $\mathcal{F}(k)$, and construct the mapping $\Phi(k) = -1 - \kappa r_\alpha \mathcal{F}(k)$ as before. Again we find

$$v_\beta = \frac{1}{2}u_\alpha\left[1+\sqrt{1-\frac{8\mathsf{E}(k)}{\kappa u_\alpha^2}}\right]\,, \tag{4.25}$$

from which (4.24) follows whenever $\kappa u_\alpha^2 > 8\mathsf{E}(k)$, a condition that holds for any $\varepsilon \in [0,1]$ under the hypothesis. From (4.25) we obtain

$$\varepsilon + \frac{2\kappa r_\alpha u_\alpha}{\kappa u_\alpha^2}\mathsf{E}(k)\left[1 + \frac{8}{\kappa u_\alpha^2}\mathsf{E}(k)\right] < \Phi(k) < \varepsilon \tag{4.26}$$

and thus $\Phi(k)$ maps into $[0,1]$ if

$$\frac{2\kappa r_\alpha u_\alpha}{\kappa u_\alpha^2}\mathsf{E}(k)\left[1 + \frac{8}{\kappa u_\alpha^2}\mathsf{E}(k)\right] > -\varepsilon\,, \tag{4.27}$$

a condition which is guaranteed by the hypothesis, since $\mathsf{E}(k) \le \mathsf{E}(0) = \pi/2$. The remainder of the proof now follows that of Lemma 4.3. □

It is not an accident of the method that the hypothesis on κu_α^2 depends on ε in Lemma 4.4 but not in Lemma 4.3. One sees by simple examples that the requirements in these lemmas cannot be significantly inproved.

5. THE BUBBLE SOLUTIONS; LOCAL PROPERTIES

We proceed to use unduloids as constructed above as comparison surfaces, in order to control the behavior of a pendent drop "bubble" solution. We consider initially such a solution, near a vertical point (r_α, u_α) at which $\psi = \pi/2$, $1+\kappa r_\alpha u_\alpha = \varepsilon$, $0 < \varepsilon < 1$. By (3.4) the vertical is isolated and there is an interval $r_\alpha < r < r_\beta$, $u_\alpha < u < u_\beta$ in which the profile can be described by a solution $u(r)$ of (3.1). We shall show that if $\kappa u_\alpha^2 > 4\pi$ then r_β must be finite, that is, r_β can be chosen so that $\psi = \pi/2$ at (r_β, u_β). By Lemmas (4.2) and (4.3) we can find unduloids $v(r)$ and $w(r)$, satisfying respectively

$$(r \sin\varphi)_r = -\kappa r u_\alpha \tag{5.1}$$

$$(r \sin\omega)_r = -\kappa r w_\beta \tag{5.2}$$

such that $v(r_\alpha) = w(r_\alpha) = u_\alpha$, $\varphi(r_\alpha) = \omega(r_\alpha) = \pi/2$, and v, w become vertical again at points (ρ_β, v_β), (σ_β, w_β) as in the lemmas (see Figure 8).

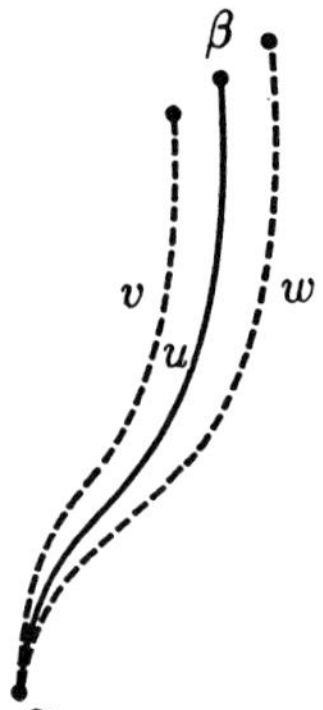

Figure 8. Comparison with unduloids; outgoing case

We have $(r\sin\psi)_r = -\kappa u$ and hence

$$r(\sin\psi - \sin\varphi) = -\int_{r_\alpha}^{r} \rho(u - u_\alpha)d\rho < 0 \tag{5.3}$$

along any common interval of continuation. Thus, the "bubble" solution is less steep, lies below $v(r)$, and necessarily extends beyond ρ_β. Similarly

$$r(\sin\psi - \sin\omega) = -\int_{r_\alpha}^{r} \rho(r - w_\beta)d\rho > 0 \tag{5.4}$$

on any interval in which $u \leq w_\beta$. There follows by (4.17) that when $r = \rho_\beta$ on the solution curve,

$$\cos\psi < \varepsilon + \frac{(1-\varepsilon)\pi}{\kappa u_\alpha^2 - \pi - \frac{4\pi^2}{\kappa u_\alpha^2}} < \varepsilon + \frac{\pi}{\kappa u_\alpha^2} \tag{5.5}$$

for sufficiently large $|u_\alpha|$, depending on ε.

We consider the interval $r > \rho_\beta$ on the solution curve, in which $u < v_\beta - \dfrac{2}{\kappa u_\alpha}$. In this interval we have

$$
(5.6)\qquad \begin{aligned}
(\cos\psi)_u &= \kappa u + \frac{\sin\psi}{r} \\
&< \kappa u_\alpha - \frac{2\mathsf{E}(\varepsilon)}{u_\alpha} + \frac{1}{\rho_\beta} - \frac{2}{u_\alpha} \\
&= \frac{\varepsilon}{1+\varepsilon}\kappa u_\alpha - \frac{2\mathsf{E}(\varepsilon)+2}{u_\alpha}
\end{aligned}
$$

and thus, setting $u_\beta = u(\rho_\beta)$,

$$
\begin{aligned}
\cos\psi &< \varepsilon + \frac{\pi}{\kappa u_\alpha^2} + \frac{\varepsilon}{1+\varepsilon}\kappa u_\alpha(u-u_\beta) - \frac{2\mathsf{E}+2}{u_\alpha}(u-u_\beta) \\
&< \varepsilon + \frac{\pi}{\kappa u_\alpha^2} + \frac{\varepsilon}{1+\varepsilon}\kappa u_\alpha(u-v_\beta) - 2\frac{\mathsf{E}(\varepsilon)+1}{u_\alpha}(u-v_\beta)
\end{aligned}
$$

for large enough $|u_\alpha|$. It follows that $\cos\psi$ must vanish (that is, the solution becomes vertical) at a value u satisfying

$$
(5.7)\qquad u < v_\beta - \frac{1+\varepsilon}{\kappa u_\alpha}\left\{\frac{1+\dfrac{\pi}{\varepsilon\kappa u_\alpha^2}}{1-2\dfrac{\mathsf{E}+1}{\varepsilon\kappa u_\alpha^2}(1+\varepsilon)}\right\}
$$

which lies within the required range when $|u_\alpha|$ is large.

The above discussion yields conditions under which an "inner" vertical (r_α, u_α) must be followed by an "outer" vertical, at a larger u and at a distance $O\left(\dfrac{-1}{\kappa u_\alpha}\right)$. In going from an "outer" to an "inner" vertical, special consideration has to be given to the case in which ε is near unity, in order to account for the non-uniformity in behavior (formation of sharp necks) near the vertical axis in that case. We distinguish the case $\varepsilon > 0.9$, and suppose that $1 + \kappa r_\alpha u_\alpha = -\varepsilon$. By Lemma 4.4, if κu_α^2 is large there is an unduloid $v(r)$ which has a vertical at (ρ_β, v_β) satisfying (4.21) and (4.22). We have

$$
(5.8)\qquad \rho_\beta = \frac{-1+\varepsilon}{\kappa u_\alpha} + O\left(\frac{1}{\kappa^2 |u_\alpha|^3}\right).
$$

Also

$$
(5.9)\qquad r(\sin\psi - \sin\varphi) = \kappa \int_r^{r_\alpha} \rho(u - v_\beta)\,dr < 0
$$

in the interval for which $u < v_\beta$, from which follows that the solution curve extends back at least to ρ_β, see Figure 9.

The curve is bounded below by the unduloid of Lemma 4.2, and cannot extend to the left beyond the vertical of that unduloid, but we no longer obtain a significant lower bound on $\sin\psi(\rho_\beta)$ by that comparison, as that unduloid becomes a circular arc centered on the u-axis when $\varepsilon \to 1$. Suppose however that on the

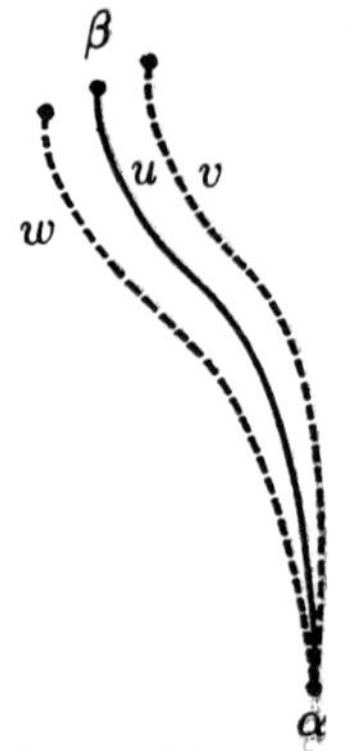

Figure 9. Comparison with unduloids; returning case

continuation to the left of ρ_β there were to hold everywhere $\sin\psi < 1/2$. Then the curve would meet the u-axis at a height

$$u_a < v_b + \sqrt{3}\rho_b \tag{5.10}$$

which is not possible. Thus, at a height $\le u_a$ there is a point with $r \le \rho_b$ at which $\sin\psi \ge 1/2$. At this point

$$\begin{aligned}\frac{\sin\psi}{r} &> -\frac{\kappa u_\alpha}{2(1-\varepsilon)} + O(1)\\ &= -5\kappa u_\alpha + O(1)\,.\end{aligned} \tag{5.11}$$

Thus

$$\begin{aligned}(\cos\psi)_u = \frac{\sin\psi}{r} + \kappa u &> -5\kappa u_\alpha + \kappa u_\alpha + O(1)\\ &= -4\kappa u_\alpha + O(1)\,.\end{aligned} \tag{5.12}$$

Since $\frac{\pi}{2} < \psi < \pi$ at the point, (5.12) implies that $\sin\psi$ is increasing there as the curve is traversed upward. But r is decreasing, thus $\frac{\sin\psi}{r}$ is increasing. Since $|u|$ is decreasing, we find that $(\cos\psi)_u$ becomes still more positive, and thus that

$$(\cos\psi)_u > -4\kappa u_\alpha + C \tag{5.13}$$

on the continuation, for some constant C depending only on ε. Integration from $u_\beta = u(\rho_\beta)$ to $u > u_\beta$ yields

$$\begin{aligned}\cos\psi &> \cos\psi_\beta - (4\kappa u_\alpha - C)(u - u_\beta)\\ &> -\frac{1}{2}\sqrt{3} - (4\kappa u_\alpha - C)(u - v_\beta)\end{aligned} \tag{5.14}$$

for sufficiently large $|u_\alpha|$. Thus at a height

$$u < -\frac{1}{2}\sqrt{3}\frac{1}{4\kappa u_\alpha - C} = -\frac{\sqrt{3}}{8\kappa u_\alpha} + O\left(\frac{1}{\kappa u_\alpha^2}\right)$$

there will occur another vertical point, at a radial distance $r < \rho_\beta$.

Finally, we suppose an arc returning from an "outer" vertical (r_α, u_α), with $1 + r_\alpha u_\alpha = -\varepsilon$, $0 < \varepsilon \leq 0.9$. The curve comes within a distance ρ_β of the origin, as given by (4.21), and its slope exceeds that of the returning unduloid of Lemma 4.2. A formal calculation based on (4.5) yields

$$\sin\psi(\rho_\beta) = 1 - O\left(\frac{1}{\kappa u_\alpha^2}\right),$$

and the further disucssion follows that of the preceding case.

We collect the above discussion in a single statement:

LEMMA 5.1. *Given $\varepsilon > 0$, suppose a solution curve is vertical at (r_α, u_α) with $1 + r_\alpha u_\alpha = \pm\varepsilon$. There exist $C > 0$ and $C_1 > 0$ depending only on ε, such that if $\kappa u_\alpha^2 > C$, then another vertical appears at (r'_α, u'_α), with $1 + r'_\alpha u'_\alpha = \mp\varepsilon + O\left(\frac{1}{\kappa u_\alpha^2}\right)$ and $u_\alpha < u'_\alpha < u_\alpha - \dfrac{C_1}{\kappa u_\alpha}$.*

We are now prepared to discuss the global behavior of the family of solution curves, as $u_0 \to -\infty$.

6. THE BUBBLE SOLUTIONS: GLOBAL PROPERTIES

LEMMA 6.1. *Let $A > 0$. Then as $u_0 \to -\infty$, the initial segment of the solution curve, joining $(0, u_0)$ to $(R, 0)$, remains uniformly bounded from the u-axis in the interval $0 \geq u \geq -A$.*

Proof. Suppose the lemma false. Since the curve crosses the r-axis with positive slope ([4], p. 319), there would either be a sequence of vertical points approaching the axis within the given interval, or else there would be a sequence of solutions in the family, for which $0 < \psi_A < \pi/2$, $r_A \to 0$. Thus there must be a point (r_α, u_α) in the interval $A > u > A - \sqrt{3}\, r_A$ at which $\sin\psi_\alpha > \frac{1}{2}$, $r_\alpha < r_A$.

We write $1 + \kappa r_\alpha u_\alpha = \varepsilon$, and we may assume $0.9 < \varepsilon < 1$. Thus, at (r_α, u_α) we have

$$(6.1) \qquad \begin{aligned} (\cos\psi)_u &\equiv \kappa u + \frac{\sin\psi}{r} \\ &> -4\kappa u_\alpha > 0, \end{aligned}$$

from which follows that $\cos\psi$ is decreasing, and $\sin\psi$ increasing as the curve is traversed downward from (r_α, u_α). Clearly r is decreasing; thus, sufficiently near (r_α, u_α) we certainly have

$$(6.2) \qquad (\cos\psi)_u > \kappa u - 5\kappa u_\alpha > 0$$

which shows that $\sin\psi$ continues to increase, and it follows that (6.2) holds at least in the range $u_\alpha > u > 5u_\alpha$.

We consider the interval $u_\alpha > u > u_\alpha + \frac{\sqrt{3}}{4\kappa u_\alpha}$. We may assume A sufficiently large that this interval lies in the preceding one just introduced. Integrating (6.2) from u to u_α we find

$$\begin{aligned} \cos\psi &< \frac{1}{2}\sqrt{3} + \kappa\left[\frac{u+u_\alpha}{2} - 5u_\alpha\right](u - u_\alpha) \\ &< \frac{1}{2}\sqrt{3} + \kappa\left[-4u_\alpha + \frac{\sqrt{3}}{8\kappa u_\alpha}\right](u - u_\alpha) \end{aligned} \tag{6.3}$$

which becomes negative at a value interior to the required interval.

Putting together what has been done till now, we see that if A is large enough, and $|u_0|$ large enough that $r_A < \frac{1}{10\kappa A}$, then a vertical must occur at a height $\geq -\left(A + \frac{\sqrt{3}}{2\kappa A}\right)$. This means we may assume a sequence of verticals approaching the axis, at heights uniformly bounded below. We consider a representative such vertical, at (r_α, u_α), $1 + r_\alpha u_\alpha = \varepsilon$, $0.9 < \varepsilon < 1$. A reasoning analogous to that of Lemma 4.3 assures us of the existence of an unduloid $w(r)$, vertical at (r_α, u_α) and at (σ_β, w_β), with

$$\sigma_\beta = -\frac{2}{\kappa w_\beta} - r_\alpha \tag{6.4}$$

$$w_\beta = u_\alpha + \frac{2\mathsf{E}(k)}{\kappa w_\beta} \tag{6.5}$$

(The restriction $\kappa u_\alpha^2 > 4\pi$ is not needed in this case.)

Since the mean curvature of this unduloid is exactly that of the solution at the height $u = w_\beta$, the unduloid lies below the solution, and its inclination angle φ satisfies $\varphi < \psi$ at each r in the range $r_\alpha < r < \sigma_\beta$ (that is, the solution is less steep). This shows, in particular, that the solution curve below u_α extends outward at least to σ_β, which in turn is bounded below for fixed A. Further, as $r_\alpha \to 0$, a subsequence of the family of unduloids thus constructed tends, uniformly with all derivatives on any interval $a < r < \sigma_\beta - a$, to a lower quarter circle. Denoting the corresponding portions of the original solution surfaces by S_a and the annuli onto which they project on the base plane by Ω_a, we obtain

$$\begin{aligned} \int_{S_a} \frac{1}{W^2}\,dS &= \int_{\Omega_a} \frac{1}{W}\,d\Omega = \int_{\Omega_a} |\cos\psi|\,d\Omega \\ &> \int_{\Omega_a} |\cos\varphi|\,d\Omega = C > 0 \end{aligned} \tag{6.6}$$

where C depends only on A, as $u_0 \to -\infty$. But if S_α denotes the portion of S up to the "neck" (r_α, u_α), then certainly

$$\int_{S_\alpha} \frac{1}{W^2}\,dS > \int_{S_a} \frac{1}{W^2}\,dS\,. \tag{6.7}$$

On the other hand, we obtain immediately from (1.13) that as $r_\alpha \to 0$

$$\int_{S_\alpha} \frac{1}{W^2}\,dS \to 0\,. \tag{6.8}$$

This contradiction establishes the lemma. □

LEMMA 6.2. *There exists* $\delta(A) > 0$ *such that on the segment of the solution curves considered in Lemma* 6.1 *there holds* $|u_r| > \delta$, *as* $u_0 \to -\infty$.

Proof. Consider any "neck" (r_α, u_α) in the interval. We introduce a catenoid (mean curvature zero) whose neck coincides with the given one. Then, on the continuations of the solution to the next verticals above and below, the solution curve lies outside the catenoid (see the discussion in §5) and has, for any given r, a larger inclination. The inclination of the catenoid is bounded from zero since r_α is bounded from zero by Lemma 6.1. Since every point of the solution curve is accessible in this way, we obtain the result. □

By Lemma 6.2, the portion of the solution curve in the interval considered can be viewed as a function $r(u)$, satisfying the equation

$$\frac{r}{(1+r_u^2)^{3/2}} r_{uu} - \left(\frac{1}{\sqrt{1+r_u^2}} + \kappa r u \right) = 0 . \tag{6.9}$$

Since, by Lemma 6.1, r is bounded from zero, and by Lemma 6.2 $|r_u|$ bounded from infinity, we obtain from (6.9) also a bound on $|r_{uu}|$, and, subsequently, of all higher derivatives, depending only on A, as $u_0 \to -\infty$. Thus we obtain:

LEMMA 6.3. *There is a subsequence of solutions of* (6.9) *that converges to a solution* $R(u)$.

Our task is to show that $R'(u) \neq 0$ and that $R(u)$ has the expected asymptotic properties. In what follows, we restrict attention to such a subsequence. It will suffice for the first step to show that with increasing $|u_0|$, verticals eventually disappear from any solution segment as considered in Lemma 6.1.

LEMMA 6.4. *On segments as considered in Lemma* 6.1, *for given* $\varepsilon > 0$, *all verticals satisfy* $|1 + \kappa r u| < \varepsilon$ *for all sufficiently large* $|u_0|$, *depending only on* ε.

Proof. Suppose that for some $\varepsilon > 0$ there were a sequence of values $u_0 \to -\infty$, for each of which a vertical (r_α, u_α) would appear, with $|1 + r_\alpha u_\alpha| > \varepsilon$, and $0 > u_\alpha \geq -A$. By Lemma 3.4, $|1 + ru| > \varepsilon$ at every vertical for which $u < u_\alpha$.

Choose $|u_0|$ so large that $\sqrt{\kappa}|u_0| > C$, where C is the constant of Lemma 5.1. Then that lemma applies to every vertical in the interval $\left[-\sqrt{|u_0|}, u_0\right]$. We wish to apply to this configuration the Green's Identity (1.13); to do so we need an estimate for the surface integrals that appear:

LEMMA 6.5. *Let* $v(r)$ *be an unduloid as defined in Lemma* 4.2, *with* $\rho_\beta > r_\alpha$, *and let* φ *denote its inclination angle. Then*

$$\begin{aligned} \int_{r_\alpha}^{\rho_\beta} r \cos\varphi \, dr &= (\rho_\beta - r_\alpha)^2 \int_0^{\pi/2} \sin^2\tau \sqrt{1 - k^2 \sin^2\tau} \, d\tau \\ &= \frac{4\varepsilon^2}{(\kappa u_\alpha)^2} \int_0^{\pi/2} \sin^2\tau \sqrt{1 - k^2 \sin^2\tau} \, d\tau \end{aligned} \tag{6.10}$$

with $k = \varepsilon$. Let $v(r)$ be an unduloid as defined in Lemma 4.4. Then

$$\left| \int_{\rho_\beta}^{r_\alpha} r \cos\varphi dr - \frac{4\varepsilon^2}{(\kappa u_\alpha)^2} \int_0^{\pi/2} \sin^2\tau \sqrt{1 - k^2 \sin^2\tau}\, d\tau \right| < \frac{C}{\kappa^3 u_\alpha^4} \tag{6.11}$$

for some constant C and $0 < k < 1$, for all $|u_\alpha|$ large enough, depending on ε.

The proof is obtained easily from the representations for the unduloids given in §4.

Returning to the proof of Lemma 6.3, we estimate the surface integral of $1/W^2$ on the solution surface between two successive verticals at r_α, r_β in $\left[-\sqrt{|u_0|},\, u_0\right]$. To do so, we observe that

$$\int_{S_{\alpha\beta}} \frac{1}{W^2} dS = 2\pi \left| \int_{r_\alpha}^{r_\beta} \cos\psi\, r dr \right| > 2\pi \left| \int_{r_\alpha}^{\rho_\beta} r \cos\varphi dr \right| \tag{6.12}$$

where φ corresponds to an appropriately chosen unduloid comparison curve. We conclude

$$\kappa \int_{S_{\alpha\beta}} \frac{1}{W^2} dS > C \frac{\varepsilon^2}{\kappa u_\alpha^2} \tag{6.13}$$

for some constant C.

We estimate from below the sum of these terms in the indicated interval. To do so, we first note from Lemma 5.1 that in an iterval of (small) size δu, there appear at least $C\sqrt{\kappa}|u|$ successive verticals, to each of which (6.13) applies. Thus the part S_0 of the surface up to $-\sqrt{|u_0|}$ yields

$$\kappa \int_{S_0} \frac{1}{W^2} dS > C(\varepsilon) \int_{u_0}^{-\sqrt{|u_0|}} \frac{-1}{u} du > C \ln|u_0| \tag{6.14}$$

$C > 0$ depending only on ε, and hence for the portion S_α of S up to (r_α, u_α) we find

$$\kappa \int_{S_\alpha} \frac{1}{W^2} dS > \kappa \int_{S_0} \frac{1}{W^2} dS > C \ln|u_0| \tag{6.15}$$

as $u_0 \to -\infty$. But Lemma 3.6 yields

$$\begin{aligned} \kappa \int_{S_\alpha} \frac{1}{W^2} dS &= -\frac{\pi}{2} \kappa r_\alpha u_\alpha (\kappa r_\alpha u_\alpha + 2) \\ &\leq \frac{\pi}{2}(1 - \varepsilon^2) \end{aligned} \tag{6.16}$$

according to the hypothesis. This contradiction completes the proof. □

From Lemma 6.4 we see that if verticals persist in any fixed compact region as $u_0 \to -\infty$, then they must all tend to the hyperbola $\kappa r u = -1$, and the limiting configuration of Lemma 6.3 would have one or more verticals on that hyperbola. By Corollary 3.4, at most one such vertical can occur.

In fact, it is possible for a vertical to appear on the hyperbola; that happens when $u_0 = u_{0c}$, the largest value of u_0 for which any verticals can appear. But it cannot happen for the limiting configuration.

To see that, we observe first that since at most one vertical can appear, the limiting configuration defines in a neighborhood of $r = 0$ a surface $u(r)$ satisfying (3.1), with an isolated singularity at the origin and which becomes vertical at (r_α, u_α), $\kappa r_\alpha u_\alpha = -1$. By [7], p. 149 we find

$$u(r) = -\frac{1}{\kappa r} + O(r) \tag{6.17}$$

at $r \to 0$. Also, $\sin\psi \to 1$ as $r \to 0$. Writing (1.13) for members of the approximating sequence with base plane at u_α, we find

$$\kappa r_\alpha^2 \cos\gamma_\alpha = \frac{\kappa}{\pi}\int_{S_\alpha} \frac{1}{W^2}dS + \frac{1}{2}\left(\kappa^2 r_\alpha^2 u_\alpha^2 + 2\kappa r_\alpha u_\alpha \sin\gamma_\alpha\right) \tag{6.18}$$

For the same u_0 we choose the base plane at $u < u_\alpha$ and obtain

$$\kappa r^2 \cos\gamma = \frac{\kappa}{\pi}\int_{S_u} \frac{1}{W^2}dS + \frac{1}{2}\left(\kappa^2 r^2 u^2 + 2\kappa r u \sin\gamma\right). \tag{6.19}$$

Subtracting (6.19) from (6.18) and denoting the tubular piece joining the base planes by $S_{\alpha u}$, we have

$$\begin{aligned}\kappa\left(r_\alpha^2\cos\gamma_\alpha - r^2\cos\gamma\right) = \frac{\kappa}{\pi}\int_{S_{\alpha u}} \frac{1}{W^2}dS &+ \frac{1}{2}\left(\kappa^2 r_\alpha^2 u_\alpha^2 + 2\kappa r_\alpha u_\alpha \sin\gamma_\alpha\right)\\ &- \frac{1}{2}\left(\kappa^2 r^2 u^2 + 2\kappa r u \sin\gamma\right).\end{aligned} \tag{6.20}$$

We let now $u_0 \to -\infty$. By the choice of u_α there follows $\cos\gamma_\alpha \to 0$, $\sin\gamma_\alpha \to 1$, $\kappa r_\alpha u_\alpha \to -1$. Thus for the limiting surface we obtain

$$-\kappa r^2 \cos\gamma = \frac{\kappa}{\pi}\int_{S_{\alpha u}} \frac{1}{W^2}dS - \frac{1}{2}\left(1 + \kappa^2 r^2 u^2 + 2\kappa r u \sin\gamma\right). \tag{6.21}$$

Now we let $u \to -\infty$ and use (6.17). We are led to

$$\kappa \int_{S_{\alpha\infty}} \frac{1}{W^2}dS = 0$$

a contradiction that shows that in fact no vertical can appear in the limit. Finally we have proved:

THEOREM. *There exists a global solution $U(r)$ of (3.1), defined for all $r > 0$, with an isolated singularity at $r = 0$. In any compactum, $U(r)$ is the uniform limit (together with all derivatives) of "bubble solutions" whose existence is proved in [4], as $u_0 \to -\infty$.*

Remarks. 1) $U(r)$ admits an asymptotic expansion in powers of r that is divergent for every r, as indicated in [5].

2) The above theorem proves Conjecture 3 on p. 101 of [3], up to the uniqueness which remains open. Conjecture 2 was proved by Bidaut-Veron in [8].

I wish to thank Zheng Chao Han for a number of helpful conversations. Han pointed out to me that the Green's Identity used by Bidaut-Veron in [8] is a consequence of (1.13). Bidaut-Veron's identity is in fact equivalent to an identity used by Wente in [9]; that identity would have sufficed for the needs of the present paper, however it does not apply to asymmetric configurations as does (1.13). I derived (1.13) originally with asymmetric problems in mind, and noticed only fortuitously its utility for the problem treated here. It seems clear that the present work has not exhausted the potential for the identity, and I plan to return in the near future to the original program of studying asymmetric equilibrium configurations.

This work was supported in part by a grant from the National Aeronautics and Space Administration, and in part by a grant from the National Science Foundation. Figures 2, 3, 5, 6, 8 and 9 are taken from reference [4] and are reproduced here by permission of the Royal Society of London. Figure 4 is taken from reference [5] and is reproduced by permission of Springer-Verlag.

REFERENCES

1. H.C. Wente, *The symmetry of sessile and pendent drops,* Pac. J. Math. **88** (1980), 387–397.

2. R. Finn, *Green's identities and pendent liquid drops II,* Proc. Symp. Cont. Mech. and Related Probs. Anal., Tbilisi, 1991, to appear.

3. R. Finn, *Equilibrium Capillary Surfaces,* Springer-Verlag, 1986.

4. P. Concus and R. Finn, *The shape of a pendent liquid drop,* Phil. Trans. Roy. Soc. Lon., series A, **292** (1979), 307–340.

5. P. Concus and R. Finn, *A singular solution of the capillary equation I: existence,* Invent. Math. **29** (1975), 143–148.

6. C.E. Delaunay, *Sur la surface de révolution dont la courbure moyenne est constante,* J. Math. Pures Appl. **6** (1841), 309–315.

7. P. Concus and R. Finn, *A singular solution of the capillary equation II: uniqueness,* Invent. Math. **29** (1975), 149–160.

8. M.F. Bidaut-Veron, *Global existence and uniqueness results for singular solutions of the capillary equation,* Pac. J. Math. **125** (1986), 317–334.

9. H.C. Wente, *The stability of the axially symmetric pendent drop.* Pac. J. Math. **88** (1980), 421–470.

REMARKS ON INCOMPREHENSIBLE MODELS OF FLUID MECHANICS

P.L. Lions

Ceremade, Université Paris-Dauphine
Place de Lattre de Tassigny
75775 Paris Cedex 16

Dedicated to the memory of R.J. DiPerna

1 Introduction

We review here some recent results concerning the Cauchy problem for various incompressible models of Fluid Mechanics. Let us first recall the basic models : in order to avoid technicalities with boundary conditions, we shall consider the simple case of periodic boundary conditions. In other words, all unknown functions will be periodic on $\mathbf{R}^2$ or on $\mathbf{R}^3$ with a period T_i in each variable x_i ($i = 1, 2$ or $i = 1, 2, 3$). Next, if $u = u(x,t)$ denote the vectorfield corresponding to the velocity field, the incompressibility condition reads

$$\operatorname{div} u = 0 .$$

The Euler equation corresponds to the additionnal equation

$$\frac{\partial u}{\partial t} + (u \cdot \nabla)u + \nabla p = 0$$

where p is the pressure field, another unknown (scalar) function that can be thought as a Lagrange multiplier corresponding to the constraint (1).

The Navier-Stokes equation is

$$\frac{\partial u}{\partial t} + (u \cdot \nabla)u - \nu\Delta u + \nabla p = 0$$

where $\nu > 0$ is the so-called viscosity or more precisely the inverse of the Reynolds number. We shall also discuss density-dependent models in which case we deal with another scalar unknown function ρ (the density) which is nonnegative. And, in the case of the density-dependent Euler equation, we add to (1)

Developments in Partial Differential Equations and Applications to Mathematical Physics, Edited by G. Buttazzo *et al.*, Plenum Press, New York, 1992

$$\frac{\partial \rho}{\partial t} + \operatorname{div}(\rho u) = 0$$

$$\frac{\partial}{\partial t}(\rho u) + \operatorname{div}(\rho u \otimes u) + \nabla p = 0 .$$

Finally, in the case of the density-dependent Navier-Stokes equation, we add to (1) and (4) the equation

$$\frac{\partial}{\partial t}(\rho u) + \operatorname{div}(\rho u \otimes u) - \operatorname{div}(\nu(\rho)\nabla u) + \nabla p = 0$$

where ν is a bounded from above and from below, continuous function on $[0, \infty)$.

We discuss in section II the state of the art on the analysis of the Cauchy problem for Euler equations (both the classical and the density-dependent ones), while section III is devoted to Navier-Stokes equations.

2 Euler equations

We begin with the classical model namely equations (1)-(2). As it is well-known the situation is extremely different when $N = 2$ or $N = 3$. The reason is that, if one introduces the vorticity field $\omega = \operatorname{curl} u$, when $N = 2$ the vorticity ω is a scalar which is passively transported along the flow lines i.e. ω satisfies the equation

$$\frac{\partial \omega}{\partial t} + \operatorname{div}(u\omega) = 0 .$$

This is why we begin our discussion in the easier case of two dimensions. In that case, we observe immediately that (1) and (7) yield, at least formally,

$$\int_Q \beta(\omega)\, dx \qquad \text{is independent of} \quad t \geq 0 ,$$

for all $\beta \in C(\mathbf{R})$, where Q is the "periodic box" $(0, T_1) \times (O, T_2)$. In particular, L^∞ bounds on the vorticity are available and this explains why global smooth solutions are known to exist uniquely : for instance if u_0 (or ω_0) are the initial conditions i.e. u_0 is a periodic vector-field on $\mathbf{R}^2$ (resp. ω_0 is a periodic function on $\mathbf{R}^2$ with zero mean) and if we impose

$$u|_{t=0} = u_0$$

or

$$\omega|_{t=0} = \omega_0 ,$$

then there exists a unique solution $u \in C([0,\infty); X^2)$ if $X = H^s_{\text{per}}$ or $C^{1,\alpha}_{\text{per}}$ with $s > 2$ or $\alpha \in (0,1)$ and the subscript "per" recalls that we deal with periodic functions. The existence and uniqueness of a global solution can be pushed to the situations where $\omega_0 \in L^\infty(Q)$ (with a periodic extension to $\mathbf{R}^2$) - a result due to V.I. Yudovich [25], see also J.Y. Chemin [5] for additionnal properties of these solutions, and [6] for the proof that vortex patches preserve with time a smooth boundary, a remarkable fact. At this stage, we wish to mention the excellent survey by A. Majda [21] on Euler equations.

In particular, in [21], various reasons are indicated in order to motivate the study of less regular initial conditions and we want to discuss two cases : i) $\omega_0 \in L^p(Q)$, ii)

ω_0 is a bounded measure on Q and we need to add $u_0 \in L^2(Q)$ in the case ii) and in the case i) when $p = 1$. The second case is mostly open regarding the existence of a global weak solution and the only informations available are the study of the possible losses of compactness by concentration phenomena made by R.J. DiPerna and A. Majda [13],[14], and the recent existence result by J.M. Delort [9] when ω_0 has a constant sign (of course, in the periodic setting, this condition is meaningless and has to be adapted: the most general condition being, up to a change of sign, $\omega_0^- \in L^1(Q)$), which also provides a global existence result in the case i) when $p = 1$. We conclude our discussion of the two-dimensional case by considering the case i) with $p > 1$. In that case several formulations are possible. In view of (8), L^p bounds on ω hold globally in t and thus by elliptic regularity ($\operatorname{div} u = 0$, $\operatorname{curl} u = \omega$) u is bounded in $W^{1,p}(Q)$ and thus compact in $L^2(Q)$. This is why, one immediately obtains the existence of a global solution u of (1)-(2) in $L^\infty(0,\infty; W^{1,p}(Q)^2)$ - the term $(u \cdot \nabla)u$ being interpreted as $\operatorname{div}(u \otimes u)$ in view of (1). But, of course, one can also formulate the Euler equation by using (7) and

$$-\Delta\phi = \omega \quad \text{in } Q \quad , \quad \phi \text{ is periodic} \quad , \quad u = \left(\frac{\partial\phi}{\partial x_2}, -\frac{\partial\phi}{\partial x_1}\right) .$$

And, using elliptic regularity and Sobolev embeddings, one obtains the existence of a global weak solution of (7) and (11) $\omega \in L^\infty(0,\infty; L^p(Q))$ provided $p \geq \frac{4}{3}$ - in fact, the borderline case $p = \frac{4}{3}$ requires some additionnal work... At this stage, we may also invoke the results of R.J. DiPerna and P.L. Lions [10] on divergence free ordinary differential equations and linear first-order equations to deduce $\omega \in C([0,\infty); L^p(Q))$. A third formulation uses the idea of renormalized solutions due to R.J. DiPerna and P.L. Lions [12], used in [10], and consists in replacing (7) by (for example)

$$\frac{\partial}{\partial t}\beta(\omega) + \operatorname{div}(u(\beta(\omega)) = 0 \quad , \quad \text{for all } \beta \in C_0^\infty(\mathbf{R}) .$$

Then, one can show (see R.J. DiPerna and P.L. Lions [11]) the following facts

i) If $\omega_0 \in L^p(Q)$ with $p > 1$ (or even if $|\omega_0| \log |\omega_0| \in L^1(Q)$), there exists a global solution $\omega \in L^\infty(0,\infty; L^p(Q))$ of (11)-(12).

ii) Any such solution satisfies : $\omega \in C([0,\infty); L^p(Q))$ and thus $u \in C([0,\infty); W^{1,p}(Q)^2)$, $p \in C([0,\infty); \tilde{W}^{1,p/2}(Q))$ where $\tilde{W}^{1,r}(Q) = \left\{\varphi \in L^1(Q) \, , \, \frac{\partial\varphi}{\partial x_i} \in \mathcal{H}^r(Q)\right\}$ if $r \leq 1$ and $\mathcal{H}^r(Q)$ denotes the Hardy space (see [7]).

iii) If $u \in L^\infty(0,\infty; W^{1,p}(Q))$ is a solution of (1)-(2) then ω is a solution of (11)-(12) provided $p \geq 2$ and thus $u \in C([0,\infty); W^{1,p}(Q)^2)$.

iv) If $\omega \in L^\infty(0,\infty; L^p(Q))$ is a solution of (7) and (11) then ω is a solution of (11)-(12) provided $p \geq \frac{4}{3}$ and thus $\omega \in C([0,\infty); L^p(Q))$.

The main remaining open question is clearly the uniqueness of any of those global weak solutions. The only partial result in that direction - that we are aware of - is the following fact that we proved recently : let $p \in (1,\infty)$, there exists a unique solution $\omega \in C([0,\infty); L^p(Q))$ of (11)-(12) for generic initial conditions $\omega_0 \in L^p(Q)$ i.e. for a countable intersection of dense open sets in $L^p(Q)$ (with zero mean...).

Concerning density-dependent Euler equations, even in two dimensions, no global results are known nor examples of breakdowns of smooth solutions - a phenomenon which seems to be expected from physical and numerical experiments. We can only build examples of global weak solutions (shear flows), that remain smooth if smooth initially, but which show that a priori estimates in $W^{1,p}$ for (ρ, u) do not hold for any $p \in (1, \infty)$.

The situation for the classical Euler equations in 3 dimensions ($N = 3$) is equally mysterious. And, except for local in time existence of unique smooth solutions and for numerical evidences of singular phenomena, nothing is known. A refined criterion for the possible breakdown of smooth solutions is available (see T. Beale, T. Kato and A. Majda [3], G. Ponce [22]) but the actual occurence of such a breakdown is an outstanding open problem. We have been able - see R.J. DiPerna and P.L. Lions [11] - to produce simple examples which show that a priori estimates in $W^{1,p}$ for $1 < p < \infty$ are not possible.

3 Navier-Stokes equations

We shall discuss first the Navier-Stokes equations (1) and (3). By the fundamental work of J. Leray [18],[19],[20] (see also E. Hopf [16]), one knows that if $u^0 \in L^2(\Omega)$ there exists a global weak solution $u \in L^2(0,T; H^1(Q)^3) \cap L^\infty(0,\infty; L^2(Q)^3)$ (for all $T < \infty$) but the persistence of smooth solutions and the uniqueness of weak solutions are fundamental open problems. Various attempts have been made to improve the available regularity which can be found in R. Temam [24], C. Guillopé, C. Foias and R. Temam [15]. For instance, L. Tartar (see [15]) showed that weak solutions satisfy $u \in L^1(0,T; C(\mathbf{R}^3)^3)$ ($\forall\, T < \infty$). Also, since the early work by J. Leray, various authors have analysed the possible singular set of weak solutions and the best available result in that direction is due to L. Caffarelli, R.V. Kohn and L. Nirenberg [4]. Recently, P. Constantin [8] has shown, among other properties of solutions, that $D_x^2 u \in L^q(Q \times (0,T))$ for all $T < \infty$ and for all $q < 4/3$ - with a little extra effort, one can show that $D_x^2 u \in L^{\frac{4}{3},\infty}(Q \times (0,T))$ for all $T < \infty$. Also recently, R. Coifman, P.L. Lions, Y. Meyer and S. Semmes [7], using Harmonic Analysis arguments, have shown the following facts for all $T < \infty$

i) $D_x^2 p \in L^1(0,T; \mathcal{H}^1(Q))$, and thus $\nabla p \in L^1(0,T; L^{3/2}(Q)^3)$, $p \in L^1(0,T; L^3(Q))$,
ii) $u \in L^\infty(0,T; W^{1,1}(Q))$ - a small extension of a result by P. Constantin [8] -,
iii) $u \cdot \nabla u$, $\nabla p \in L^2(0,T; \mathcal{H}^1(Q))$ where $\mathcal{H}^1$ denotes the Hardy space.

Clearly enough, much more remains to be discovered or at least understood !

We conclude this brief review with density-dependent Navier-Stokes equations still in 3 dimensions ($N = 3$) i.e. equations (1),(4) and (6). First, we have to prescribe initial conditions and since we want to consider the possibility of a vacuum i.e. densities ρ which can vanish on some set of positive measure, we prescribe

$$\rho|_{t=0} = \rho_0 \quad \text{in } Q \quad , \quad \rho u|_{t=0} = m_0 \quad \text{in } Q$$

where $\rho_0 \in L^\infty(Q)$, $\rho_0 \geq 0$ a.e., $m_0 \in L^2(Q)$.

It is shown in R.J. DiPerna and P.L. Lions [11] that there exists a global weak solution (ρ, u) of (1),(4),(6) and (8) which satisfies $\rho \in L^\infty(Q \times (0,\infty))$, $\rho u^2 \in L^\infty(0,\infty; L^1(Q))$,

$u \in L^2(0,\infty; H^1(Q))$. In addition, any such solution satisfies $\rho \in C([0,\infty); L^p(Q))$ for all $1 \leq p < \infty$ and

$$\int_Q \beta(\rho)\, dx \quad \text{is independent of} \quad t \geq 0 \quad , \quad \text{for all } \beta \in C(\mathbf{R})\, .$$

Again, this result relies upon the ODE's and first-order equations results in R.J. DiPerna and P.L. Lions [10]. It extends previous results due to S.N. Antontsev and A.V. Kazhikov [1], A.V. Kazhikov [17], S.N. Antontsev, A.V. Kazhikov and V.N. Manakhov [2]. Of course, the uniqueness and the regularity of solutions are open problems since if $\rho_0 \equiv 1$, (14) shows that for any solution we have $\rho \equiv 1$ on $Q \times [0,\infty)$ and u is then a solution of the classical Navier-Stokes equations.

References

1. S.N. Antontsev and A.V. Kazhikov, Mathematical study of flows of nonhomogeneous fluids. Lecture Notes, Novosikirsk State University, 1973 (in Russian).

2. S.N. Antontsev, A.V. Kazhikov and V.N. Monakhev, Boundary values problems in mechanics of nonhomogeneous fluids. North-Holland Amsterdam, 1990.

3. T. Beale, T. Kato and A. Majda, *Remarks on the breakdown of smooth solutions for the 3D Euler equations.* Comm. Math. Phys., 94 (1984), p. 61-66.

4. L. Caffarelli, R.V. Kohn and L. Nirenberg, *On the regularity of the solutions of Navier-Stokes equations.* Comm. Pure Appl. Math., 35 (1982), p. 771-831.

5. J.Y. Chemin, *Sur le mouvement des particules d'un fluide parfait incompressible bidimensionnel.* Invent., 103 (1991), p. 599-629.

6. J.Y. Chemin, *Persistance de structures géométriques dans les fluides incompressibles bidimensionnels.* In Séminaire EDP, 1990-1991, Ecole Polytechnique, Palaiseau, 1991.

7. R. Coifman, P.L. Lions, Y. Meyer and S. Semmes, *Compensated compactness and Hardy spaces.* To appear in J. Math. Pures Appl.

8. P. Constantin, *Remarks on the Navier-Stokes equations.* J. A.M.S.

9. J.M. Delort, *Existence de nappes de tourbillon en dimension deux.* To appear in J. A.M.S.

10. R.J. DiPerna and P.L. Lions, *Ordinary differential equations, transport theory and Sobolev spaces.* Invent., 98 (1989), p. 511-547.

11. R.J. DiPerna and P.L. Lions, in preparation, see also in Séminaire EDP, 1988-1989, Ecole Polytechnique, Palaiseau, 1989.

12. R.J. DiPerna and P.L. Lions, *On the global existence for Boltzmann equations : global existence and weak stability.* Ann. of Math., 130 (1989), p. 321-366.

13. R.J. DiPerna and A. Majda, *Concentrations in regularizations for 2D incompressible flow.* Comm. Pure Appl. Math., 40 (1987), p. 301-345.

14. R.J. DiPerna and A. Majda, *Reduced Hausdorff dimension and concentration-cancellation for two dimensional incompressible flow.* J. A.M.S., 1 (1988), p. 59-95.

15. C. Guillopé, C. Foias and R. Temam, *Lagrangian representation of a flow.* J. Diff. Eq. ; 57 (1985), p. 440-449.

16. E. Hopf, *Uber die Anfangswertaufgabe für die hydrodynamischen Grundgleichangen.* Math. Nacht., 4 (1951), p. 213-231.

17. A. V. Kazhikov, *Resolution of boundary value problems for nonhomogeneous viscous fluids.* Dokl. Akad. Nauk., 216 (1974), p. 1008-1010 (in Russian).

18. J. Leray, *Etude de diverses équations intégrales non linéaires et de quelques problèmes que pose l'hydrodynamique.* J. Math. Pures Appl., 12 (1933), p. 1-82.

19. J. Leray, *Essai sur les mouvements plans d'un liquide visqueux que limitent des parois.* J. Math. Pures Appl., 13 (1934), p. 331-418.

20. J. Leray, *Essai sur le mouvement d'un liquide visqueux emplissant l'espace.* Acta Math., 63 (1934), p. 193-248.

21. A. Majda, *Vorticity and the mathematical theory of incompressible fluid flow.* Comm. Pure Appl. Math., 39 (1986), p. 187-220.

22. G. Ponce, *Remarks on a paper by J.T. Beale, T. Kato and A. Majda.* Comm. Math. Phys., 98 (1985), p. 349-353.

23. L. Tartar, In Macroscopic Modelling of turbulent flows. Lecture Notes in Physics, ♯ 230, Springer, Berlin, 1985.

24. R. Temam, Navier-Stokes equations. North-Holland, Amsterdam, 1977.

25. V.I. Yudovich, *Non-stationary flow of an ideal incompressible liquid.* Zh. Vych. Mat., 3 (1963), p. 1032-1066 (en Russe).

ON SOLUTIONS WITH BLOW-UP AT THE BOUNDARY FOR A CLASS OF SEMILINEAR ELLIPTIC EQUATIONS

Moshe Marcus

Department of Mathematics
Technion

1. INTRODUCTION

In this paper we consider positive solutions of the equation,

$$\Delta u = h(x)f(u) \quad \text{in } D, \tag{1.1}$$

where D is a domain in R^N $(N \geq 3)$ and

$(A)_0$ $\quad f \in C^1([0,\infty))$, $f' \geq 0$, $f(0)=0$ and $f(t)>0$ for $t>0$,

$(B)_0$ $\quad h \in C^\alpha(\bar{D})$ for some $\alpha \in (0,1)$ and $h \geq 0$.

We are interested in *large solutions* of (1.1), i.e. positive solutions which blow up uniformly on the boundary. Throughout the paper we shall assume that D is a *standard* domain, i.e. ∂D is compact of class C^2. When D is unbounded, we shall say that a large solution is *regular* if it tends to zero at infinity.

The existence of large solutions of (1.1), under quite general conditions on h and f, is well-known. In fact, by Osserman [7] and Keller [5], such solutions exist if h and 1/h are bounded and f satisfies the additional condition,

$(A)_1$ $\quad \psi(a) := \int_a^\infty (2F(t))^{-1/2} dt < \infty, \quad \forall a>0,$

where F is the primitive of f such that $F(0)=0$. Furthermore, assuming $(A)_0$, condition $(A)_1$ is necessary.

Here we shall be interested mainly in the question of uniqueness and the (asymptotic) boundary behaviour of large solutions. These questions were first

Developments in Partial Differential Equations and Applications to Mathematical Physics, Edited by G. Buttazzo *et al.*, Plenum Press, New York, 1992

considered by Loewner and Nirenberg [6], with respect to the equation

$$\Delta u = u^{N^*}, \quad \text{where } N^* = (N+2)/(N-2) \tag{1.2}$$

which is related to a problem in differential geometry. They described the precise asymptotic behaviour of large solutions of (1.2) at the boundary and used this result in order to establish uniqueness. Recently these questions were treated, for various classes of equations, in [1-3] and [8].

A solution of (1.1) is *maximal* if it dominates every positive solution of (1.1). A unique large solution is necessarily a maximal solution. Recently Dynkin [4] showed that there exist certain relations between hitting probabilities for superdiffusions and maximal solutions of (1.1) with $f(u) = u^p$, $(1<p\leq 2)$. In this connection the question of uniqueness is of special interest.

2. PRELIMINARY RESULTS

We start by listing some basic facts concerning equation (1.1) always assuming that $(A)_0$ and $(B)_0$ hold.

2.1

If D is a bounded domain and φ is a non-negative, continuous function on ∂D, then problem

$$\Delta u = h(x)f(u) \quad \text{in } D, \quad u=\varphi \text{ on } \partial D, \tag{2.1}$$

possesses a unique non-negative solution. If φ is not identically zero the solution is positive in D.

2.2

Let D be a bounded domain. If u, v are non-negative functions in $C(\bar{D})$ such that

$$\Delta u \leq h(x)f(u) \quad \text{and} \quad \Delta v \geq h(x)f(v) \quad \text{in } D,$$

and $u \geq v$ on ∂D then $u \geq v$ in D.

2.3 Theorem ([5],[7])

Let D be an arbitrary domain in R^N and let f be a function satisfying conditions $(A)_0$ and $(A)_1$. Then every positive function u in $C^2(D)$ such that

$$\Delta u \geq f(u) \text{ in } D, \tag{2.2}$$

satisfies the inequality,

$$u(x) \leq \mu(\delta(x)), \quad \forall x \in D, \tag{2.3}$$

where $\delta(x) = \text{dist}(x,\partial D)$ and $\mu: R_+ \to R_+$ is a continuous, non-increasing function (independent of D) such that

$$\mu(\delta) \to \begin{cases} \infty \text{ when } \delta \to 0 \\ 0 \text{ when } \delta \to \infty. \end{cases} \tag{2.4}$$

2.4 Corollary

If D is an unbounded standard domain and u satisfies (2.2) then $u \to 0$ at infinity.

2.5

Let D be an unbounded (standard) domain and φ a non-negative, continuous function on ∂D such that $\varphi \not\equiv 0$. Assume that conditions $(A)_0$, $(A)_1$ and $(B)_0$ are satisfied and that

$$\liminf_{|x| \to \infty} h(x) > 0. \tag{2.5}$$

Then problem (2.1) possesses a unique positive solution. The solution tends to zero at infinity.

Proof If a positive solution exists, then (by (2.5) and corollary 2.4) it must tend to zero at infinity. Hence, by the maximum principle, it is unique. Now let R be sufficiently large so that $\partial D \subset B_R := \{x \in R^N : |x| < R\}$ and let u_R be the solution of the problem,

$$\Delta u = h(x)f(u) \text{ in } D \cap B_R, \quad u = \varphi \text{ on } \partial D \quad \text{and} \quad u = 0 \text{ on } \partial B_R. \tag{2.6}$$

By the maximum principle, the set of solutions $\{u_R\}$ is uniformly bounded and u_R increases with R. Hence $u := \lim_{R \to \infty} u_R$ is a solution of (2.1).

Remark Note that assumption (2.5) is needed for uniqueness but not for the existence of a solution.

3. EXISTENCE RESULTS

In this section we present a result concerning the existence of large solutions of (1.1) relaxing the conditions on the coefficient h, so that h may vanish on compact subsets of D or at infinity. This is an extension of a result of [3] in which equation (1.1) was considered with $f(u) = u^p$ and with h vanishing at most at a finite number of points in D or at infinity.

3.1 Theorem

Let D be a standard domain. Suppose that f satisfies conditions $(A)_0, (A)_1$ and that h satisfies

$(B)_1$ $\quad h \in C^{\alpha}(\bar{D})$ for some $\alpha \in (0,1)$, $h \geq 0$ in D and $h>0$ on ∂D.

In addition, if D is unbounded, assume that either (2.5) holds or there exist $p>1$ and $\beta>0$ such that,

$$\liminf_{t \to 0} f(t)\, t^{-p} > 0, \tag{3.1}$$

$$\liminf_{|x| \to \infty} h(x)\, |x|^{\beta} > 0. \tag{3.2}$$

Then there exists a large solution v of (1.1) which (if D is unbounded) tends to zero at infinity. More precisely, if D is unbounded,

$$v(x) = O(|x|^{-\rho}) \qquad \text{as } |x| \to \infty, \tag{3.3}$$

where $\rho := \max(N-2, (2-\beta)(p-1))$ when (3.1), (3.2) hold and $\rho := N-2$ when (2.5) holds.

Proof Let δ_0 be a positive number such that $h>0$ in the set $E_{\delta_0} := \{x \in \bar{D}: \delta(x) \leq \delta_0\}$. (Recall that $\delta(x) = \text{dist.}(x, \partial D)$.)

If D is bounded and γ is a positive number, let v_γ denote the solution of problem (2.1) with $\varphi \equiv \gamma$. By theorem 2.3, if $0<\delta_2<\delta_1<\delta_0$ then the set $\{v_\gamma: \gamma>0\}$ is uniformly bounded in the strip $E_{\delta_1} \cap E^c_{\delta_2}$. Consequently (by the maximum principle) this set of functions is uniformly bouded in every compact subset of D. In addition v_γ is monotone increasing with respect to γ. Thus the limit $v := \lim_{\gamma \to \infty} v_\gamma$ exists and v is a large solution of (1.1).

If D is unbounded let R_0 be sufficiently large so that $E_{\delta_0} \subset B_{R_0}$ and $h>0$ in $B^c_{R_0}$. For $\gamma > R_0$ let v_γ be the solution of problem (2.6) with $R=\gamma$ and $\varphi \equiv \gamma$. By theorem 2.3, the set of functions $\{v_\gamma: \gamma \geq 2R_0\}$ is uniformly bounded on the sphere ∂B_{R_1} where $R_1 = 3R_0/2$. Hence, by the maximum principle, this set of functions is uniformly bounded in $B^c_{R_1}$. (Initially v_γ is defined only in $D \cap B_\gamma$ but we extend it to D by setting $v_\gamma = 0$ in B^c_γ.) In view of this fact, the argument used in the first part of the proof shows that the set of functions $\{v_\gamma: \gamma \geq 2R_0\}$ is uniformly bounded in every closed subset of D. By the maximum principle, v_γ is monotone increasing with respect to γ. Hence $v := \lim_{\gamma \to \infty} v_\gamma$ is a large solution of (1.1).

If (2.5) holds, then by corollary 2.4, $v \to 0$ at infinity. Let

$$\mu := R_0^{N-2} \sup\{v(x): |x| = R_0\} \quad \text{and} \quad w_\epsilon(x) := \mu |x|^{2-N+\epsilon}.$$

Then $\Delta w_\epsilon = 0$ in $B^c_{R_0}$ and for every $\epsilon > 0$, $v \leq w_\epsilon$ on ∂B_R for all sufficiently large R. By definition, $v \leq w_\epsilon$ on ∂B_{R_0}. Hence by the comparison principle 2.2 it follows that $v \leq w_0$ in $B^c_{R_0}$.

Now suppose that (3.1) and (3.2) hold. Let M be a bound of $\{v_\gamma: \gamma \geq 2R_0\}$ in $B^c_{R_1}$. By (3.1) and $(A)_0$ there exists a positive constant c such that

$$f(t) \geq ct^p \quad \text{for } 0 \leq t \leq M. \tag{3.4}$$

By (3.2) and the choice of R_1, if $c>0$ is sufficiently small we have also,

$$h(x) \geq c|x|^{-\beta} \text{ for } |x| \geq R_1. \tag{3.5}$$

It follows that, for $\gamma \geq 2R_0$,

$$\Delta v_\gamma = h(x)f(v_\gamma) \geq c^2|x|^{-\beta}v_\gamma^p \quad \text{for } |x| \geq R_1. \tag{3.6}$$

By [3; th.2.1] the equation

$$\Delta w = c^2|x|^{-\beta}w^p \ , \tag{3.7}$$

possesses a large solution W in $B^c_{R_1}$ such that (with $\rho := \max(N-2, (2-\beta)(p-1))$,

$$W(x) \leq \text{const.}|x|^{-\rho} \quad \text{for } |x| \geq 2R_1.$$

By (3.6) and the comparison principle 2.2,

$$v_\gamma \leq W \quad \text{in } B^c_{R_1}, \text{ for every } \gamma \geq 2R_0.$$

Hence, $v \leq W$ in $B^c_{R_1}$ so that v satisfies (3.3). This completes the proof of the theorem.

3.2 Remarks

The large solution constructed in the above proof is obviously the *smallest large solution* of problem (1.1). We shall refer to it briefly as the s.l.s. of the problem.

If D is a radially symmetric domain and if h is radially symmetric then the same is true with respect to v_γ and therefore with respect to the s.l.s.

4. BOUNDARY BEHAVIOUR AND UNIQUENESS

In this section we discuss the question of uniqueness and the asymptotic boundary behaviour of large solutions of problem (1.1). We start with two basic lemmas.

4.1 Lemma

Consider equation (1.1) with $h \equiv 1$ assuming that f satisfies $(A)_0$, $(A)_1$ and

$(A)_2$ $\exists \mu>0$ and $t_0 \geq 1$ such that $f(t)/t^{1+\mu}$ is monotone increasing for $t \geq t_0$.

Let U be the s.l.s. of (1.1) with $D=B_R$. Then U is radially symmetric and

$$U(R-\delta)/\phi(\delta) \to 1 \quad \text{as} \quad \delta \to 0, \tag{4.1}$$

where ϕ is the inverse of the function ψ defined in $(A)_1$.

For the proof of the lemma see [2].

4.2 Lemma

Consider equation (1.1) under the same assumptions as in lemma 4.1. Let W be the smallest solution of (1.1) in $B_R \cap B^c_{R_0}$ $(R>R_0>0)$ such that

$$W=0 \text{ on } \partial B_R \quad \text{and} \quad W(x) \to \infty \text{ uniformly as } x \to \partial B_{R_0}. \tag{4.2}$$

Then W is radially symmetric and

$$W(R_0+\delta)/\phi(\delta) \to 1 \text{ as } \delta \to 0. \tag{4.1'}$$

Remark The function W is obtained as the limit of a sequence of solutions $\{v_j\}$ of (1.1) in the annulus such that $v_j=0$ on ∂B_R and $v_j=j$ on ∂B_{R_0}. The proof given below is based on an argument of [6], which was applied also in [2] (see proof of lemma 2.2 there).

Proof Given $0<\epsilon$, set $w_\epsilon(x)=\phi(r-R_0+\epsilon)$, where $R_0<r=|x|$. Then

$$\Delta w_\epsilon = -\sqrt{2F(w_\epsilon)}\,\frac{N-1}{r} + f(w_\epsilon) \quad \text{for } R_0<r. \tag{4.3}$$

By $(A)_2$, $\sqrt{F(t)}/f(t) \to 0$ as $t \to \infty$. Given $b \in (0,1)$ choose t_b sufficiently large so that

$$\sqrt{2F(t)}\,\frac{N-1}{R_0} \leq bf(t) \text{ for } t \geq t_b. \tag{4.4}$$

Now let $\delta_b := \psi(t_b)/2$, i.e. $t_b=\phi(2\delta_b)$, and let $\epsilon \in (0,\delta_b)$. Then $w_\epsilon > t_b$ in the annulus $R_0<|x|<R_0+\delta_b$ and consequently, by (4.3) and (4.4)

$$\Delta w_\epsilon \geq (1-b)f(w_\epsilon) \quad \text{for } 0<r-R_0<\delta_b. \tag{4.5}$$

Denote $\beta := (1-b)^{1/\mu}$ and $u_\epsilon := \beta(w_\epsilon - \phi(\delta_b+\epsilon))$. Assume that t_b is sufficiently large so that $\delta_b \leq (R-R_0)/2$ and $t_b \geq t_0/\beta$. By $(A)_2$,

$$f(\beta t) \leq \beta^{1+\mu} f(t) \quad \text{for } t \geq t_0/\beta.$$

Hence by (4.5),

$$\Delta u_\epsilon = \beta\Delta w_\epsilon \geq \beta^{1+\mu} f(w_\epsilon) \geq f(\beta w_\epsilon) \geq f(u_\epsilon) \quad \text{for } 0<r-R_0<\delta_b. \tag{4.6}$$

Now let $\{\epsilon_j\}$ be a sequence of positive numbers tending to zero such that $\frac{1}{2}\phi(\epsilon_j) \leq j$ for $j=1,2,\ldots$. Then u_{ϵ_j} is a subsolution of the problem,

$$\begin{aligned} &\Delta u = f(u) \text{ in } R_0<|x|<R_0+\delta_b, \\ &u=j \quad \text{for } |x|=R_0, \; u=0 \quad \text{for } |x|=R_0+\delta_b. \end{aligned} \tag{4.7}$$

Consequently, $u_{\epsilon_j} \leq W$ in the above annulus. Since $\epsilon_j \to 0$ it follows that

$$b\,\phi(r-R_0) \leq W(r) \quad \text{for} \quad 0<r-R_0<\delta_b.$$

Letting b tend to 1 we obtain

$$\liminf_{r\to R_0} W(r)/\phi(r-R_0) \geq 1.$$

On the other hand, by a standard argument, lemma 4.1 implies that

$$\limsup_{r\to R_0} W(r)/\phi(r-R_0) \leq 1.$$

This completes the proof of the lemma.

The result stated below was obtained in [2]. A similar result was independently obtained in [8] where the author considered equation (1.1) with more general elliptic operators but with a more restricted class of nonlinearities (namely, $f(u)=u^p$). The proof given here (using lemmas 4.1 and 4.2) is a modification of the proof of [2]. It clarifies the local nature of the result and it enables us to apply the same arguments to other related problems for which the locality is crucial (see, for instance, 5.1 in next section).

4.3 Theorem

Let D be a standard domain and suppose that f satisfies $(A)_0$-$(A)_2$ and that h satisfies $(B)_1$. If U is a large solution of (1.1) in D then it satisfies,

$$\lim_{x\to\partial D} U(x) \,/\, \phi(\delta(x)\sqrt{h^*(x)}) = 1, \tag{4.8}$$

where $h^*(x) := \max\{h(y) \colon y\in\partial D,\ |y-x|=\delta(x)\}$.

Proof In this proof we shall assume (as we may) that h is defined and continuous in a neighborhood of $\bar{D}$. Let δ_0 be sufficiently small so that ,

(i) $h(x)>0$ whenever $\delta(x)<\delta_0$;

(ii) if $\delta(x)<\delta_0$, there exists a unique point y on ∂D such that $|x-y| = \delta(x)$. (This point will be denoted by P(x).)

Let $E := \{x \in D: \delta(x) < \delta_0/4\}$. For $x_0 \in E$ denote by $B(x_0)$ the ball of radius $\delta(x_0)$ centered at x_0 and set $\alpha(x_0) := \inf_{B(x_0)} h$. If U_0 is the s.l.s. of

$$\Delta u = \alpha(x_0) f(u) \quad \text{in } B(x_0), \tag{4.9}$$

then $U \le U_0$ in $B(x_0)$. Consequently, by lemma 4.1,

$$U(x) \le \phi(\delta(x)\sqrt{\alpha(x_0)})(1+o(1)), \tag{4.10}$$

for every x on the segment $[x_0, P(x_0)]$, where o(1) is a quantity which tends to zero as x tends to $P(x_0)$. This quantity depends on $\delta(x_0)$ but can be chosen to be independent of x_0.

Now from $(A)_2$ we obtain,

$$\phi(\beta\tau) \le (1+o(1))\beta^{-2/\mu}\phi(\tau) \qquad \forall \beta \in (0,1), \tag{4.11}$$

where o(1) is a quantity (independent of β) which tends to zero as $\tau \to 0$. Since $h^*(x_0) \ge \alpha(x_0)$, (4.9) and (4.11) yield,

$$U(x) \le \phi(\delta(x)\sqrt{h^*(x)})(\alpha(x_0)/h^*(x))^{-1/\mu}(1+o(1)), \qquad \forall x \in [x_0, P(x_0)]. \tag{4.12}$$

(Note that for x on this segment $h^*(x) = h^*(x_0)$.) As the inequality holds for every x_0 on the surface $\{x \in D: \delta(x) = \delta_0/4\}$ we conclude that,

$$\limsup_{x \to \partial D} U(x)/\phi(\delta(x)\sqrt{h^*(x)}) \le 1. \tag{4.13}$$

Next, let $E' := \{x \in \bar{D}^c: \delta(x) < \delta_0/4\}$. For $x_0 \in E'$, denote by $K(x_0)$ the annulus $\delta(x_0) < |x - x_0| < 2\delta(x_0)$ and set $\gamma(x_0) := \sup_{K(x_0)} h$. Let W be as in lemma 4.2 with respect to the equation $\Delta w = \gamma(x_0) f(w)$ in $K(x_0)$. Then $W \le U$ in $K(x_0) \cap D$. Therefore, by lemma 4.2,

$$U(x) \ge \phi(\delta(x)\sqrt{\gamma(x_0)})(1+o(1)), \quad \forall x \in [x_0, P(x_0)], \tag{4.14}$$

with o(1) as before. Since $h^*(x_0) \le \gamma(x_0)$, (4.14) and (4.11) imply that

$$\liminf_{x \to \partial D} \frac{U(x)}{\phi(\delta(x)\sqrt{h^*(x)})} \ge 1. \tag{4.15}$$

This completes the proof of the theorem.

Combining theorems 3.1 and 4.3 one obtains the following result,

<u>4.4 Theorem</u>

Suppose that f satisfies $(A)_0$–$(A)_2$ and

$(A)_3$ $\quad f(\sigma t) \le \sigma f(t)$ for $\forall \sigma \in (0,1)$ and $\forall t>0$.

Suppose also that h satisfies $(B)_1$ and (if D is unbounded) either (2.5) or (3.1) and (3.2). Then there exists a unique large solution of (1.1) which tends to zero at infinity (when D is unbounded).

Condition $(A)_3$ is used in order to establish the uniqueness which is proved as follows. Let u_1, u_2 be two large solutions which tend to zero at infinity if D is unbounded. Then by theorem 4.3, for every $\gamma \in (0,1)$, $u_2 > \gamma u_1$ in a neighborhood of ∂D. By $(A)_3$, $\Delta(\gamma u_1) \ge h(x) f(\gamma u_1)$. Therefore, in view of the fact that the solutions tend to zero at infinity, the maximum principle implies that $u_2 > \gamma u_1$ in D for each γ as above. Consequently $u_2 \ge u_1$ and hence (by symmetry) $u_2 = u_1$.

5. SOME RELATED PROBLEMS

In this section we discuss certain questions concerning solutions of (1.1) with (partial) blow-up on the boundary and large solutions of a related equation.

5.1

Consider equation (1.1) in a standard domain D. Let Γ be a closed subset of ∂D and let φ be a non-negative function on ∂D which is continuous in the complement of Γ. Under the assumptions of theorem 3.1, there exists a solution u_Γ of (1.1) such that

$$u_\Gamma = \varphi \text{ on } \partial D \backslash \Gamma, \qquad u_\Gamma(x) \to \infty \quad \text{as } x \to \Gamma, \tag{5.1}$$

and $u_\Gamma \to 0$ at infinity (if D is unbounded). Indeed such a solution can be obtained as the limit of a sequence of solutions $\{u_{j,\Gamma}\}$ of (1.1) such that $u_{j,\Gamma} = \varphi_j$ on ∂D where $\{\varphi_j\}$ is an increasing sequence of continuous functions on ∂D satisfying,

$$0 \le \varphi_j \le j, \; \varphi_j = j \text{ on } \Gamma \text{ and } \varphi_j \to \varphi \text{ in } \partial D \backslash \Gamma. \tag{5.2}$$

Suppose that Γ^0 (=the interior of Γ) is not empty and that in addition to the previous assumptions, f satisfies condition $(A)_2$. Then the arguments used in the proof of theorem 4.3 show that

$$\lim_{x \to \Gamma^0} u_\Gamma(x) / \phi(\delta(x)\sqrt{h^*(x)}) = 1, \tag{5.3}$$

with uniform convergence with respect to compact subsets of Γ^0. (Here one makes use of the local nature of these arguments.) Of course this result does not imply the uniqueness of u_Γ. To settle this question, precise information on the asymtotic behaviour of u_Γ at the relative boundary of Γ is needed.

Next we consider the equation

$$\Delta u = h(x) f(u) - g(x) \qquad x \in D, \tag{5.4}$$

where g is a non-negative function in C(D) and D is a standard domain. The previous results are readily extended to this equation under appropriate conditions on g.

5.2 Lemma

Suppose that f satisfies conditions $(A)_0$-$(A)_2$, that h satisfies $(B)_1$ and that g is a non-negative function in C(D) such that

$$g(x)/f(\phi(\delta(x))) \to 0 \quad \text{as} \quad x \to \partial D. \tag{5.5}$$

In addition suppose that

$$\limsup_{t \to \infty} f(\beta t)/f(t) > 0, \quad \forall \beta > 0. \tag{5.6}$$

Then every large solution of (5.4) in D satisfies (4.8).

Proof Suppose that U is a large solution of (5.4) in D. Then $\Delta U \le h(x)f(U)$ and by theorem 4.3 and the comparison principle 2.2,

$$U(x) \ge \phi(\delta(x)\sqrt{h^*(x)})\,(1+o(1)), \tag{5.7}$$

with h^* as in (4.8). Here o(1) is a quantity which tends to zero as $x \to \partial D$. From (5.7), (4.11) and the monotonicity of ϕ we obtain,

$$U(x) \ge c_1(1+o(1))\phi(\delta(x)), \tag{5.8}$$

where

$$c_1 := \min\,(1, M_h^{-2/\mu}) \quad \text{and} \quad M_h := \max_{\partial D} h.$$

Denote $E(\delta_0) := \{x \in D: \delta(x) < \delta_0\}$. If δ_0 is a sufficiently small positive number, then by (5.6), (5.8) and $(B)_1$ there exists a constant $c_2 > 0$ such that,

$$h(x)f(U(x)) \ge c_2 f(\phi(\delta(x)), \quad \text{for } x \in E(\delta_0). \tag{5.9}$$

From (5.5) and (5.9) it follows that for every $\epsilon > 0$ there exists $\delta_0 > 0$ such that,

$$\Delta U(x) \ge (1-\epsilon)h(x)f(U(x)), \quad \text{for } x \in E(\delta_0). \tag{5.10}$$

Hence, as in the proof of theorem 4.3, we obtain,

$$U \le \phi(\delta(x)\sqrt{h^*(x)})\,(1+o(1)), \tag{5.11}$$

with o(1) as in (5.7). From (5.7) and (5.11) we obtain (4.8).

5.3 Lemma

Suppose that g satisfies the assumptions of lemma 5.2, that h satisfies $(B)_1$

and that f satisfies conditions $(A)_0$-$(A)_2$ and (5.5). If D is unbounded we assume in addition that h satisfies (2.5) and that

$$g(x) \to 0 \quad \text{as } |x| \to \infty. \tag{5.12}$$

Then there exists a large solution of (5.4) in D which (if D is unbounded) tends to zero at infinity.

<u>Proof</u> First assume that D is bounded. Let u_j (resp. v_j) be the solution of (5.4) (resp. (1.1)) which is identically equal to j on ∂D. Then $u_j \geq v_j$. The sequence $\{v_j\}$ converges (uniformly in compact subsets of D) to a large solution V of (1.1). Therefore the argument used in the first part of the proof of lemma 5.2 shows that if δ_0 is a sufficiently small positive number then,

$$\Delta u_j(x) \geq \tfrac{1}{2}h(x)f(u_j(x)), \qquad \text{for } x \in E(\delta_0). \tag{5.13}$$

By theorem 2.3, (5.13) implies that $\{u_j\}$ is uniformly bounded in every compact subset of $E(\delta_0)$. Hence (by the maximum principle) it is uniformly bounded in every compact subset of D. Therefore $U := \lim u_j$ is a large solution of (5.4).

Next assume that D is unbounded. Choose R_0 sufficiently large so that $\partial D \subset B_{R_0}$ and $m_0 := \inf\{h(x): |x| \geq R_0\} > 0$. (Here we use (2.5).) Consider the equation

$$\Delta u(x) = m_0 f(u(x)) - \hat{g}(|x|) \quad \text{for } |x| > R_0, \tag{5.14}$$

where

$$\hat{g}(r) := \sup\{g(x): |x| = r\}.$$

Let u_j be the solution of equation (5.14) in the annulus $R_0 < |x| < j$ such that

$$u_j(x) = j \text{ on } |x| = R_0 \text{ and } u_j(x) = 0 \text{ on } |x| = j.$$

Then u_j is radially symmetric and the sequence $\{u_j\}$ is monotone increasing and bounded (by the argument of the first part of the proof). Therefore $U := \lim u_j$ is a radially symmetric large solution of (5.14). We claim that $U \to 0$ at infinity. Indeed if $\liminf_{r \to \infty} U(r) > 0$ then for sufficiently large R we obtain (from (5.12))

$$\Delta U \geq \tfrac{1}{2} m_0 f(U) \quad \text{in } |x| > R. \tag{5.15}$$

This implies that $U \to 0$ at infinity and thus leads to a contradiction. Therefore $\liminf_{r \to \infty} U(r) = 0$. Now suppose that $\limsup_{r \to \infty} U(r) > 0$. Then there exists a positive ϵ and a sequence of disjoint intervals (a_n, b_n) such that $a_n \to \infty$, $U(a_n) = U(b_n) = \epsilon$ and $U > \epsilon$ in each interval. It follows that the inequality (5.15) holds in (a_n, b_n) for large n. But this is impossible by the maximum principle. Therefore $U \to 0$ at infinity.

Now let w_j be a solution of equation (5.4) in $D \cap B_j$ satisfying

$$w_j(x)=j \text{ on } \partial D \text{ and } w_j(x)=0 \text{ on } |x|=j.$$

Since $\Delta U(x) \leq h(x)f(U(x))-g(x)$ for $|x|>R_0$ it follows that w_j is dominated by U for $|x|>R_0$. Using this fact and the argument of the first part of the proof we conclude that the sequence $\{w_j\}$ converges in D to a large solution of (5.4) which tends to zero at infinity. This completes the proof.

The following theorem is a consequence of the previous two lemmas. In the case $f(u)=u^p$ it is due to [8] where more general elliptic operators were also considered.

5.4 Theorem

Suppose that f satisfies $(A)_0$-$(A)_3$ and (5.6), that h satisfies $(B)_1$ and that g is a continuous non-negative function in D satisfying (5.5). If D is unbounded assume also that h satisfies (2.5) and that g satisfies (5.12). Then there exists a unique large solution of (5.4) which (if D is unbounded) tends to zero at infinity. This solution satisfies (4.8).

The uniqueness of the large solution follows by the same argument that was used in the proof of theorem 4.4.

REFERENCES

[1] Bandle C. and Marcus M., *Sur les solutions maximales de problèmes elliptiques nonlinéaires: bornes isopérimetriques et comportement asymptotique*, C.R.Acad.Sci.Paris, **311** Série I (1990), 91-93.

[2] Bandle C. and Marcus M., *Large solutions of semilinear elliptic equations: existence, uniqueness and asymptotic behaviour*, (to appear in J. D'Anal. Math.)

[3] Bandle C. and Marcus M., *Large solutions of semilinear elliptic equations with 'singular' coefficients*, (to appear in the proceedings of the conference on "Optimization and Nonlinear Analysis" held in Haifa, 1990).

[4] Dynkin E.B., *A probabilistic approach to a class of nonlinear differential equations*, (to appear).

[5] Keller J.B., *On solutions of $\Delta u=f(u)$*, Comm. Pure Appl. Math. **10** (1957), 503-510.

[6] Loewner C. and Nirenberg L., *Partial differential equations invariant under conformal or projective transformations*, in Contributions to Analysis (L.Ahlfors ed.) Academic Press N.Y. (1974), 245-272.

[7] Ossermann R., *On the inequality* $\Delta u \geq f(u)$, Pac. J. Math. 7 (1957), 1641-1647.

[8] Veron L., *Semilinear elliptic equations with uniform blow-up on the boundary*, (to appear).

ON SOME SINGULAR BOUNDARY VALUE PROBLEM FOR HEAT OPERATOR

Sigeru Mizohata

Osaka Electro-Communication University
Osaka, Japan

1 Introduction

In [2], S. Itô treated the following initial-boundary value problem for heat operator :

$$(I.B.P.)_0 \begin{cases} (1) \ \frac{\partial}{\partial t}u = \Delta u, \ \text{in } \Omega \subset R^n , \\ (2) \ Bu = a(x)\frac{\partial u}{\partial n} + b(x)u = 0, \ x \in \partial\Omega , \\ \quad \text{where } \frac{\partial u}{\partial n} \text{ is the derivative in the direction of} \\ \quad \text{outernormal, and } a(x) \geq 0, \ b(x) \geq 0, \ a(x)+b(x)=1, \\ (3) \ u|_{t=0} = u_0(x). \end{cases}$$

He proved that this problem is well-posed by constructing the fundamental solution. Later several authors (cf. [1], [3], [5], [8], [9]), treated the *elliptic boundary value problem* relaxing the boundary condition (2), by the method of functional analysis. Recently T.Terakado retreated (I.B.P.) using Laplace transformation [10], and Y.Kannai treated the problem for more general setting by using the theory of pseudodifferential operators [4].

Our problem is to consider the necessary condition concerning a(x) and b(x) in order that (I.B.P.) be well-posed in a simple case. To be more precise,

$$(I.B.P.) \begin{cases} (1) \ \frac{\partial}{\partial t}u = \Delta u, \ \text{in } \Omega, t \geq 0, \Omega = \{(x,y) \mid y < 0\}, \\ (2) \ Bu = a(x)\frac{\partial u}{\partial y} + b(x)u \,|_{y=0} = 0, \ x \in R , \\ \quad \text{where a(x), b(x)} \in \mathcal{B}, \text{ namely they are in } C^\infty \text{ and bounded} \\ \quad \text{with all their derivatives, and are real-valued, we assume that} \\ \quad \text{they satisfy } a(x)^2 + b(x)^2 = 1, \\ (3) \ u|_{t=0} = u_0(x). \end{cases}$$

Let us make precise the definition of H^∞-wellposedness. For our purpose, it would be better to define it in a weak sense.

Definition 1 *We say that (I.B.P.) is H^∞-wellposed, if there exists N_0 such that for any $u_0(x) \in H^\infty(\Omega)$ satisfying the compatiblity conditions*

(C) $B(\Delta^j u_0(x,y)) = 0$, *for j=0,1,2,*$\cdots, N_0 - 1$,

there exists a unique solution for $t \in [0,T]$

$$u(x,y,t) \in C_t^0([0,T]; H^2(\Omega)) \cap C_t^1([0,T]; L^2(\Omega)),$$

satisfying $B(u(x,y,t)) = 0$ *for any* $t \in [0,T]$.

Our result is the following. We put the assumption on a(x):

(A) At the point x_0 where $a(x_0) = 0$, the vanishing order of a(x) is finite.

Theorem 1 *Under the assumption (A), in order that the above (I.B.P.) be* H^∞*-wellposed, the following conditions are necessary:*
(1) $a(x)$ *does not change the sign. We assume therefore* $a(x) \geq 0$.
(2) On the set $\{x \,;\, a(x) = 0\}$, $b(x)=1$.

Remark. We put the assumption (A). It is desirable to prove the Theorem without it.

Let us describe a little the outline of the proof. First for the convenience of the proof, *we write hereafter the boundary condition in the form*

$$Bu = a(x)\frac{\partial u}{\partial y} - b(x)u = 0 \tag{1.1}$$

replacing b(x) by -b(x).
We assume that a(x) vanishes at the origin with vanishing order n. We assume that

$$(B) \quad \begin{cases} 1) & \text{n is odd, or} \\ 2) & \text{n is even, a(x)>0 (x} \neq 0\text{) and b(0)=1.} \end{cases}$$

Then the Theorem can be restated as follows:

Theorem 2 *Under the assumptions (A) and (B), problem (I.B.P.) is never* H^∞*-wellposed.*

Hereafter we assume that (I.B.P.) is H^∞-wellposed and (B), and show that these two assumptions lead to a contradiction.

First observe that if we put $u(\cdot,t) = T_t u_0$, T_t is a semi-group. T. Komura obtained the necessary and sufficient condition to the infinitesimal generator in Fréchet spaces [6]. However it seems difficult to apply her result to actual problem. Instead of that, suggested by it, we use the truncated Laplace transform:

$$\hat{u}(x,y;\lambda) = \int_0^1 e^{-\lambda t} u(x,y,t)dt, \tag{1.2}$$

assuming without loss of generality $T \geq 1$. Here λ is a positive parameter tending to infinity.

On the other hand, by Banach's closed graph theorem, there exists a constant C and some positive integer q (≥ 2) such that

$$\max_{t\in[0,T]} \| u(\cdot,t) \|_{H^2(\Omega)} \leq C\| u_0(x,y) \|_{H^q(\Omega)}. \tag{1.3}$$

Now we have

$$(\lambda - \Delta)\hat{u}(x,y;\lambda) = u_0(x,y) - e^{-\lambda}u(x,y,1). \tag{1.4}$$

Then (1.2) and (1.3) imply

$$\text{(1.5)} \qquad \begin{cases} \| \hat{u}(\cdot;\lambda) \|_2 \leq const. \| u_0 \|_{q}, \\ \| u(x,y,1) \|_2 \leq const. \| u_0 \|_{q}. \end{cases}$$

Choice of initial data. Suggested by (1.4) and in view of compatibility condition (C), we choose, as initial data, functions of specific form. Let $f(x) \in H^\infty$. First introduce

$$\text{(1.6)} \qquad \begin{aligned} & u_1(x,y) \\ &= f(x)\alpha(y) + (\lambda - \frac{d^2}{dx^2})f(x)\cdot\frac{y^2}{2!}\alpha(y) + \cdots + \left(\lambda - \frac{d^2}{dx^2}\right)^{N_0} f(x)\cdot\frac{y^{2N_0}}{(2N_0)!}\alpha(y) \\ &= \left(\sum_{j=0}^{N_0}\left(\lambda - \frac{d^2}{dx^2}\right)^j f(x)\cdot\frac{y^{2j}}{(2j)!}\right)\alpha(y), \end{aligned}$$

where $\alpha(y)$ (cut-off function) is a fixed C^∞-function satisfying $\alpha(y) = 1$ for $-1/2 \leq y \leq 0$, and $= 0$ for $y \leq -1$. Then the initial data $u_0(x,y)$ is defined by

$$\text{(1.7)} \qquad u_0(x,y) = (\lambda - \Delta)u_1(x,y).$$

It is easy to see that

$$\text{(1.8)} \qquad u_0(x,y) = \left(\lambda - \frac{d^2}{dx^2}\right)^{N_0+1} f(x)\cdot\frac{y^{2N_0}}{(2N_0)!}\alpha(y) + \cdots,$$

where $\cdots$ is a sum of functions of the form

$$c_i\left(\lambda - \frac{d^2}{dx^2}\right)^{i_1} f(x)\cdot y^{i_2}\alpha^{(i_3)}(y),$$

with $i_1 \leq N_0$, $i_2 \leq 2N_0$, and $i_3 = 1, 2$. Hence it has its y-support only in $y \in [-1, -1/2]$. Observe that $u_0(x,y)$ satisfies compatibility condition (C) for any f(x), and the mapping from $f(x) \in H^\infty$ to u_1, u_0, and also to $\hat{u}(\cdot;\lambda)$ are linear. The core part of our argument consists of the choice of f(x). As we shall show in §3, our f(x) is connected with some almost null solution ψ of a functional equation defined on the boundary $y = 0$. Let us remark that

$$\text{(1.9)} \qquad Bu_1(x,y) = -b(x)f(x).$$

This subject was treated in [7] with very short proof. This article describes fairly in detail its proof.

2 Functional equation on the boundary

Taking account of (1.4) and (1.7), we express u(x,y,1) in the form

$$\text{(2.1)} \qquad \begin{cases} (\lambda - \Delta)u_D(x,y) = u(x,y,1) \\ u_D(x,0) = 0\,. \end{cases}$$

Since $u(x,y,1) \in H^2$, we can find a unique solution in H^4. The suffix D of u_D means that u_D is the solution of such a Dirichlet problem.

Hence

$$\text{(2.2)} \qquad v(x,y) = \hat{u}(x,y;\lambda) - u_1(x,y) + e^{-\lambda}u_D(x,y)$$

satisfies

(2.3) $$(\lambda - \Delta)v(x,y) = 0.$$

Since $v(x,y) \in H^2$ and satisfies (2.3), we can express v(x,y) by Poisson's formula:

$$v(v,y) = \frac{1}{2\pi}\int_{-\infty}^{\infty} e^{ix\xi}e^{y\sqrt{\lambda+\xi^2}}\hat{v}(\xi,0)d\xi \,.$$

Denoting $v(x,0) = v(x)$, in view of (1.9), we see that

(2.4) $$(\,a(x)\Lambda \;-\; b(x)\,)v(x) \;=\; b(x)f(x) \;+\; e^{-\lambda}g(x)\,,$$

where

$$g(x) \;=\; a(x)\frac{\partial}{\partial y}u_D\big|_{y=0}\,, \text{ and } \sigma(\Lambda) \;=\; (\lambda+\xi^2)^{\frac{1}{2}} \;.$$

Then denoting $\Lambda v(x) \;=\; w_0(x)$, we arrive at *the fundamental equation*

(2.5) $$Aw_0 \equiv (a(x) \;-\; b(x)\Lambda^{-1})w_0(x) \;=\; b(x)f(x) \;+\; e^{-\lambda}g(x)\;,$$

where $w_0 \in H^{1/2}$ and $f(x) \in H^\infty$.

We should note that for any $f(x) \in H^\infty$, there exists $(w_0, g) \in H^{1/2} \times H^{5/2}$ such that the correspondence $f \to (w_0, g)$ is a linear mapping. Let us make precise this mapping. (1.5) implies

$$\| \hat{u}(\cdot;\lambda) \|_2 \le const\cdot \| u_0 \|_q \le const\cdot \| (\lambda+\xi^2)^{N_0+1}(1+\xi^2)^{\frac{q}{2}}\hat{f}(\xi) \|_0 \;.$$

The same inequality holds for $\| u(x,y,1) \|_2$.

For our analysis it is convenient and even necessary to use another norm in H^s:

(2.6) $$\| f(x) \|_s^2 = \int_{-\infty}^{\infty}(\lambda+\xi^2)^s|\hat{f}(\xi)|^2 d\xi \,.$$

Let us notice if we use

$$\| f(x,y) \|_2^2 = \| \frac{\partial^2 f}{\partial y^2} \|_0^2 + \| \Lambda\frac{\partial f}{\partial y} \|_0^2 + \| \Lambda^2 f(x,y) \|_0^2 \,,$$

we have

(2.7) $$\| f(x,0) \|_{\frac{3}{2}} \le C \| f(x,y) \|_2$$

where C is a constant independent of λ. We use hereafter the notation constant. *This means always a constant independent of* λ. Now we denote

$$N \;=\; 2N_0 \;+\; 2 \;+\; q \;+\; 2.$$

Then in the sense of new norm defined by (2.6), we obtain

$$\| \hat{u}(x,y;\lambda) \|_2 \,,\; \| u_1(x,y) \|_2 \,,\; \| u(x,y,1) \|_2 \le C \| f(x) \|_N \,,$$

where C is a constant independent of λ.

Hence taking account of the definitions of $v(x)$, $w_0(x)$, and of (2.7), we obtain

(2.8) $$\| w_0 \|_{\frac{1}{2}} \,, \| g(x) \|_{\frac{3}{2}} \le C_1 \| f(x) \|_N \,,$$

where C_1 is a constant independent of λ. Concerning g(x), we can show it directly. First

$$\frac{\partial}{\partial y}\hat{u}_D(\xi,0) \;=\; -\int_{-\infty}^{0} e^{\alpha y}\hat{h}(\xi,y)dy\,,$$

where $\alpha = (\lambda + \xi^2)^{1/2}$, $h(x,y) = u(x,y,1)$, and $\hat{h}$ is Fourier transform of h(x,y) with respect to x. Hence

$$| \frac{\partial}{\partial y} \hat{u}_D(\xi, 0) |^2 \leq \frac{1}{2\alpha} \int_{-\infty}^{0} | \hat{h}(\xi, y) |^2 \, dy \, .$$

This implies that

$$\int_{-\infty}^{\infty} (\lambda + \xi^2)^{5/2} \, | \frac{\partial}{\partial y} \hat{u}_D(\xi, 0) |^2 \, d\xi \ \leq \ \frac{1}{2} \int \int (\lambda + \xi^2)^2 \, | \hat{h}(\xi, y) |^2 \, dy d\xi \ \leq \ \frac{1}{2} \parallel h(x,y) \parallel_2^2 .$$

Hence $\parallel g(x) \parallel_{\frac{5}{2}} \leq \parallel u(x,y,1) \parallel_2 \leq const \cdot \parallel f \parallel_N$. A fortiori, (2.8) is true.

3 Almost null solution ψ

Let us introduce a cut-off function $\beta(x) \in C_0^\infty$ with small support contained in a neighborhood V of $x = 0$, and taking the value 1 for $| x |$ small. Let us recall the assumption

(3.1) $$a(x) = x^n \alpha(x) \, , \quad \alpha(0) \neq 0 \, .$$

we modify $\alpha(x)$ outside V, and we can assume $\alpha(x) \in \mathcal{B}$ and bounded away from 0. We are concerned with the operator $A^* = a(x) - \Lambda^{-1} b(x)$. Then

(3.2) $$(a(x) - \Lambda^{-1} b(x))(\alpha(x)^{-1} \beta(x) \psi) = (x^n - \Lambda^{-1} \tilde{b}(x))(\beta(x) \psi),$$

where $\tilde{b}(x) = b(x)/\alpha(x)$.

Neglecting β in (3.2), we are concerned with a good approximate solution of the equation

(3.3) $$(x^n - \Lambda^{-1} \tilde{b}(x)) \psi(x) = 0 \, .$$

Expecting the existence of ψ in $(\mathcal{S}')$, We solve this equation by converting it to Fourier image. Let

$$\tilde{b}(x) = b_0 + b_1 x + \cdots + b_l x^l + b_{l+1}(x) x^{l+1} \quad (\text{ Taylor expansion }).$$

Then (3.3) becomes

(3.4) $$\left[x^n - \Lambda^{-1} (\sum_{j=0}^{n-1} b_j(x) x^j) \right] \psi = \left(\Lambda^{-1} (\sum_{j=n}^{l} b_j(x) x^j) \right) \psi + \Lambda^{-1} b_{l+1}(x) x^{l+1} \psi \, .$$

Neglecting the last term, we consider its Fourier transform

(3.5) $$\begin{aligned} &\left[\left(\frac{d}{d\xi} \right)^n - i^{-n} (\lambda + \xi^2)^{-\frac{1}{2}} \left(\sum_{j=0}^{n-1} b_j \left(i \frac{d}{d\xi} \right)^j \right) \right] \hat{\psi}(\xi) \\ &= \left[i^{-n} (\lambda + \xi^2)^{-\frac{1}{2}} \sum_{j=n}^{l} b_j \left(i \frac{d}{d\xi} \right)^j \right] \hat{\psi}(\xi) \, . \end{aligned}$$

Let us denote it by

(3.6) $$L_0 \hat{\psi} = Q \hat{\psi} \, .$$

We solve this equation by

(3.7) $$L_0(u_0) = 0 \, , \ L_0(u_1) = Q u_0 \, , \cdots , \ L_0(u_m) = Q u_{m-1} \, , \cdots$$

We have the following proposition, which is the core part of our argument.

Proposition 1 *Under the assumption (B), we can find a sequence of solutions u_j satisfying*

$$(3.8)\qquad |\, u_j^{(p)}(\xi)\,| \leq \left(\lambda(\lambda+\xi^2)\right)^s A^{j+1}(2c_0\varphi^{-\frac{1}{2n}})^p \lambda^{-\frac{j}{2n}} \Psi(\lambda)^{j+1}\,,\quad p=0,1,2,\cdots$$

where A, c_0 are positive constants and

$$(3.9)\qquad s = \frac{n-1}{4n}\,,\ \varphi = \lambda + \xi^2\,,\ \Psi(\lambda) = \sum_{m=0}^{\infty}(c_R\, c(n)\lambda^{-2s})^m\,.$$

Moreover it holds that

$$|\, u_0(\xi)\,|\,,\ |\,\hat{\psi}_J(\xi)\,| \sim (\lambda(\lambda+\xi^2))^s \ \text{for } \lambda \text{ large,}$$

where

$$\hat{\psi}_J(\xi) = u_0 + u_1 + \cdots + u_J\,.$$

Then

$$(\, L_0 - Q\,)\hat{\psi}_J = -Q(u_J)\,.$$

Hence

$$(3.10)\qquad \|\,(x^n - \Lambda^{-1}\tilde{b}(x))\psi_J\,\|_{-\frac{1}{2}} \leq \|\,Q(u_J)\,\|_{-\frac{1}{2}} + \|\,b_{l+1}(x)x^{l+1}\psi_J\,\|_{-\frac{1}{2}}\,.$$

The right-hand side is estimated by

$$const\cdot \|\,\varphi^{-\frac{1}{2}}\lambda^s(\lambda+\xi^2)^s\,\|_{-\frac{1}{2}}\,\lambda^{-\frac{J}{2n}} + const\cdot\|\,\varphi^{-\frac{l+1}{2n}}\lambda^s(\lambda+\xi^2)^s\,\|_{-\frac{1}{2}}\,.$$

Later, we use a quantity equivalent to $\|\,\Lambda^{-2}\psi_J\,\|_1$ which is again equivalent to

$$\|\,\varphi^{-\frac{1}{2}}\lambda^s(\lambda+\xi^2)^s\,\|_0 = \|\,\varphi^{-\frac{1}{4}}\lambda^s(\lambda+\xi^2)^s\,\|_{-\frac{1}{2}}\,.$$

Then, since $\varphi^{-(l+1)/2n+1/4} \leq \lambda^{-(l+1)/2n+1/4}$, We have

$$(3.11)\qquad \|\,(x^n - \Lambda\tilde{b}(x))\psi_J\,\|_{-\frac{1}{2}} \leq const\cdot\left(\lambda^{-\frac{J}{2n}-\frac{1}{4}} + \lambda^{-\frac{l+1}{2n}+\frac{1}{4}}\right)\|\,\Lambda^{-2}\psi_J\,\|_1\,.$$

This shows fairly clearly the accuracy of approximation. Namely when we choose J and l large, then the decreasing order in λ increases, when $\lambda \to \infty$.

Now we show that why we can neglect $\beta(x)$ in (3.2). To explain it, it is convenient to introduce

Definition 2 *We say that $f(x;\lambda)$ is negligible, and denote it by $f \approx 0$, when $\|\,f(x;\lambda)\,\|_0$ $(= (\int |\,f(x;\lambda)\,|^2\,dx)^{1/2})$ is rapidly decreasing function of λ when $\lambda \to \infty$, namely when $\|\,f(x;\lambda)\,\|_0$ is estimated by any negative power of λ. Next, $f \approx g$ means that $f - g \approx 0$.*

Now we see that

$$(3.12)\qquad (\,1 - \beta(x)\,)\,\psi_J \approx 0\,.$$

In fact, since $1-\beta(x) = 0$ in a neighborhood of the origin,

$$(\,1 - \beta(x)\,)\,\psi_J = \frac{1}{x^k}(\,1 - \beta(x)\,)(\,x^k\psi_J(x)\,)\,.$$

Hence its L^2-norm is estimated by $const.\ \|\,(\frac{d}{d\xi})^k\hat{\psi}_J(\xi)\,\|$ by virtue of Proposition 1, this is estimated in the form: $const\cdot\|\,\varphi^{-k/2n}\lambda^s(\lambda+\xi^2)^s\,\|_0 \leq const\cdot\lambda^{-k/2n+1/2}\ \|\,\Lambda^{-2}\psi_J\,\|_1$; and here we can choose k as large as we wish. Namely $\| (1-\beta)\psi_J \|_0$ is rapidly decreasing function of λ. Hence

$$(3.13) \qquad (x^n - \Lambda^{-1}\tilde{b}(x))(\beta\psi_J) \approx (x^n - \Lambda^{-1}\tilde{b}(x))\psi_J .$$

The existence theorem of ψ is postponed to the last part of this article. However, let us remark the following fact. The characteristic equation of L_0 (cf. (3.5)) is

$$(3.14) \qquad (i\mu)^n - (\lambda+\xi^2)^{-\frac{1}{2}}(b_0 + \sum_{j=1}^{n-1} b_j(i\mu)^j) = 0 .$$

Observe that $b_0 = b(0)\alpha(0)^{-1}$. The principle part of characteristic roots are obtained by

$$(3.15) \qquad (i\mu)^n - a^n b_0 = 0 , \text{ where } a = (\lambda+\xi^2)^{-\frac{1}{2n}} .$$

It can be shown that in order that $L_0(u_0(\xi)) = 0$ have a non-trivial solution in $(\mathcal{S}')_\xi$, it is necessary and sufficient that the characteristic equation (3.14) has *at least one pure imaginary root.* And it can be seen that it is equivalent to that (3.15) has such a root. Evidently if n is odd this is realized for any real non-zero b_0, and for n even this is realized when and only when $b_0 > 0$, namely $\alpha(0)$ and b(0) have the same sign. This condition is nothing but (B).

4 Final step

Now we return to the fundamental equation (2.5):

$$(4.1) \qquad Aw_0 = b(x)f(x) + e^{-\lambda}g(x) .$$

This suggests that , when $\lambda \to \infty$, Aw_0 and b(x)f(x) are nearly equal. Our idea is to use as f(x) some function closely related to $\psi(x)$. Namely, roughly speaking, we choose

$$f(x) = b(x)^{-1}\Lambda^{-2}\alpha(x)^{-1}\psi(x) .$$

However this function is not at all in H^∞ ($f \in H^1$, but$\notin H^2$), hence we cannot consider w_0 in (4.1). So, we are forced to use the following artifice. We choose

$$(4.2) \qquad f(x) = \beta(x)b(x)^{-1}\Lambda^{-2}\beta(x)\alpha(x)^{-1}\psi(x) .$$

where ψ is ψ_J just considered in §3. $\beta(x)$ is the cut-off function introduced at the beginning of §3. $b(x)^{-1}$ is the modification of $b(x)^{-1}$ just in the same way as $\alpha(x)$. Now instead of f(x) itself, we consider

$$(4.3) \qquad \tilde{f}(x) = \rho_\delta * f(x) \in H^\infty ,$$

where $\rho_\delta *$ is mollifier of size δ. Then, since $\tilde{f} \in H^\infty$, there exists $w_0 \in H^{1/2}$, and g(x) of the equation

$$(4.4) \qquad Aw_0 = b(x)\tilde{f} + e^{-\lambda}g .$$

This relation is not enough for our purpose. We use another mollifier of size ϵ

$$(4.5) \qquad w = \rho_\epsilon * w_0(x) \in H^\infty ,$$

and consider

$$(4.6) \qquad (Aw, b(x)f)_1 .$$

This is equal to

$$(\Lambda A w,\ \Lambda b(x) f\ )_0 = \left\langle A w,\ \overline{\Lambda^2 b(x) f} \right\rangle .$$

The right-hand side makes sense, since $w \in H^\infty$.

Now in view of (4.2), and Definition 2 and the argument used in the proof of (3.12),

$$\Lambda^2 b(x) f = \Lambda^2 \beta \Lambda^{-2} \beta \alpha(x)^{-1} \psi \approx \beta \alpha(x)^{-1} \psi .$$

Hence

$$Re\,(A w,\ b(x) f\ )_1 = Re \left\langle w,\ \overline{A^*(\beta \alpha^{-1} \psi)} \right\rangle + (rapidly\ decreasing)$$

$$\leq \| w \|_{\frac{1}{2}} \| A^*(\beta \alpha^{-1} \psi) \|_{-\frac{1}{2}} + (rapidly\ decreasing) .$$

In view of (3.2) and (3.13), we see that

$$\| A^*(\beta \alpha^{-1} \psi) \|_{-\frac{1}{2}} = \| (x^n - \Lambda^{-1} \tilde{b}(x)) \psi \|_{-\frac{1}{2}} + (rapidly\ decreasing) .$$

Concerning $\| w \|_{1/2}$, in view of (2.8) and Lemma 2 of next section, it holds

$$\| w \|_{\frac{1}{2}} \leq \| w_0 \|_{\frac{1}{2}} \leq const \cdot \| \tilde{f} \|_N \leq const \cdot \lambda^{\frac{N-1}{2}} \| f \|_1 .$$

Hence

$$(4.7) \qquad Re\,(A w,\ b(x) f\ )_1 \leq const \cdot \lambda^{\frac{N-1}{2}} \| (x^n - \Lambda^{-1} \tilde{b}(x)) \psi \|_{-\frac{1}{2}} \| f \|_1 + (rapidly\ decreasing) .$$

Now from (3.11),

$$(4.8) \qquad Re\,(A w,\ b(x) f\ )_1 \leq const \cdot \lambda^{\frac{N-1}{2}} (\lambda^{-\frac{J}{2n} - \frac{1}{4}} + \lambda^{-\frac{l+1}{2n} + \frac{1}{4}}) \| f \|_1^2 + (rapidly\ decreasing) .$$

In fact, we have

$$(4.9) \qquad \| f \|_1 \geq \frac{1}{2}\, inf \mid b(x)^{-1} \alpha(x)^{-1} \mid \| \Lambda^{-1} \psi_J \|_0 , \quad \text{for } \lambda \text{ large.}$$

Now we fix J and l as follows

$$(4.10) \qquad J = l = nN .$$

Then

$$(4.11) \qquad Re\,(A w,\ b(x) f\ )_1 \leq \epsilon(\lambda) \| f \|_1^2 ,$$

where $\epsilon(\lambda) \to 0$, when $\lambda \to \infty$.

On the other hand, in next section we show that

Proposition 2 *If we choose the sizes of mollifiers ρ_δ, ρ_ϵ suitably, it holds that*

$$(4.12) \qquad Re\,(A w,\ b(x) f\ )_1 \geq \frac{1}{2} \| b(x) f \|_1^2 \sim \frac{1}{2} \| f \|_1^2 .$$

(4.11) and (4.12) are not compatible, hence we proved Theorem 2.

5 On mollifiers

In previous section we used two mollifiers $\rho_\delta *$ and $\rho_\epsilon *$. Its property is fairly delicate, since the sizes δ and ϵ depend on parameter λ. First we estimate $\| \tilde{f} - f \|_1$.

Lemma 1 *We choose δ in the form : $\delta = \theta\lambda^{-1/2}$. Then $\| \tilde{f} - f \|_1 / \| f \|_1$ becomes as small as we wish by choosing θ (> 0) small.*

Proof. Recall $f(x) = \beta b(x)^{-1}\Lambda^{-2}\beta a(x)^{-1}\psi(x)$. Thus, by the property of ψ,

$$f \sim (b(0)a(0))^{-1}\Lambda^{-2}\psi .$$

It is enough to prove Lemma in the case $| \hat{f}(\xi) | = \varphi^{s-1}$, $\varphi = \lambda + \xi^2$ (see Appendix E). Now

$$\| \tilde{f} - f \|_1^2 = \int | \hat{\rho}(\delta\xi) - 1 |^2 | \hat{f}(\xi) |^2 \varphi d\xi .$$

We divide the domain of integral in two parts $| \xi | \leq M\sqrt{\lambda}$, and $| \xi | \geq M\sqrt{\lambda}$, and denote the corresponding integral by I_1 and I_2 respectively.

Estimate of I_1. $| \hat{\rho}(\delta\xi) - 1 |^2 \leq \delta^2\xi^2 \cdot sup_\xi | \hat{\rho}'(\xi) |^2$. Hence

$$I_1 \leq \delta^2 \sup_\xi | \hat{\rho}'(\xi) |^2 \int_{|\xi| \leq M\sqrt{\lambda}} \xi^2 | \hat{f}(\xi) |^2 \varphi d\xi \leq \delta^2 (M\sqrt{\lambda})^2 \sup_\xi | \hat{\rho}'(\xi) |^2 \int | \hat{f}(\xi) |^2 \varphi d\xi$$
$$= \theta^2 M^2 \sup_\xi | \hat{\rho}'(\xi) |^2 \int | \hat{f}(\xi) |^2 \varphi d\xi .$$

Estimate of I_2. We need to use the form of $| \hat{f}(\xi) |$. Since $2s - 1 < 0$, $\varphi^{2s-1} < | \xi |^{4s-2}$.

$$I_2 = 2\int_{M\sqrt{\lambda}}^\infty \varphi^{2(s-1)}\varphi d\xi \leq 2\int_{M\sqrt{\lambda}}^\infty \xi^{4s-2} d\xi = \frac{2}{1-4s}(M\sqrt{\lambda})^{4s-1} .$$

On the other hand

$$\| f \|_1^2 = \int \varphi^{2s-1} d\xi = const \cdot \lambda^{2s-\frac{1}{2}} .$$

Hence

$$\| \tilde{f} - f \|_1^2 \leq (\theta^2 M^2 sup | \hat{\rho}'(\xi) |^2 + const \cdot M^{4s-1}) \| f \|_1^2 .$$

Observe that $4s - 1 < 0$. First we choose M large, and next we choose θ small.

Note. θ was chosen independently of λ. The advantage depends on the specific form of f.

Hereafter we use the notation "const". This means always some constant independent of λ. However, in the statement of Lemmas, we state it explicitly.

Lemma 2 *Under the same assumption of Lemma 1, we have for $N > 1$,*

$$\| \tilde{f} \|_N \leq \theta^{-(N-1)}\lambda^{\frac{N-1}{2}} const \cdot \| f \|_1 \equiv const \cdot \lambda^{\frac{N-1}{2}} \| f \|_1 ,$$

where const. is independent of λ.

Proof.

$$\| \tilde{f} \|_N^2 = \| \rho_\delta * f \|_N^2 = \int | \hat{\rho}(\delta\xi) |^2 | \hat{f}(\xi) |^2 \varphi^N d\xi$$
$$\leq sup\left(| \hat{\rho}(\delta\xi) |^2 \varphi^{N-1}\right) \| f \|_1^2 \quad \text{Now putting } \delta\xi = \xi',$$

$$| \hat{\rho}(\delta\xi) |^2 \varphi^{N-1} = \frac{| \hat{\rho}(\xi') |^2 (\lambda\delta^2 + \xi'^2)^{N-1}}{\delta^{2(N-1)}} \qquad \text{since } \lambda\delta^2 = \theta^2 ,$$
$$= \theta^{-2(N-1)}\lambda^{N-1} sup\left(| \hat{\rho}(\xi) |^2 (\theta^2 + \xi^2)^{N-1}\right) .$$

Corollary. $\| \tilde{f} \|_{3/2} \leq const \cdot \lambda^{1/4} \| f \|_1$.

Lemma 3 *Let $k(x) \in H^{3/2}$. Then*

$$\| \rho_\epsilon * k - k \|_1 \leq \sqrt{\epsilon}\, const \cdot \| k \|_{\frac{3}{2}} ,$$

where const. is independent of λ.

Proof. As Lemma 1, we consider

$$\| \rho_\epsilon * k - k \|_1^2 = \int | \hat{\rho}(\epsilon\xi) - 1 |^2 | \hat{k}(\xi) |^2 \varphi d\xi .$$

We divide the integral in two parts $| \xi | \leq 1/\epsilon$, and $| \xi | \geq 1/\epsilon$, and denote its integral by J_1 and J_2 respectively. For J_1,

$$| \hat{\rho}(\epsilon\xi) - 1 |^2 \leq \epsilon^2 \xi^2 \cdot sup | \hat{\rho}'(\xi) |^2 \text{ . Hence}$$

$$\begin{aligned} J_1 &\leq \epsilon\ sup | \hat{\rho}'(\xi) |^2 \int_{|\xi| \leq 1/\epsilon} \epsilon | \xi | \frac{| \xi |}{\sqrt{\lambda + \xi^2}} | \hat{k}(\xi) |^2 (\lambda + \xi^2)^{\frac{3}{2}} d\xi \\ &\leq \epsilon\ sup | \hat{\rho}'(\xi) |^2 \int_{|\xi| \leq 1/\epsilon} | \hat{k}(\xi) |^2 (\lambda + \xi^2)^{\frac{3}{2}} d\xi . \\ J_2 &\leq 4 \int_{|\xi| \geq 1/\epsilon} | \hat{k}(\xi) |^2 (\lambda + \xi^2)^{-\frac{1}{2}} (\lambda + \xi^2)^{\frac{3}{2}} d\xi . \end{aligned}$$

Since $(\lambda + \xi^2)^{-1/2} \leq | \xi |^{-1} \leq \epsilon$, we obtain $J_1 + J_2 \leq \epsilon\ max(sup\ | \hat{\rho}'(\xi) |^2 ,\ 4) \| k \|_{3/2}^2$. Therefore

$$\| \rho_\epsilon * k - k \|_1 \leq \sqrt{\epsilon}\ max(sup\ | \hat{\rho}'(\xi) | ,\ 2) \| k \|_{3/2} .$$

Lemma 4 *Provided that ϵ satisfies $\lambda\epsilon^2 \leq 1$, we have*

$$\| [A , \rho_\epsilon *]\ w_0 \|_1 \leq const \cdot \sqrt{\epsilon} \| w_0 \|_{\frac{1}{2}}$$

where $w_0 \in H^{1/2}$ and const. is independent of λ and w_0.

Proof.

$$\Big[a(x) - b(x)\Lambda^{-1} , \rho_\epsilon * \Big]\ w_0 = (a(x)\rho_\epsilon * w_0 - \rho_\epsilon * a(x) w_0) - (b(x)\rho_\epsilon * - \rho_\epsilon * b(x))\Lambda^{-1} w_0 .$$

We consider the first term. This can be written

$$- (\hat{\rho}(\epsilon D) a(x) - a(x) \hat{\rho}(\epsilon D)) w_0 .$$

Let us consider only the first term of the asymptotic expansion:

$$\| a'(x) \epsilon \hat{\rho}'(\epsilon D) w_0 \|_1 \leq const \cdot \epsilon \| \hat{\rho}'(\epsilon D) w_0 \|_1 .$$

Now

$$\int | \hat{\rho}'(\epsilon\xi) |^2 | \hat{w}_0(\xi) |^2 \varphi d\xi = \int | \hat{\rho}'(\epsilon\xi) |^2 | \hat{w}_0(\xi) |^2 \varphi^{\frac{1}{2}} \varphi^{\frac{1}{2}} d\xi \leq \left(sup | \hat{\rho}'(\epsilon\xi) |^2 \varphi^{\frac{1}{2}} \right) \| w_0 \|_{\frac{1}{2}}^2 .$$

The factor can be written

$$sup\ | \hat{\rho}'(\xi) |^2 \left(\lambda + \frac{\xi^2}{\epsilon^2} \right)^{\frac{1}{2}} = \frac{1}{\epsilon} sup\ | \hat{\rho}'(\xi) |^2 \left(\lambda\epsilon^2 + \xi^2 \right)^{\frac{1}{2}} .$$

Hence we obtain the desired inequality, provided that $\lambda\epsilon^2 \leq 1$. The estimates of other terms are easier.

Proof of Proposition 2.

It is enough to prove

$$\| Aw - b(x)f \|_1 \leq \frac{1}{2} \| b(x)f \|_1 . \tag{5.1}$$

In fact

$$Re\,(Aw , b(x)f)_1 = Re\,(Aw - b(x)f , b(x)f)_1 + \| b(x)f \|_1^2$$
$$\geq \| b(x)f \|_1^2 - \| Aw - b(x)f \|_1 \| b(x)f \|_1 .$$

First let us observe

$$Aw_0 = b(x)\tilde{f} + e^{-\lambda}g(x) \in H^{\frac{3}{2}}$$

It follows that

$$\| Aw_0 \|_{\frac{3}{2}} \leq const \cdot \lambda^{\frac{1}{4}} \| f \|_1 . \tag{5.2}$$

In fact, $\| Aw_0 \|_{\frac{3}{2}} \leq \| b(x)\tilde{f} \|_{\frac{3}{2}} + e^{-\lambda} \| g(x) \|_{\frac{3}{2}}$

$$\| g(x) \|_{\frac{3}{2}} \leq \| \tilde{f} \|_N \leq const \cdot \lambda^{\frac{N-1}{2}} \| f \|_1 \quad \text{(Lemma 2).} \tag{5.3}$$

Taking account of Corollary of Lemma 2, we obtain (5.2) for λ large.

Now let us consider

$$Aw - b(x)f = (Aw - Aw_0) + (Aw_0 - b(x)\tilde{f}) - b(x)(f - \tilde{f}) . \tag{5.4}$$

First $\| Aw_0 - b(x)\tilde{f} \|_1 = e^{-\lambda} \| g(x) \|_1 \leq const \cdot e^{-\lambda}\lambda^{\frac{N-1}{2}} \| f \|_1$ (by (5.3)). Next, $\| b(x)(\tilde{f} - f) \|_1 \leq 2\, max \,| b(x) | \| \tilde{f} - f \|_1$.

We fix θ hence δ (in Lemma 1) in such a way that the right-hand side is less than $\| f \|_1 /4$.

Let us show that it holds

$$\| Aw - Aw_0 \|_1 \leq const \cdot \sqrt{\epsilon}\lambda^{\frac{N-1}{2}} \| f \|_1 . \tag{5.5}$$

$Aw - Aw_0 = (\rho_\epsilon * Aw_0 - Aw_0) + [A , \rho_\epsilon *]\, w_0$.

$$\| (\rho_\epsilon * -1)Aw_0 \|_1 \leq const \cdot \sqrt{\epsilon} \| Aw_0 \|_{\frac{3}{2}} \quad \text{(by Lemma 3)}$$
$$\leq const \cdot \sqrt{\epsilon}\, \lambda^{\frac{1}{4}} \| f \|_1 \quad \text{(by (5.2));}$$

$$\| [A , \rho_\epsilon *]w_0 \|_1 \leq const \cdot \sqrt{\epsilon} \| w_0 \|_{\frac{1}{2}} \quad \text{(by Lemma 4)}$$
$$\leq const \cdot \sqrt{\epsilon} \| \tilde{f} \|_N \quad \text{(by (2.8))} \leq const \cdot \sqrt{\epsilon}\, \lambda^{\frac{N-1}{2}} \| f \|_1 \quad \text{(by Lemma 2)}.$$

Thus (5.5) was confirmed. Summing up:

$$\| Aw - b(x)f \|_1 \leq const \cdot \sqrt{\epsilon}\lambda^{\frac{N-1}{2}} \| f \|_1 + const \cdot e^{-\lambda}\lambda^{\frac{N-1}{2}} \| f \|_1 + \frac{1}{4} \| f \|_1 . \tag{5.6}$$

Finally we choose

$$\epsilon = \lambda^{-N} \tag{5.7}$$

Thus we proved (5.1) for λ large. $\| b(x)f \|_1 \sim \| f \|_1$ is proved in Appendix E.

6 On construction of ψ (I).

This part is fairly lengthy. Let us consider (see (3.6))

$$L_0 u_0 = \left[\left(\frac{d}{d\xi}\right)^n - i^{-n}(\lambda+\xi^2)^{-\frac{1}{2}} \sum_{j=0}^{n-1} b_j \left(i\frac{d}{d\xi}\right)^j \right] u_0 = 0\,.$$

The characteristic equation is

$$(\,i\mu\,)^n - (\lambda+\xi^2)^{-\frac{1}{2}} \left(b_0 + \sum_{j=1}^{n-1} b_j (i\mu)^j \right) = 0\,.$$

Denote

(6.1) $$a(\xi;\lambda) = (\lambda+\xi^2)^{-\frac{1}{2n}}\,.$$

If we put $i\mu/a = \nu$, the above equation becomes

$$\nu^n - \left(b_0 + \sum_{j=1}^{n-1} b_j (a\nu)^j \right) = 0\,.$$

Namely, ν is the function of a defined by

(6.2) $$\nu = \sqrt[n]{b_0 + b_1 a\nu + b_2 a^2\nu^2 + \cdots + b_{n-1}a^{n-1}\nu^{n-1}}\,,$$

where a is a small parameter. Observe that b_0 is real and $\neq 0$. By implicit function theorem in complex sense, ν are holomorphic function of a. We are concerned with in particular real-valued function ν. Sincr b_j are all real, if n is odd, there exists one and only one real-valued root for a real and small, and if n is even, when $b_0 > 0$, there exists two real roots. Hence we see that

(6.3) $$\begin{cases} \mu_j = i^{-1}\omega^{j-1} b_0' a(1+\varphi_j(a))\,,\ 1 \le j \le n\,,\ b_0' = \sqrt[n]{b_0}\ \text{(real)} \\ \omega = e^{\frac{2\pi i}{n}}\ ;\ \varphi_j(a) \text{ are holomorphic functions in a neighborhood} \\ \text{of } a = 0 \text{ and } \varphi_j(0) = 0. \end{cases}$$

Let us notice that μ_1 is pure imaginary root. Then μ_j are divided in three cases: $Re\mu_j \equiv 0$, $Re\,\mu_j > 0$, and $Re\,\mu_j < 0$. In order to see clearly the estimates of derivatives of u_0, putting $(\frac{d}{d\xi})^j u_0 = u_j$, ($j = 0, 1, \cdots, n-1$), and $U = {}^t(u_0, u_1, \cdots, u_{n-1})$, we treat $L_0(u_0) = 0$ in matrix form

(6.4) $$\frac{d}{d\xi}U = AU\,.$$

We choose the diagonalizer N of A in such a way that the entries of the last column of N are all 1 (see §8). Putting $NU = V$, we obtain

(6.5) $$\frac{d}{d\xi}V = (D+C)V$$

where $C = N_\xi N^{-1}$, and D is the diagonal matrix, whose (j,j)-entry is μ_j.

We go further. Introducing M, whose diagonal entries are all 0, put $(I+M)V = W$, (5.5) can be written,

$$\frac{d}{d\xi}W = (D + D_0 + R_1)W, \tag{6.6}$$

where D_0 is also diagonal, whose (j,j)-entry is

$$d_{jj} = \frac{n-1}{2}a_\xi a^{-1} + c_{jj}(a)a_\xi, \tag{6.7}$$

where $c_{jj}(a)$ is holomorphic in a neighborhood of $a = 0$. To define u_0 satisfying (3.7) is equivalent to define W_H satisfying

$$\left[\frac{d}{d\xi} - (D + D_0)\right] W_H = R_1 W_H. \tag{6.8}$$

We construct W_H by successive approximations :

$$W_H = W_0 + W_1 + \cdots + W_m + \cdots$$

$$\begin{aligned} \left[\frac{d}{d\xi} - (D + D_0)\right] W_0 &= 0 \\ \left[\frac{d}{d\xi} - (D + D_0)\right] W_m &= R_1 W_{m-1} \end{aligned} \tag{6.9}$$

Let $W_0 = {}^t(w_{0,1}, w_{0,2}, \cdots, w_{0,n})$. Then the first equation of (6.9) is

$$\left(\frac{d}{d\xi} - d_j(\xi)\right) w_{0,j} = 0, \quad (1 \le j \le n). \tag{6.10}$$

The fundamental solution is

$$\begin{aligned} e_j(\xi,\eta) &= exp\left(\int_\eta^\xi d_j(\eta')d\eta'\right) \\ &= exp\left(-\sqrt{-1}\,\omega^{j-1}b_0'\int_\eta^\xi a(\eta')(1+\varphi_j(a))d\eta'\right)\left(\frac{a(\xi)}{a(\eta)}\right)^{\frac{n-1}{2}} \\ &\times exp\left(\int_\eta^\xi c_{jj}(a)a_\xi(\eta')d\eta'\right). \end{aligned} \tag{6.11}$$

In particular, for $j = 1$, the first factor is 1 in absolute values, and the third factor is negligible. More precisely

$$(1 - const\cdot\lambda^{-\frac{1}{2n}})\left(\frac{\lambda+\eta^2}{\lambda+\xi^2}\right)^s \le |\, e_1(\xi,\eta)\,| \le (1 + const\cdot\lambda^{-\frac{1}{2n}})\left(\frac{\lambda+\eta^2}{\lambda+\xi^2}\right)^s. \tag{6.12}$$

where

$$s = \frac{n-1}{4n} \tag{6.13}$$

Finally we define

$$\begin{cases} w_{0,1}(\xi) = e_1(\xi, 0), \\ w_{0,j}(\xi) \equiv 0 \quad (2 \le j \le n), \end{cases} \tag{6.14}$$

Hence

$$|\, W_0(\xi)\,| \sim \left(\frac{\lambda}{\lambda+\xi^2}\right)^s, \quad \text{when } \lambda \to \infty. \tag{6.15}$$

7 Construction of W_H.

Let us proceed to the definition of W_1, $W_2,\cdots$. First it is shown that (see Appendix)

$$|\, R_1(\xi,\lambda)\,| \le C_R(\lambda+\xi^2)^{-1+\frac{1}{2n}} \equiv C_R\varphi^{-1+\frac{1}{2n}} \,. \tag{7.1}$$

Denote

$$S = S(\xi,\lambda) = \left(\frac{\lambda}{\lambda+\xi^2}\right)^s \quad (= e_1(\xi,0)\,)\,. \tag{7.2}$$

Put $R_1W_0 = {}^t(f_1,\cdots,f_n)$. We define $W_1 = {}^t(w_{1,1},w_{1,2},\cdots,w_{1,n})$ by

$$w_{1,1}(\xi) = \int_0^\xi e_1(\xi,\eta)f_1(\eta)d\eta\,,$$

$$w_{1,j}(\xi) = \begin{cases} \displaystyle\int_{-\infty}^\xi e_j(\xi,\eta)f_j(\eta)d\eta\,, & \text{when } Re\ d_j < 0\,,\\ \displaystyle -\int_\xi^\infty e_j(\xi,\eta)f_j(\eta)d\eta\,, & \text{when } Re\ d_j > 0\,. \end{cases}$$

Further, when n is even, there exists j (>1) such that $Re\ d_j(\xi) \equiv 0$. In this case we define in the same way as $w_{1,1}(\xi)$. It is shown that

$$|\, W_1\,| \equiv \sum_{j=1}^n |\, w_{1,j}(\xi)\,| \le C_Rc(n)\lambda^{-2s}2S\,, \tag{7.3}$$

where c(n) is a constant depending only on n. We repeat this consideration. We get

$$|\, W_m(\xi)\,| \le (C_Rc(n)\lambda^{-2s})^m 2S \quad (\, m = 1,2,\cdots)$$

Hence

$$|\, W_H(\xi)\,| \le (1+\sum_{m=1}^\infty (C_Rc(n)\lambda^{-2s})^m)2S \equiv \Psi(\lambda)2S\,. \tag{7.4}$$

The estimates of $(\frac{d}{d\xi})^pW_H = W_H^{(p)}$ ($p = 1,2,\cdots$) is fairly delicate. For this purpose first we establish the following estimates using Cauchy's integral representation (see Appendix).

Lemma 5

$$\begin{matrix} |\, (D+D_0)^{(p)}\,| \le c_0C^pp!\varphi^{-\frac{1}{2n}-\frac{p}{2}} \\ |\, R_1^{(p)}\,| \le C_RC^pp!\varphi^{-1+\frac{1}{2n}-\frac{p}{2}} \end{matrix} \qquad p = 0,1,2,\cdots,$$

where c_0, C_R, C are positive constants ≥ 1 independent of λ, ξ, and if necessary, taking C large we can assume $C/2c_0 \ge 1$.

Hence we have

$$|\, W_H^{(p)}(\xi)\,| \le (2c_0\varphi^{-\frac{1}{2n}})^p\Psi(\lambda)2S\,,\quad p = 1,2,\cdots\,, \tag{7.5}$$

provided that λ satisfies $\lambda \ge (p\cdot 2C/c_0)^{1/2s}$.

The proof of (7.5) is the following. We consider first $W_0^{(p)}$. Starting from

$$W_0{}' = (D+D_0)W_0\,,\quad |\, W_0\,| \le 2S\,, \tag{7.6}$$

we get

$$|\, W_0^{(p)}(\xi)\,| \le (2c_0\varphi^{-\frac{1}{2n}})^p\,|\, W_0\,|\,. \tag{7.7}$$

We show this by induction on p. From (7.6) we have

$$W_0^{(p+1)} = (D+D_0)W_0^{(p)} + \sum_{j\geq 1} \binom{p}{j} (D+D_0)^{(j)} W_0^{(p-j)} .$$

Now we assume that (7.7) is true for $p = 0, 1, 2, \cdots, p$.
Then, general term of the right-hand side is estimated by

$$\frac{p!}{(p-j)!} c_0 C^j \varphi^{-\frac{1}{2n}} \varphi^{-\frac{j}{2}} (2c_0 \varphi^{-\frac{1}{2n}})^{p-j} \mid W_0 \mid .$$

This is estimated again by

$$p^j C^j (c_0 \varphi^{-\frac{1}{2n}})(2c_0\varphi^{-\frac{1}{2n}})^p (2c_0)^{-j} \varphi^{(\frac{1}{2n}-\frac{1}{2})j} \mid W_0 \mid = \frac{1}{2}(2c_0\varphi^{-\frac{1}{2n}})^{p+1} (p\frac{C}{2c_0}\varphi^{-2s})^j \mid W_0 \mid .$$

Thus, if

(7.8) $$p \frac{C}{2c_0} \lambda^{-2s} \leq \frac{1}{4} ,$$

$$\sum_{j\geq 1} \binom{p}{j} (D+D_0)^{(j)} W_0^{(p-j)} \leq \frac{1}{2} (2c_0\varphi^{-\frac{1}{2n}})^{p+1} \left(\sum_{j=1}^{p} \frac{1}{2^2} + \frac{1}{2^4} + \cdots \right) \mid W_0 \mid$$
$$\leq \frac{1}{4} (2c_0\varphi^{-\frac{1}{2n}})^{p+1} \mid W_0 \mid .$$

Hence we get (7.7) for $p+1$.

In order to estimate $W_1^{(p)}$, $W_2^{(p)}, \cdots$ we use the following lemmas, (proofs are given in Appendix).

Lemma 6 *Let $f(\xi;\lambda)$ satisfy*

$$\mid f^{(p)}(\xi;\lambda) \mid \leq (2c_0\varphi^{-\frac{1}{2n}})^p S \quad \textit{for } p \textit{ satisfying}$$

(*) $$p \cdot \frac{C}{2c_0} \leq \frac{1}{4}\lambda^{2s} ,$$

and let $g(\xi;\lambda)$ satisfy

$$\mid g^{(p)}(\xi;\lambda) \mid \leq AC^p p! \, \varphi^{q-\frac{p}{2}} \quad p = 0, 1, 2, \cdots$$

Then

$$\mid (\frac{d}{d\xi})^p (gf) \mid \leq \frac{3}{2} A \varphi^q (2c_0\varphi^{-\frac{1}{2n}})^p S ,$$

provided that p satisfies (*).

Lemma 7 *Let $V(\xi;\lambda)$ satisfy*

$$\mid V^{(p)} \mid \leq (2c_0\varphi^{-\frac{1}{2n}})^p S \quad \textit{for } p \textit{ satisfying } (*).$$

Then the solution W of the equation

$$\left[\frac{d}{d\xi} - (D+D_0) \right] W = R_1 V$$

satisfies

$$\mid W^{(p)} \mid \leq (2c_0\varphi^{-\frac{1}{2n}})^p (C_R \, c(n) \lambda^{-2s}) S$$

for p satisfying (∗), where W satisfies the initial condition explained at the beginning of this section.

Let us remark that Lemma 6 is used to prove Lemma 7. Now, by repeated use of Lemma 7, we can show that

$$| W_m^{(p)} | \leq (2c_0\varphi^{-\frac{1}{2n}})^p (C_R\, c(n)\lambda^{-2s})^m 2S \,. \tag{7.9}$$

Thus we proved (7.5).

8 Construction of ψ (II).

We return to the construction of ψ.

$$\hat{\psi}_J = u_0 + u_1 + \cdots + u_J \,, \tag{8.1}$$

where u_0, u_1 , $\cdots$ are defined by (3.7):

$$L_0(u_0) = 0,\ L_0(u_1) = Q(u_0),\ L_0(u_2) = Q(u_1),\ \cdots \,. \tag{8.2}$$

In preceding section we constructed W_H and obtained the estimate of its derivatives (7.5) :

$$| W_H^{(p)}(\xi) | \leq (2c_0\varphi^{-\frac{1}{2n}})^p \Psi(\lambda) 2S \,, \quad p = 1, 2, \cdots \,, \tag{8.3}$$

Now, in order to obtain the estimate of u_1 and its derivatives, first we need to know the estimates of $(\frac{d}{d\xi})^p Q(u_0)$. We explain how to derive it from (8.3). Let

$$U_0 = {}^t(u_0,\, u_0',\cdots,\, u_0^{(n-1)}) \,. \tag{8.4}$$

Then

$$U_0 = N^{-1}(I + M)^{-1} W_H = N^{-1}(I + \tilde{M}) W_H \,.$$

Let, $N^{-1} = (\tilde{n}_{ij})_{1 \leq i,j \leq n}$, we see that (see the last part of this section, concerning the structure of N).

$$| \left(\frac{d}{d\xi}\right)^p \tilde{n}_{ij} | \leq L\, C^p\, p!\, \varphi^{\frac{n-i}{2n} - \frac{p}{2}} \,, \quad (1 \leq i \leq n) \,, \quad p = 0, 1, 2, \cdots \tag{8.5}$$

where C is the same one as in Lemma 5. Also, Let $\tilde{M} = (\tilde{m}_{ij})$, we have

$$| \left(\frac{d}{d\xi}\right)^p \tilde{m}_{ij} | \leq L'\, C^p\, p!\, \varphi^{-2s - \frac{p}{2}} \,, \quad p = 0, 1, 2, \cdots \tag{8.6}$$

By applying Lemma 6, we obtain

$$| \left(\frac{d}{d\xi}\right)^p u_0^{(i)} | \leq \frac{3}{2}(L + \epsilon'(\lambda))\, \varphi^{\frac{n-1-i}{2n}} (2c_0\varphi^{-\frac{1}{2n}})^p\, \Psi(\lambda) 2S \tag{8.7}$$

where $\epsilon'(\lambda) \to 0$ when $\lambda \to \infty$ provided that p satisfies (∗).

In particular for $i = n-1$, we obtain

$$(8.8) \qquad \left| \left(\frac{d}{d\xi}\right)^p u_0^{(n-1)} \right| \le \frac{3}{2}(L+\epsilon'(\lambda))\,(2c_0\varphi^{-\frac{1}{2n}})^p\,\Psi(\lambda)2S\,.$$

Now we return to $Q(u_0)$.

$$(8.9) \qquad \begin{aligned} Q(u_0) &= \varphi^{-\frac{1}{2}}\tilde{Q}(i\frac{d}{d\xi})\frac{d}{d\xi}u_0^{(n-1)}(\xi)\,, \\ \text{where} \quad \tilde{Q} &= b_n + b_{n+1}(i\frac{d}{d\xi}) \cdots + b_l(i\frac{d}{d\xi})^{l-n}\,. \end{aligned}$$

From (8.8) we obtain

$$(8.10) \qquad \left| \left(\frac{d}{d\xi}\right)^p Q(u_0) \right| \le \frac{3}{2}(L+\epsilon'(\lambda))\,K\,\varphi^{-\frac{1}{2}}(2c_0\varphi^{-\frac{1}{2n}})^{p+1}\,\Psi(\lambda)2S\,,$$

where K is a constant independent of λ.

Now denoting $U_1 = (u_1, u_1', \cdots, u_1^{(n-1)})$, $L(u_1) = Q(u_0)$ is equivalent to

$$\left(\frac{d}{d\xi} - A\right) U_1 = {}^t(0, 0, \cdots, 0, Q(u_0))\,.$$

This is again equivalent to

$$(8.11) \qquad \left[\frac{d}{d\xi} - (D + D_0 + R_1)\right] (I+M)NU_1 = (I+M)N\,{}^t(0,0,\cdots,0,Q(u_0))\,.$$

Observe that M has also the same type estimate as (8.6).

We can prove (see Appendix),

Lemma 8 *For the equation*

$$\left[\frac{d}{d\xi} - (D + D_0 + R_1)\right] W = F\,,$$

suppose that

$$| F^{(p)}(\xi) | \le A\varphi^{-\frac{1}{2}}(2c_0\varphi^{-\frac{1}{2n}})^{p+1}\,S, \quad p = 0, 1, 2, \cdots, p_0\,.$$

Then there exists a solution W satisfying

$$| W^{(p)} | \le A\,(2c_0c'(n))\lambda^{-\frac{1}{2n}}(2c_0\varphi^{-\frac{1}{2n}})^p\,\Psi(\lambda)\,S, \text{ for } p = 0, 1, 2, \cdots, p_0$$

where $c'(n) = 2\int_{-\infty}^{\infty}(1+\eta^2)^{-\frac{1}{2}-\frac{1}{2n}}d\eta$, *and* p_0 *is the integer satisfying* $(*)$.

Apply this lemma to (8.11), taking $F = (I+M)\,{}^t(Q(u_0), \cdots, Q(u_0))$. Then denoting $W_1 = (I+M)NU_1$, we get

$$(8.12) \qquad | W_1^{(p)} | \le 2(L+\epsilon')\,(2c_0c'(n))K\,\lambda^{-\frac{1}{2n}}\,(2c_0\varphi^{-\frac{1}{2n}})^p\,\Psi(\lambda)^2\,2S\,.$$

By the above argument (see (8.8)), putting $K' = 2c_0c'(n)K$, we obtain

$$(8.13) \qquad \left| \left(\frac{d}{d\xi}\right)^p u_1^{(n-1)} \right| \le 2^2(L+\epsilon')^2\,K'\,\lambda^{-\frac{1}{2n}}\,(2c_0\varphi^{-\frac{1}{2n}})^p\,\Psi(\lambda)^2\,2S\,.$$

Observe that, *comparing this with (8.8), we gained the factor* $\lambda^{-\frac{1}{2n}}$. Repeating this argument, we see that we can obtain

$$(8.14) \qquad \left| \left(\frac{d}{d\xi}\right)^p u_j^{(n-1)}(\xi) \right| \le (2(L+\epsilon'))^{j+1} K'^j \lambda^{-\frac{j}{2n}} (2c_0\varphi^{-\frac{1}{2n}})^p \Psi(\lambda)^{j+1} 2S .$$

Proof of Proposition 1.

We explain briefly about diagonalizer N. First consider the case

$$A = \begin{bmatrix} 0 & 1 & & & 0 \\ & & 1 & & \\ 0 & & & \ddots & \\ & & & \ddots & 1 \\ 1 & 0 & 0 & \cdots \cdots & 0 \end{bmatrix}$$

The characteristic roots are $\mu = 1, \omega, \omega^2, \cdots, \omega^{n-1}$, where $\omega = e^{2\pi i/n}$. Let μ be one of those, then each row vector of N is a null vector of ${}^tA - \mu I$. Hence we choose as diagonalizer

$$N_0 = \begin{bmatrix} 1 & 1 & \cdots \cdots \cdots & 1 \\ \omega^{n-1} & \omega^{n-2} & \cdots \cdots \cdots & 1 \\ & \cdots & \cdots \cdots \cdots & \\ (\omega^{n-1})^{n-1} & (\omega^{n-1})^{n-2} & \cdots \cdots \cdots & 1 \end{bmatrix} = \left[(\omega^{i-1})^{n-j}\right]_{1\le i,j\le n} .$$

Hence

$$N_0^{-1} = \frac{1}{n} \begin{bmatrix} 1 & (\omega^{-1})^{n-1} & (\omega^{-2})^{n-1} & \cdots \cdots \\ 1 & (\omega^{-1})^{n-2} & (\omega^{-2})^{n-2} & \cdots \cdots \\ \cdots & \cdots & \cdots & \cdots \cdots \\ \cdots & \cdots & \cdots & \cdots \cdots \\ 1 & 1 & 1 & \cdots \cdots \end{bmatrix} = \left[(\omega^{-(j-1)})^{n-i}\right]_{1\le i,j\le n} .$$

Next in the case

$$A = \begin{bmatrix} 0 & 1 & & & 0 \\ & & 1 & & \\ 0 & & & \ddots & \\ & & & \ddots & 1 \\ c^n & 0 & 0 & \cdots \cdots & 0 \end{bmatrix}$$

the characteristic roots are $\mu = c, c\omega, c\omega^2, \cdots, c\omega^{n-1}$. The corresponding diagonalizer N is

$$N = N_0 \begin{bmatrix} c^{n-1} & & & 0 \\ & c^{n-2} & & \\ & & \ddots & \\ 0 & & & 1 \end{bmatrix} .$$

Hence

$$N^{-1} = \begin{bmatrix} c^{-(n-1)} & & & 0 \\ & c^{-(n-2)} & & \\ & & \cdots & \\ & & \cdots & \\ 0 & & & 1 \end{bmatrix} N_0^{-1}\,, \quad c = i^{-1}b_0{}'a\,.$$

In the actual case

$$A = \begin{bmatrix} 0 & 1 & & & & 0 \\ & & 1 & & & \\ & & & \cdots & & \\ 0 & & & & \cdots & \\ & & & & & 1 \\ b_0 a^n & \cdots & \cdots & \cdots & \cdots & \cdots \end{bmatrix},$$

here the entries of the last row entail the change of N_0, however the change occurs only by the addition of terms of first order in a.

To see the principal part of u_0, recall that, in the definition of $W_H = W_0 + W_1 + \cdots$, $W_0 = {}^t(e_1(\xi,0), 0, \cdots, 0)$. Now taking account of (8.5) and the structure of N, we see that the principal part of u_0 is just

$$n^{-1}(i^{-1}b_0{}'\,a)^{-(n-1)}e_1(\xi,0) \;=\; n^{-1}(ib_0{}'^{-1})^{n-1}\varphi^{\frac{n-1}{2n}}e_1(\xi,0)\,.$$

Hence in view of (6.14),

$$|\,u_0\,| \sim n^{-1}\,|\,b_0{}'\,|^{-(n-1)}\,\lambda^s(\lambda+\xi^2)^s\,.$$

Thus, if we change the definition of $w_{0,1}(\xi)$ to $n(i^{-1}b_0{}')^{n-1}e_1(\xi,0)$, we have

$$|\,u_0(\xi)\,| \sim (\lambda(\lambda+\xi^2))^s\,.$$

Next, we see that $|\,u_0^{(j)}(\xi)\,|$ is estimated by (3.8) in view of (8.7). In fact since $i=0$ in this case, $(n-1-i)/2n = (n-1)/2n = 2s$, and $\varphi^{2s}S = (\lambda(\lambda+\xi^2))^s$. Concerning u_1, instead of (8.13) we see that $|\,(\frac{d}{d\xi})^p u_1(\xi)\,|$ is estimated by the right-hand side multiplied by $\varphi^{(n-1)/2n}$ by the same reason as u_0. Thus we get (3.8) for $j=1$. For $j = 2, 3, \cdots$, the situation is the same. Thus we proved Proposition 1.

Appendix

We give the proofs of Lemma 5 $\sim$ 8, and some comments on the property of $f(x)$.

A Proof of Lemma 5.

The key point is the following fact. Let $\lambda > 0$, and consider the complex disc D of center ξ ($\in \mathbf{R}$) of radius R : $D = \{z \mid\, |\,z-\xi\,| \le R\}$, Then if we choose $R = \sqrt{\lambda+\xi^2}/4$, there exist positive constants ρ_0, ρ_1 such that

(A.1) $$\rho_0 \le \frac{|\,\lambda+z^2\,|}{\lambda+\xi^2} \le \rho_1\,,$$

where z runs through D, and ρ_0 and ρ_1 are independent of $(\lambda,\xi) \in \mathbf{R}_+ \times \mathbf{R}$.

To prove this, it is enough to prove it assuming $\xi \geq 0$. Then we confirm the assertion by dividing, for instance, into the following cases.

$$1)\ \xi \leq \frac{1}{2}\sqrt{\lambda}\,,\quad 2)\ \frac{1}{2}\sqrt{\lambda} \leq \xi \leq \sqrt{\lambda}\,,\quad 3)\ \sqrt{\lambda} \leq \xi \leq 2\sqrt{\lambda}\,,\quad 4)\ \xi \geq 2\sqrt{\lambda}\,,$$

To estimate, for instance $(\frac{d}{d\xi})^p(\lambda+\xi^2)^{-\sigma}$ $(\sigma > 0)$, we start from

$$(\lambda+\xi^2)^{-\sigma} = \frac{1}{2\pi i}\int_{|z-\xi|=R}\frac{1}{z-\xi}(\lambda+z^2)^{-\sigma}dz\,.$$

Hence

$$|\,(\frac{d}{d\xi})^p(\lambda+\xi^2)^{-\sigma}\,| \leq \frac{\max_{z\in D}|\lambda+z^2|^{-\sigma}\cdot p!}{R^p} \leq \rho_0^{-\sigma}\frac{(\lambda+\xi^2)^{-\sigma}p!}{(\lambda+\xi^2)^{\frac{p}{2}}4^{-p}}$$
$$= \rho_0^{-\sigma}4^p(\lambda+\xi^2)^{-\sigma-\frac{2}{p}}p!\,,\ p=0,1,2,\cdots\,.$$

In the same way as above we see easily that

$$\text{(A.2)}\qquad \sup_{z\in D}|a_\xi(\lambda,z)| \leq \rho_2(\lambda+\xi^2)^{-\frac{1}{2}-\frac{1}{2n}} \equiv \rho_2\varphi^{-\frac{1}{2}-\frac{1}{2n}}\,,$$

where ρ_2 is also independent of λ, ξ.

Let us remark that $MD - DM = D_0 - C$. The matrices C and M have the following properties

1) $C = N_\xi N^{-1} = (c_{ij})\,,\ c_{ij} = c_{ij}^{(0)}a_\xi a^{-1} + c_{ij}^{(I)}(a)a_\xi,$

where $c_{ij}^{(0)}$ is a constant and $c_{ij}^{(I)}(a)$ is a holomorphic in a neighborhood of $a = 0$. Observe that $c_{ii}^{(0)} = (n-1)/2$.

2) Let $M = (m_{ij})$. $m_{ij} = m_{ij}^{(0)}a_\xi a^{-2} + m_{ij}^{(I)}(a)a_\xi + m_{ij}^{(II)}(a)a_\xi,$

where $m_{ij}^{(0)}$ is a constant and $m_{ij}^{(I)}$, $m_{ij}^{(II)}$ are holomorphic in a neighborhood of the origin.

Finally

$$\text{(A.3)}\qquad R = MC - D_0M + M_\xi\,.$$

Symbolically we can express the above properties in the following way

$$C \in a_\xi a^{-1}\,,\ M \in a_\xi a^{-2}\,,\ D_0 \in a_\xi a^{-1}\,.$$

Hence we have

$$\text{(A.4)}\qquad \max_{z\in D}|R_1(\lambda,z)| \leq C_R\,\varphi^{-1+\frac{1}{2n}}\,.$$

B Proof of Lemma 6.

$$(gf)^{(p)} = gf^{(p)} + \sum_{i=1}^{p}\binom{p}{j}g^{(j)}f^{(p-j)}\,.$$

General term in the sum is estimated by

$$\frac{p!}{(p-j)!}AC^j\varphi^{q-\frac{1}{2}j}(2c_0\varphi^{-\frac{1}{2n}})^{p-j}S$$
$$\leq A\,\varphi^q(2c_0\varphi^{-\frac{1}{2n}})^p\left(p\frac{C}{2c_0}\right)^j\varphi^{-j(\frac{1}{2}-\frac{1}{2n})}S$$
$$\leq A\,\varphi^q(2c_0\varphi^{-\frac{1}{2n}})^p\left(p\frac{C}{2c_0}\lambda^{-2s}\right)^j S\,.$$

Hence $(gf)^{(p)}$ is estimated, under assumption (*), by

$$A\varphi^q(2c_0\varphi^{-\frac{1}{2n}})^p\left[1+\sum_{j=1}^{p}\left(\frac{1}{4}\right)^j\right]\,.$$

C Proof of Lemma 7.

For $p = 0$, we explained the construction of W at the beginning of §7. (see (7.3)). we use the estimate of $R_1^{(p)}$ in Lemma 5. The number c(n) is

$$c(n) = 2\int_{-\infty}^{\infty}(1+\xi^2)^{-1+\frac{1}{2n}}d\xi \,. \tag{C.1}$$

Hence

$$|W| \leq C_R\, c(n)\lambda^{-2s}S \,.$$

For $p = 1$

$$W' = (D + D_0)W + R_1 V \tag{C.2}$$

yields the estimate

$$|W'| \leq c_0\, \varphi^{-\frac{1}{2n}}(C_R\, c(n)\lambda^{-2s})S + C_R\varphi^{-1+\frac{1}{2n}}S \,.$$

Observe that

$$\begin{aligned} C_R\varphi^{-1+\frac{1}{2n}} &\leq C_R\lambda^{-4s}\varphi^{-\frac{1}{2n}} \\ &\leq (C_Rc(n)\lambda^{-2s})(2c_0\varphi^{-\frac{1}{2n}})\left(\frac{\lambda^{-2s}}{2c_0c(n)}\right) . \end{aligned}$$

Now $(*)$ implies $\lambda^{-2s}/2c_0 \leq 1/4$, and since $c(n) > 2$, hence last factor $\leq 1/8$, we get the desired estimate for $p = 1$.

For general p, we prove it by induction on p.
First $((D + D_0)W)^{(p)}$ is estimated, using Lemma 6, by

$$\frac{3}{2}(C_R\, c(n)\lambda^{-2s})(2c_0\, \varphi^{-\frac{1}{2n}})^p(c_0\, \varphi^{-\frac{1}{2n}})S \,,$$

and $(R_1V)^{(p)}$ is estimated by

$$\frac{3}{2}(C_R\varphi^{-1+\frac{1}{2n}})(2c_0\varphi^{-\frac{1}{2n}})^pS \,,$$

which is estimated again, using the above estimate of $C_R\, \varphi^{-1+\frac{1}{2n}}$, by

$$\frac{3}{2}(C_Rc(n)\lambda^{-2s})(2c_0\varphi^{-\frac{1}{2n}})^{p+1}\left(\frac{\lambda^{-2s}}{2c_0c(n)}\right)S \,.$$

Observe that the last factor is less than 1/8. Since $3/4+3/2\times 1/8 < 1$, we get the desired estimate of $W^{(p+1)}$.

D Proof of Lemma 8.

We assume $A = 1$. We solve this eqiuation in the following way.

$$\begin{aligned} \left[\frac{d}{d\xi} - (D + D_0)\right] W_0 &= F \,, \\ \left[\frac{d}{d\xi} - (D + D_0)\right] W_1 &= R_1 W_0 \,, \\ \left[\frac{d}{d\xi} - (D + D_0)\right] W_m &= R_1 W_{m-1} \,, \quad m = 1, 2, 3, \cdots \,. \end{aligned}$$

The solution is obtained by

$$W = W_0 + W_1 + \cdots + W_m + \cdots$$

where all W_m $(m \geq 0)$ are solved with initial condition 0 explained at the beginning of §7.

First let us show that

$$\text{(D.1)} \qquad |W_0^{(p)}| \leq 2c_0\, c'(n)\lambda^{-\frac{1}{2n}}(2c_0\, \varphi^{-\frac{1}{2n}})^p S .$$

It will be clear that

$$|W_0| \leq 2c_0\, c'(n)\lambda^{-\frac{1}{2n}} S .$$

Next from

$$\text{(D.2)} \qquad W'_0 = (D + D_0)W_0 + f ,$$

it follows that

$$|W'_0| \leq 2c_0\, c'(n)\lambda^{-\frac{1}{2n}}(c_0\, \varphi^{-\frac{1}{2n}})S + \varphi^{-\frac{1}{2}}(2c_0\, \varphi^{-\frac{1}{2n}})S .$$

Now last term $\leq 2c_0\, c'(n)\lambda^{-\frac{1}{2n}}(2c_0\, \varphi^{-\frac{1}{2n}})((2c_0\, c'(n))^{-1}\lambda^{-2s})S$. Since $(2c_0)^{-1}\lambda^{-2s} \leq 1/4$, and $c'(n) > 2$, last factor is less than $1/8$. Thus (D.1) is proved for $p = 1$.

For general p, we prove it by induction on p. Now

$$|((D + D_0)W_0)^{(p)}| \leq \frac{3}{2}(2c_0\, c'(n))\lambda^{-\frac{1}{2n}}(c_0\, \varphi^{-\frac{1}{2n}})\, (2c_0\, \varphi^{-\frac{1}{2n}})^p S ,$$

$$|F^{(p)}| \leq \varphi^{-\frac{1}{2}}(2c_0\, \varphi^{-\frac{1}{2n}})^{p+1} S .$$

Since $\varphi^{-\frac{1}{2}} \leq \lambda^{-\frac{1}{2n}}\lambda^{-2s} \leq \lambda^{-\frac{1}{2n}}(2c_0\, c'(n))\, (2c_0\, c'(n))^{-1}\lambda^{-2s} \leq \frac{1}{8}\lambda^{-\frac{1}{2n}}(2c_0\, c'(n))$. Since $3/4 + 1/8 < 1$, the proof is completed.

Next, Let us estimate $W_1^{(p)}$. We can apply Lemma 7. Hence

$$\text{(D.3)} \qquad |W_1^{(p)}| \leq (2c_0\, c'(n))\lambda^{-\frac{1}{2n}}(2c_0\, \varphi^{-\frac{1}{2n}})^p(C_R c(n)\lambda^{-2s})S .$$

Repeated application of Lemma 7 implies

$$\text{(D.4)} \qquad |W_m^{(p)}| \leq (2c_0\, c'(n))\lambda^{-\frac{1}{2n}}(2c_0\, \varphi^{-\frac{1}{2n}})^p(C_R c(n)\lambda^{-2s})^m S .$$

Hence we obtain

$$|W^{(p)}| \leq (2c_0\, c'(n))\lambda^{-\frac{1}{2n}}(2c_0\, \varphi^{-\frac{1}{2n}})^p\Psi(\lambda)S .$$

E Properties of the function f(x)

Recall

$$\text{(E.1)} \qquad \begin{cases} \Lambda f = \Lambda\beta b(x)^{-1}\Lambda^{-2}(\beta a(x)^{-1})\psi , \\ \Lambda b(x) f = \Lambda\beta\Lambda^{-2}(\beta a(x)^{-1})\psi . \end{cases}$$

At first we claim that these functions are expressed in the form

$$\text{(E.2)} \qquad \begin{cases} \Lambda f = b(x)^{-1}a(x)^{-1}\Lambda^{-1}\psi + h_0(x), \\ \Lambda b(x) f = a(x)^{-1}\Lambda^{-1}\psi + h_1(x), \\ \text{where} \quad \| h_i \|_0 \leq const\cdot \| \Lambda^{-2}\psi \|_0 \leq const \cdot \lambda^{-\frac{1}{2}} \| \Lambda^{-1}\psi \|_0 , \\ \text{constant being independent of } \lambda. \end{cases}$$

Let us consider Λf. We use the notion "negligible" introduced in Definition 2.

$$\Lambda f = \beta b(x)^{-1}\Lambda^{-1}(\beta\alpha(x)^{-1})\psi + [\Lambda,\beta b^{-1}]\Lambda^{-2}(\beta\alpha^{-1})\psi .$$

First term $\approx \beta b^{-1}\Lambda^{-1}\alpha(x)^{-1}\psi = \beta b^{-1}\alpha^{-1}\Lambda^{-1}\psi + \beta b^{-1}[\Lambda^{-1},\alpha^{-1}]\psi$.
Since $\beta\Lambda^{-1}\psi \approx \Lambda^{-1}\psi$, we obtain

$$\text{(E.3)} \qquad \Lambda f \approx b(x)^{-1}\alpha(x)^{-1}\Lambda^{-1}\psi + k(x) ,$$

where $k(x) = [\Lambda,\beta b^{-1}]\Lambda^{-2}(\beta\alpha^{-1})\psi + \beta b^{-1}[\Lambda^{-1},\alpha^{-1}]\psi$. Now $[\Lambda,\beta b^{-1}]$ is bounded operator in L^2 whose operator norm is estimated by some constant independent of λ. Next we have

$$\| [\Lambda^{-1},\alpha^{-1}]\psi \|_0 \leq const. \| \Lambda^{-2}\psi \|_0 ,$$

where const. is independent of λ. Finally, if a function is negligible, then its L^2 norm is estimated in the form

$$const.\lambda^{-N} \| \Lambda^{-2}\psi \|_0 ,$$

where N is arbitrary positive number. From (E.3), putting

$$\Lambda f - b(x)^{-1}\alpha(x)^{-1}\Lambda^{-1}\psi = h_0 ,$$

we obtain (E.2) for Λf. For $\Lambda b(x)f$, we have

$$\begin{aligned}\Lambda b(x)f &= \Lambda\beta\Lambda^{-2}(\beta\alpha^{-1})\psi \approx \Lambda^{-1}(\beta\alpha^{-1})\psi = (\beta\alpha^{-1})\Lambda^{-1}\psi + [\Lambda^{-1},\beta\alpha^{-1}]\psi \\ &\approx \alpha^{-1}\Lambda^{-1}\psi + [\Lambda^{-1},\beta\alpha^{-1}]\psi .\end{aligned}$$

Now in (E.2) put

$$b(x)^{-1}\alpha(x)^{-1} = (b(0)\alpha(0))^{-1} + c(x)x , \quad \alpha(x)^{-1} = \alpha(0)^{-1} + d(x)x .$$

Hence

$$\begin{aligned}b(x)^{-1}\alpha(x)^{-1}\Lambda^{-1}\psi &= (b(0)\alpha(0))^{-1}\Lambda^{-1}\psi + c(x)x\Lambda^{-1}\psi . \\ \alpha(x)^{-1}\Lambda^{-1}\psi &= \alpha(0)^{-1}\Lambda^{-1}\psi + d(x)x\Lambda^{-1}\psi .\end{aligned}$$

Now, since $|\hat{\psi}'(\xi)| \leq const.(\lambda+\xi^2)^{-1/2n}|\hat{\psi}(\xi)|$, it holds that

$$\| x\Lambda^{-1}\psi \|_0 \leq const.\lambda^{-\frac{1}{2n}} \| \Lambda^{-1}\psi \|_0 .$$

Hence (E.2) implies

$$\text{(E.4)} \qquad \begin{cases} \| \Lambda f - (b(0)\alpha(0))^{-1}\Lambda^{-1}\psi \|_0 \leq const.\lambda^{-\frac{1}{2n}} \| \Lambda^{-1}\psi \|_0 , \\ \| \Lambda b(x)f - \alpha(0)^{-1}\Lambda^{-1}\psi \|_0 \leq const.\lambda^{-\frac{1}{2n}} \| \Lambda^{-1}\psi \|_0 , \end{cases}$$

Since $|b(0)| = 1$, we obtain

$$\text{(E.5)} \qquad \| \Lambda f \|_0 \sim \| \Lambda b(x)f \|_0 , \quad \text{when } \lambda \to \infty .$$

Next we add some comments to the proof of Lemma 1. Put

$$\text{(E.6)} \qquad \Lambda f = (b(0)\alpha(0))^{-1}\Lambda^{-1}\psi + k(x) .$$

Then

$$\Lambda(\rho_\delta * f) = (b(0)\alpha(0))^{-1}\Lambda^{-1}(\rho_\delta * \psi) + \rho_\delta * k(x) .$$

Hence

$$\Lambda(\rho_\delta * -1)f = (b(0)\alpha(0))^{-1}(\rho_\delta * -1)\Lambda^{-1}\psi + (\rho_\delta * -1)k(x) .$$

Since $\| (\rho_\delta * -1)k \|_0 \leq const.\lambda^{-1/2n} \| \Lambda^{-1}\psi \|_0$, it holds that

$$\text{(E.7)} \qquad \Big| \| \Lambda(\rho_\delta * -1)f \|_0 - |b(0)\alpha(0)|^{-1} \| (\rho_\delta * -1)\Lambda^{-1}\psi \|_0 \Big| \leq const.\lambda^{-\frac{1}{2n}} \| \Lambda^{-1}\psi \|_0 \ .$$

In the proof of Lemma 1, we proved essentially the following estimate

$$\text{(E.8)} \qquad \| (\rho_\delta * -1)\Lambda^{-1}\psi \|_0^2 \leq (\theta^2 M^2 \sup |\hat{\rho}'(\xi)|^2 + const.M^{4s-1}) \| \Lambda^{-1}\psi \|_0^2 \ .$$

Note that, by (E.5),

$$\| \Lambda f \|_0 \sim |b(0)\alpha(0)|^{-1} \| \Lambda^{-1}\psi \|_0 \ .$$

Hence taking account of (E.7), it holds that

$$\frac{\| \Lambda(\rho_\delta * -1)f \|_0}{\| \Lambda f \|_0} \sim \frac{\| (\rho_\delta * -1)\Lambda^{-1}\psi \|_0}{\| \Lambda^{-1}\psi \|_0} ,$$

when $\lambda \to \infty$. Thus the proof of Lemma 1 is now complete.

F Final remark

In the argument we assumed implicitly that the vanishing order n of a(x) is greater than 1. In the case n=1, hence s=0, we can argue very easily. In this case, (3.3) becomes

$$(x - \Lambda^{-1}(b_0 + b_1 x + \cdots b_l x^l))\psi \sim 0 .$$

Hence (3.5) becomes

$$\left(\frac{d}{d\xi} + i b_0(\lambda + \xi^2)^{-\frac{1}{2}} \right) \hat{\psi}(\xi) = -i(\lambda + \xi^2)^{-\frac{1}{2}} \left(\sum_{j=1}^{l} b_j \left(i\frac{d}{d\xi} \right)^j \right) \hat{\psi}(\xi) .$$

Now $u_0(\xi)$ is defined by

$$\left[\frac{d}{d\xi} + i b_0(\lambda + \xi^2)^{-\frac{1}{2}} \right] u_0(\xi) = 0 ,$$

and denote

$$e(\xi, \eta) = exp\left(-i b_0 \int_\eta^\xi (\lambda + \eta'^2)^{-\frac{1}{2}} d\eta' \right) .$$

Put $u_0(\xi) = e(\xi, 0)$. Hence $|u_0(\xi)| = 1$, and we obtain the estimates

$$\left| \left(\frac{d}{d\xi} \right)^p u_0(\xi) \right| \leq c_{0,p}(\lambda + \xi^2)^{-\frac{p}{2}} , \quad p = 1, 2, \cdots .$$

Next,

$$u_1(\xi) = -i \int_0^\xi e(\xi, \eta)(\lambda + \eta^2)^{-\frac{1}{2}} b \left(i\frac{d}{d\eta} \right) u_0(\eta) d\eta .$$

Hence

$$|u_1(\xi)| \leq const. \int_{-\infty}^{\infty} (\lambda + \eta^2)^{-\frac{1}{2}}(\lambda + \eta^2)^{-\frac{1}{2}} d\eta = c_{1,0}\lambda^{-\frac{1}{2}} .$$

Moreover we have

$$\left|\left(\frac{d}{d\xi}\right)^p u_1(\xi)\right| \leq c_{1,p}\lambda^{-\frac{1}{2}}(\lambda+\xi^2)^{-\frac{p}{2}}\ , \quad p=1,2,\cdots .$$

In general we have

$$\left|\left(\frac{d}{d\xi}\right)^p u_j(\xi)\right| \leq c_{j,p}\lambda^{-\frac{j}{2}}(\lambda+\xi^2)^{-\frac{p}{2}}\ , \quad p=0,1,2,\cdots .$$

Namely (3.8) holds for $s=0$, $\Psi(\lambda)=1$.

References

[1] K. Hayashida, On the singular boundary value problem for elliptic equations, *Trans. Amer. Math. Soc.* 184 p.205-221 (1973).

[2] S. Itô, Fundamental solutions of parabolic differential equations and boundary value problem, *Japan. J. Math.* 27 p.55-102 (1957).

[3] A. Kaji, On the degenerate oblique derivative problems, *Proc. Japan. Acad.* 50 p.1-5 (1974).

[4] Y. Kannai, Existence and smoothness for certain degenerate parabolic boundary value problems, *Osaka J. Math.* 25 p.1-18 (1988).

[5] Y. Kato, Another approach to a non-elliptic boundary problem, *Nagoya Math.J.* 66 p.13-22 (1976).

[6] T. Komura, Groups of operators in locally convex spaces, *J. Functional Anal.* 7 p.258-296 (1968).

[7] S. Mizohata, On the wellposed singular boundary value problems for heat operator, *Journées des équationa aux dérivées partielles à Saint-Jean de Monts en Juin 1983*, Conférence n° 9, p.1-10.

[8] K. Taira, On some degenerate oblique derivative problems, *J. Fac. Sc. Univ. Tokyo, Sec* IA 23 p.259-287 (1976).

[9] K. Taira, Sur le problème de dérivée oblique I, *J. Math. Pure et Appl.* 57 p.379-395 (1978).

[10] M. Terakado, On singular initial-boundary value problems for the second order parabolic equations, *J. Math. Kyoto Univ.* 25 p.145-168 (1985).

SURFACE WAVES OF VISCOUS FLUID DOWN AN INCLINED PLANE

Hirokazu Ninomiya, Takaaki Nishida[1],
Yoshiaki Teramoto and Htay Aung Win

Department of Mathematics
Kyoto University
Kyoto, 606 Japan

INTRODUCTION

We consider free surface problems of viscous fluid down an inclined plane. In the first part we solve the initial boundary value problem for the full system globally in time under appropriate smallness conditions. In the second part we treat an approximate equation which may be called Korteweg-de Vries-Kuramoto-Sivashinsky equation and can be derived under the assumption of small amplitude and long waves for the free surface problem and we obtain travelling periodic waves of the equation.

1 FORMULATION

We consider the flows of viscous incompressible fluid down an inclined plane which has the inclination angle α with the horizontal plane. The fluid motion is governed by Navier-Stokes equation under the gravity g with the fixed boundary condition on the bottom and the free surface conditions on the top surface. For the simplicity we treat two dimensional motions of the fluid with the mean depth h_0 under the constant external pressure $\bar{\pi}$. The stationary solution is given by

$$(1.1)\qquad \begin{cases} \bar{u}_1 = \dfrac{g\sin\alpha}{2\nu}(2h_0 y - y^2), \qquad \bar{u}_2 = 0, \\ \bar{p} = \bar{\pi} - \rho g \cos\alpha (y - h_0), \end{cases}$$

where y is the vertical axis to the bottom plane, ρ is the density of the fluid and ν is the viscosity coefficient. Then the perturbation (u, v, p) from the stationary solution satisfies Navier-Stokes equation

$$(1.2)\qquad \begin{cases} u_{1,x} + u_{2,y} = 0, \\ u_{1,t} + (\bar{u}_1 + u_1)u_{1,x} + u_2(\bar{u}_{1,y} + u_{1,y}) + \dfrac{p_x}{\rho} = \nu\Delta u_1, \\ u_{2,t} + (\bar{u}_1 + u_1)u_{2,x} + u_2 u_{2,y} + \dfrac{p_y}{\rho} = \nu\Delta u_2, \end{cases}$$

[1]supported in part by Monbusho's Grant-in-Aid for Scientific Research 02452007

the fixed boundary conditions on the bottom:

$$u_1 = 0, \qquad u_2 = 0 \qquad \text{on } y = 0, \tag{1.3}$$

and the free surface conditions on the top:

$$\begin{cases} p_s = 0 & \text{on } y = h(t,x), \\ \bar{\pi} + p_n - T\dfrac{h_{xx}}{(1+h_x^2)^{\frac{3}{2}}} = 0 & \text{on } y = h(t,x), \end{cases} \tag{1.4}$$

$$h_t + (\bar{u}_1 + u_1)h_x - u_2 = 0 \qquad \text{on } y = h(t,x), \tag{1.5}$$

where $y = h(t,x)$ denotes the free surface, T is the coefficient of the surface tension and the tangential and normal stress p_s, p_n are given by

$$\begin{cases} p_s = \rho\nu(u_{1,y}+u_{2,x})\dfrac{1-h_x^2}{1+h_x^2} + \rho\nu(u_{2,y}-u_{1,x})\dfrac{2h_x}{1+h_x^2}, \\ p_n = (-p+2\rho\nu u_{2,y})\dfrac{1}{1+h_x^2} + (-p+2\rho\nu u_{1,x})\dfrac{h_x^2}{1+h_x^2} - \rho\nu(u_{1,y}+u_{2,x})\dfrac{2h_x}{1+h_x^2}. \end{cases} \tag{1.6}$$

When the initial condition (u_1, u_2, h) at $t = 0$ is supplied, we have the initial value problem (1.2)–(1.5) for $t > 0$.

When $\alpha = 0$, i.e., the bottom is horizontal, these initial value problems are investigated about a local existence theorem under $T = 0$ (Beale [1]), about a global existence theorem for small initial data under $T > 0$ (Beale [2]), about a decay rate of the solution, (Beale and Nishida [3]) and about a global existence theorem for small initial data under $T = 0$ ([8]). Also local and global in time existence theorems are obtained by Solonnikov for the motion of the fluids bounded by a free surface. When $\alpha > 0$, the local existence theorems are obtained by Teramoto [9],[10]. Here we solve the initial value problem (1.2)–(1.5) under $\alpha > 0, T > 0$ globally in time for the small initial data.

For the later purpose we consider the problem in the dimensionless form. Thus we make the variable changes:

$$\begin{cases} t = \dfrac{l_0 t'}{U_0}, & x = l_0 x', & y = h_0 y', \\ u_1 = U_0 u_1', & \bar{u}_1 = U_0 \bar{u}_1', & u_2 = \dfrac{h_0 U_0 u_2'}{l_0}, \\ h = h_0 h', & p = \rho g \sin\alpha p', & \bar{p} = \rho g \sin\alpha \bar{p}', \end{cases} \tag{1.7}$$

where

$$U_0 = \frac{g h_0^2 \sin\alpha}{l_0} \tag{1.8}$$

and l_0 is the characteristic length of the perturbation.

Following Benney [4], we introduce the following dimensionless parameters:
Reynolds' number:

$$R = \frac{U_0 h_0}{\nu} = \frac{g h_0^3 \sin\alpha}{2\nu^2}, \tag{1.9}$$

Weber's number:

$$W = \frac{T}{\rho g h_0^2}, \tag{1.10}$$

Shallowness number:

$$\delta = \frac{h_0}{l_0}. \tag{1.11}$$

Then the stationary solution has the form:

$$\bar{u}_1 = 2y - y^2. \tag{1.12}$$

The dimensionless system has the following form after omitting $'$

$$u_{1,x} + u_{2,y} = 0, \tag{1.13}$$

$$u_{1,t} + (\bar{u}_1 + u_1)u_{1,x} + u_2(\bar{u}_{1,y} + u_{1,y}) + \frac{2}{R}p_x = \frac{\delta}{R}u_{1,xx} + \frac{1}{R\delta}u_{1,yy}, \tag{1.14}$$

$$\delta^2(u_{2,t} + (\bar{u}_1 + u_1)u_{2,x} + u_2u_{2,y}) + \frac{2}{R}p_y = \frac{\delta^3}{R}u_{2,xx} + \frac{\delta}{R}u_{2,yy}, \qquad \text{in } \Omega(t), \tag{1.15}$$

and

$$h_t + (\bar{u}_1 + u_1)h_x - u_2 = 0 \qquad \text{on } S_F, \tag{1.16}$$

where

$$\Omega(t) = \{(x,y) | -\infty < x < \infty, 0 < y < h(t,x)\}$$

is the fluid region and S_F is the free surface given by

$$y = h(t,x), \qquad -\infty < x < \infty. \tag{1.17}$$

The boundary conditions have the form:

$$(\bar{u}_{1,y} + u_{1,y} + \delta^2 u_{2,x})(1 - \delta^2 h_x^2) - 4\delta^2 u_{1,x}h_x = 0 \quad \text{on } S_F, \tag{1.18}$$

$$-\delta^2 W \csc\alpha \frac{h_{xx}}{(1 + \delta^2 h_x^2)^{\frac{3}{2}}} + (h-1)\cot\alpha - p - \delta u_{2,y}\frac{1 + \delta^2 h_x^2}{1 - \delta^2 h_x^2} = 0 \quad \text{on } S_F, \tag{1.19}$$

$$u_1 = 0, \qquad u_2 = 0 \qquad \text{on } y = 0. \tag{1.20}$$

2 INITIAL VALUE PROBLEM

The initial value problem for our surface waves is formulated as follows: Given the initial fluid region

$$\Omega(0) = \{(x,y) | -\infty < x < \infty, 0 < y < h(0,x)\} \tag{2.1}$$

and the velocity u in it:

$$u(0) = (u_1(0,x,y), u_2(0,x,y)), \qquad (x,y) \in \Omega(0),$$

find the solution for (1.13)-(1.20), i.e., the fluid region

$$\Omega(t) = \{(x,y) | -\infty < x < \infty, 0 < y < h(t,x)\} \tag{2.2}$$

and the velocity u and the pressure p:

$$(u(t,x,y), p(t,x,y)) \qquad \text{in } \Omega(t)$$

for $t > 0$. In this section we do not vary δ and so we fix δ, say $\delta = 1$. For a technical reason we treat the initial value problem with the periodic boundary condition in x of the period, say 2π. We use the following notations:

$H^l(S_F)$ $(l \geq 0)$, denotes the Sobolev space of periodic functions $h(x)$ with period 2π and the norm is denoted by $\| \ \|_{S_F,l}$ or simply by $\| \ \|_l$.

$H^l(\Omega_h)$ $(l \geq 0)$, denotes the Sobolev space of functions $u(x,y)$, $(x,y) \in \Omega_h$, which are periodic with respect to x with period 2π where

$$\Omega_h = \{(x,y)|0 \leq x \leq 2\pi,\ 0 \leq y \leq h(x)\}.$$

The norm is denoted by $\| \ \|_{\Omega_h,l}$ or simply by $\| \ \|_l$.

$C(I;X)$ denotes the space of continuous functions u of $t \in I$ with the value in a Banach space X.

Theorem 2.1. *Let $h(0,x) - 1 \in H^3(S_F)$ and $u(0,x,y) \in H^2(\Omega_h)$ satisfy the compatibility conditions and*

$$\int_0^{2\pi} (h(0,x) - 1)dx = 0. \tag{2.3}$$

If R and α are small and the norms of the initial data are small in the above space, then a unique solution for (1.13)–(1.20) exists for all $t > 0$ such that

$$\begin{cases} h - 1 \in C([0,\infty); H^3(S_F)) \cap L^2([0,\infty); H^3(S_F)), \\ u \in C([0,\infty); H^2(\Omega(t))) \cap L^2([0,\infty); H^3(\Omega(t))), \\ \nabla p \in C([0,\infty); L^2(\Omega(t))) \cap L^2([0,\infty); H^1(\Omega(t))), \end{cases} \tag{2.4}$$

and that the solution decays to zero as $t \to \infty$.

The proof of Theorem 2.1 is divided in several steps.

(i) The system of equations and boundary conditions on the moving domain $\Omega(t)$ are transformed to a system and boundary conditions on a fixed domain

$$\Omega = \{(x,y); -\infty < x < \infty, 0 < y < 1\}$$

by using the Beale's transformation (Beale [2]) which preserves the divergence free condition and the regularity of solutions if $T > 0$.

(ii) The system and boundary conditions on Ω is rearranged to the system and the boundary conditions in the form:

All the linear terms are gathered on the left hand side of the equations and all the nonlinear terms on the right hand side. Then the linearized system for u and $\eta = h - 1$ has the form:

$$\operatorname{div} u = 0, \tag{2.5}$$

$$u_t + (\bar{u} \cdot \nabla)u + (u \cdot \nabla)\bar{u} + \frac{2}{R}\nabla p - \frac{1}{R}\Delta u = f_0 \qquad \text{in } \Omega, \tag{2.6}$$

$$\eta_t - u_2 = f_1 \qquad \text{on } y = 1, \tag{2.7}$$

$$u_{2,x} + u_{1,y} - 2\eta = f_2 \qquad \text{on } y = 1, \tag{2.8}$$

$$p - \cot\alpha \cdot \eta + W \csc\alpha \cdot \eta_{xx} - u_{2,y} = f \qquad \text{on } y = 1, \tag{2.9}$$

$$u = 0 \qquad \text{on } y = 0. \tag{2.10}$$

The inhomogeneous terms f_1, f_2 can be reduced to zero by introducing auxiliary velocity function and included it in f_0 and f. Beale [2].

(iii) We get a priori energy estimates for the linear system (2.5)–(2.10) with $f_1 = f_2 = 0$, which will be explained below.

(iv) The energy estimates can be modified to include the nonlinear terms, the method of which is similar to Matsumura and Nishida [7]. Then it can be applied to give a global existence of solution by combining the local existence theorem.

Now we explain briefly how to obtain the energy estimates for the linearized system. We notice that the main difference from the free surface problem with $\alpha = 0$ is the term 2η in the boundary condition (2.8), but it gives the difficulty for the estimates. Multiply the linear Navier-Stokes equation (2.6) by u and integrate it in Ω using boundary conditions:

$$\begin{aligned}
&\{\cos\alpha\|\eta\|^2 + W\csc\alpha\|\eta_x\|^2 + \frac{R}{2}\|u\|^2\}_t + \frac{1}{2}\langle u,u\rangle - R\iint|u_1u_2| \\
&\le \int_0^{2\pi} 2\eta u_1 + R\iint f_0\cdot u + \int_0^{2\pi} fu_2 \qquad (2.11)\\
&\le 2\int_0^{2\pi}(\int_0^{2\pi}|\eta_x|)|u_1| + R\iint f_0\cdot u + \int_0^{2\pi} fu_2,
\end{aligned}$$

where we used

$$\int_0^{2\pi}\eta = 0 \qquad (2.12)$$

by the periodicity condition and (2.3). Here we remember the Korn's inequality:

Lemma 2.2

$$C^{-1}\|u\|_1^2 \le \langle u,u\rangle \le C\|u\|_1^2, \qquad (2.13)$$

for some positive constant C, *where*

$$\langle v,u\rangle = \iint(\partial_i v_j + \partial_j v_i)(\partial_i u_j + \partial_j u_i), \qquad \partial_1 = \partial_x, \qquad \partial_2 = \partial_y.$$

The tangential derivatives $\partial_x, \partial_x^2, \partial_t$ of the solution have similar estimates. For example the time derivative has the estimate:

$$\begin{aligned}
&\{\cos\alpha\|\eta_t\|^2 + W\csc\alpha\|\eta_{tx}\|^2 + \frac{R}{2}\|u_t\|^2\}_t + \frac{1}{2}(1-2CR)\langle u_t,u_t\rangle \\
&\le \int_0^{2\pi} 2\eta_t u_{1,t} + R\iint f_{0,t}\cdot u_t + \int_0^{2\pi} f_t u_{2,t} \qquad (2.14)\\
&\le 2\int_0^{2\pi}(u_2+f_1)u_{1,t} + R\iint f_{0,t}\cdot u_t + \int_0^{2\pi} f_t u_{2,t},
\end{aligned}$$

The second x-derivative has the estimate:

$$\begin{aligned}
&\{\cos\alpha\|\partial_x^2\eta\|^2 + W\csc\alpha\|\partial_x^3\eta\|^2 + \frac{R}{2}\|\partial_x^2 u\|^2\}_t + \frac{1}{2}(1-2CR)\langle\partial_x^2 u,\partial_x^2 u\rangle \qquad (2.15)\\
&\le 2\int_0^{2\pi}\partial_x^2\eta\,\partial_x^2 u_1 + R\iint \partial_x^2 f_0\cdot\partial_x^2 u + \int_0^{2\pi}\partial_x^2 f\,\partial_x^2 u_2.
\end{aligned}$$

In order to estimate the terms containing η and its derivatives on the right hand side of the these inequalities, we use the x-derivative of the equation (2.9), i.e., multiply it by $\partial_x\eta$ and integrate it on $[0,2\pi]$.

$$\begin{aligned}
&\cot\alpha\|\partial_x\eta\|^2 + W\csc\alpha\|\partial_x^2\eta\|^2 \\
&\le \int(\partial_x p - \partial_x u_{2,y} - \partial_x f)\partial_x\eta \qquad (2.16)\\
&\le \{C_0(\|\partial_x p\|_1 + \|\partial_x^2 u_1\|_1) + \|\partial_x f\|\}\|\partial_x\eta\|.
\end{aligned}$$

Thus we have

$$\|\partial_x\eta\| \le \tan\alpha\{RC_0(\|f_0\|_1 + \|\partial_t u\|_1 + \|\partial_x u\|_1) + C_0\|\partial_x^2 u\|_1 + \|\partial_x f\|\}, \tag{2.17}$$

and so

$$\|\partial_x^2\eta\| \le \frac{\sin\alpha}{\sqrt{W}\cos\alpha}\{RC_0(\|f_0\|_1 + \|\partial_t u\|_1 + \|\partial_x u\|_1) + C_0\|\partial_x^2 u\|_1 + \|\partial_x f\|\}, \tag{2.18}$$

where we used the estimate of Stokes equation for the pressure.

Lemma 2.3 *If v, q satisfy*

$$\begin{cases} -\frac{1}{R}\Delta v + \frac{2}{R}\nabla q &= f_0 \quad \text{in } \Omega, \\ \nabla\cdot v &= 0 \quad \text{in } \Omega, \\ v &= 0 \quad \text{on } y=0, \\ v &= \phi \quad \text{on } y=1, \\ \int_0^{2\pi}\phi &= 0, \end{cases} \tag{2.19}$$

then it holds for all $l \ge 0$

$$\|v\|_{l+2} + \|\nabla q\|_l \quad \le \quad C(\|f_0\|_l + \|\phi\|_{l+\frac{3}{2}}). \tag{2.20}$$

Adding the above tangential estimates for small constants κ and θ, we have

$$\begin{aligned} &\{\cot\alpha(\|\eta\|_2^2 + \kappa\|\eta_t\|^2) + W\csc\alpha(\|\partial_x\eta\|_2^2 + \kappa\|\partial_x\eta_t\|^2) \\ &+\frac{R}{2}(\|u\|^2 + \|\partial_x u\|^2 + \|\partial_x^2 u\|^2 + \kappa\|u_t\|^2)\}_t \\ &+\frac{1}{4}(\langle u,u\rangle + \langle\partial_x u,\partial_x u\rangle + \langle\partial_x^2 u,\partial_x^2 u\rangle + \kappa\langle u_t,u_t\rangle) \\ &\le \ C\{\frac{R^2}{\theta}(\|f_0\|_2^2 + \kappa\|\partial_t f_0\|^2) + \frac{C_0^2}{\theta}(\|f\|_2^2 + \kappa\|\partial_t f\|^2)\}, \end{aligned} \tag{2.21}$$

where we took κ and α small using (2.17), (2.18). All terms containing η in (2.21) can be estimated by the dissipative terms containing $\langle,\rangle$ if we use the equations (2.7), (2.8), (2.9) (2.17) and (2.18). Namely we have the differential inequality of the form:

$$\frac{dE(t)}{dt} + \beta E(t) \le \frac{dE(t)}{dt} + \frac{1}{4}F(t) \le G(t), \tag{2.22}$$

where β is a small positive constant,

$$\begin{cases} E(t) = \cot\alpha(\|\eta\|_2^2 + \kappa\|\eta_t\|^2) + W\csc\alpha(\|\partial_x\eta\|_2^2 + \kappa\|\partial_x\eta_t\|^2) \\ \qquad +\frac{R}{2}(\|u\|^2 + \|\partial_x u\|^2 + \|\partial_x^2 u\|^2 + \kappa\|u_t\|^2), \\ F(t) = \langle u,u\rangle + \langle\partial_x u,\partial_x u\rangle + \langle\partial_x^2 u,\partial_x^2 u\rangle + \kappa\langle u_t,u_t\rangle, \end{cases} \tag{2.23}$$

and $G(t)$ denotes the terms on the right hand side of inequality (2.21).

Therefore we can conclude that if $\|f_0\|_2$, $\|\partial_t f_0\|$, $\|f\|_2$ and $\|\partial_t f\|$ decay exponentially as $t\to\infty$, all the tangential derivatives in $E(t)$ decay exponentially also as $t\to\infty$ and of course the dissipative terms $F(t)$ is integrable in $t\in[0,\infty)$. Then all the normal derivatives up to the same order as tangential derivatives decay exponentially as $t\to\infty$ because of Lemma 2.3 and Lemma 2.4.

Lemma 2.4 *If v, q satisfy*

$$(2.24)\qquad \begin{cases} -\frac{1}{R}\Delta v + \frac{2}{R}\nabla q &= f_0 \quad \text{in } \Omega, \\ \nabla \cdot v &= 0 \quad \text{in } \Omega, \\ v &= 0 \quad \text{on } y = 0, \\ \partial_1 v_2 + \partial_2 v_1 &= \phi_1 \quad \text{on } y = 1, \\ v_2 &= \phi_2 \quad \text{on } y = 1, \\ \int_0^{2\pi} \phi &= 0, \end{cases}$$

then it holds for all $l \geq 0$

$$(2.25)\qquad \|v\|_{l+2} + \|\nabla q\|_l \quad \leq \quad C(\|f_0\|_l + \|\phi_1\|_{l+\frac{1}{2}} + \|\phi_2\|_{l+\frac{3}{2}}).$$

Remark When the initial data are not periodic with respect to x, the estimates above are not sufficient to give a global existence theorem for the solution of the initial value problem (1.13)–(1.20). In order to treat this initial value problem we need additional arguments such as the spectral analysis for the linearized equation to get decay estimates of solution as $t \to \infty$. The spectral analysis for the linearized system (2.5)–(2.10) is not well known because it has the variable coefficients. However if R is small, the spectral analysis for the linearized system with constant coefficients is sufficient to get the decay estimate of solutions as $t \to \infty$. In fact we can obtain the eigenvalue for the Fourier transform of the linear system (2.5)–(2.10) with $\bar{u} = 1$ as

$$(2.26)\qquad \lambda = -2i\xi + (-\frac{2}{3}\cot\alpha + \frac{5}{12}R)\xi^2 + \cdots ,$$

where ξ is the variable of Fourier transform. Thus we can get the decay estimates of the linearized solution and a global existence of solution for the initial value problem by combining the energy method above. The detailed paper will be published elsewhere.

3 APPROXIMATE EQUATION AND TRAVELING WAVE SOLUTIONS

As seen in the previous section the stationary solution (zero solution) for the system (1.13)–(1.20) is asymptotically stable as $t \to \infty$ for small R. However if R increases, then the traveling wave solutions such as solitary waves and periodic waves will appear and then many complicated waves will appear. As the first step we will show the occurence of periodic traveling waves for an approximate equation of the system (1.13)–(1.20) under the assumption of small amplitude and long wave. The approximate equation can be obtained by Benney's asymptotic expansion [4] for δ, $0 < \delta \ll 1$,

$$(3.1)\qquad \begin{cases} u = u^{(0)} + \delta u^{(1)} + \delta^2 u^{(2)} + \cdots, \\ v = v^{(0)} + \delta v^{(1)} + \delta^2 v^{(2)} + \cdots, \\ p = p^{(0)} + \delta p^{(1)} + \delta^2 p^{(2)} + \cdots, \end{cases}$$

and by the small amplitude assumption

$$(3.2)\qquad h = 1 + \epsilon\eta, \qquad \epsilon \approx \delta^2.$$

The expansion of the equation (1.16) for η has the form:

$$
\begin{aligned}
&\eta_t + (2+4\eta)\eta_x + \delta(-\frac{2}{3}\cot\alpha + \frac{8}{15}R)\eta_{xx} \\
&+\delta^2(2 - \frac{40}{63}R\cot\alpha + \frac{32}{63}R^2)\eta_{xxx} + \delta^2\frac{2}{3}(\delta W)\csc\alpha\eta_{xxxx} = O(\delta^3).
\end{aligned} \tag{3.3}
$$

Thus if we neglect all the terms $O(\delta^3)$ except the term

$$\delta^2\frac{2}{3}(\delta W)\csc\alpha\eta_{xxxx},$$

where we assume $\delta W \approx O(1)$, we can obtain the following equation:

$$
\begin{aligned}
&\eta_t + (2+4\eta)\eta_x + \delta(-\frac{2}{3}\cot\alpha + \frac{8}{15}R)\eta_{xx} + \delta^2(2 - \frac{40}{63}R\cot\alpha + \frac{32}{63}R^2)\eta_{xxx} \\
&+\delta^2\frac{2}{3}\delta W\csc\alpha\eta_{xxxx} = 0.
\end{aligned} \tag{3.4}
$$

We notice that the eigenvalue of the Fourier transform of the linearized equation of (3.4) has the form:

$$\lambda = -2i\xi + \delta(-\frac{2}{3}\cot\alpha + \frac{8}{15}R)\xi^2 + \cdots . \tag{3.5}$$

We can compare it with (2.26) for $\delta = 1$. After rescaling the variables for (3.4) we obtain the approximate equation:

$$u_t + uu_x + Ru_{xx} + u_{xxx} + u_{xxxx} = 0, \tag{3.6}$$

where R is another constant but essentially corresponds to the Reynold's number and hereafter it will be treated as the bifurcation parameter. This may be called Korteweg-de Vries-Kuramoto-Sivashinsky equation because of the mixture of them. We consider the solutions of (3.6) which are periodic with respect to $x \in \mathbf{R}$ with period, say 2π, as in the previous section, and satisfy

$$\int_0^{2\pi} u(t,x)dx = 0. \tag{3.7}$$

Then we have the following theorem:

Theorem 3.1. *If $R < 1$, then $u = 0$ is asymptotically stable. If $R > 1$, then $u = 0$ becomes unstable and a family of periodic solutions with respect to t bifurcates from zero solution for small $R - 1 > 0$.*

This can be proved by applying a standard Hopf bifurcation theory for nonlinear partial differential equations. See for example Crandall and Rabinowitz [6]. The eigenvalues of the linearized equation have the form:

$$\lambda = n^2(R - n^2) + in^3, \qquad n = \pm 1, \pm 2, \cdots . \tag{3.8}$$

Next we consider traveling wave solutions of (3.6) in the form:

$$u = u(x - ct). \tag{3.9}$$

The corresponding ordinary differential equation for $u = u(z)$ has the form:

$$-cu + \frac{u^2}{2} + Ru' + u'' + u''' = c_0, \tag{3.10}$$

where c_0 is the integration constant. We rewrite the equation (3.10) in the first order system:

$$\begin{cases} u' = v, \\ v' = w, \\ w' = -w - Rv - (\frac{u^2}{2} - cu - c_0). \end{cases} \tag{3.11}$$

The system (3.11) has two equilibrium points

$$u = c - \sqrt{c^2 + 2c_0}, \qquad v = 0, \qquad w = 0, \tag{3.12}$$

and

$$u = c + \sqrt{c^2 + 2c_0}, \qquad v = 0, \qquad w = 0. \tag{3.13}$$

Nothing interesting happens for the first equilibrium point (3.12) but for the second equilibrium point (3.13) we have the following theorem.

Theorem 3.2. *The equilibrium point (3.13) is stable for $R > \sqrt{c^2 + 2c_0}$. However if $R < \sqrt{c^2 + 2c_0}$, a family of periodic solutions of (3.11) with respect to $z = x - ct$ bifurcates from the equilibrium point.*

This can be also proved by applying a standard Hopf bifurcation theory for nonlinear ordinary differential equations. See for example Carr [5].

Now we can compare two Hopf bifurcations for KdV-KS equation (3.6) as a partial differential equation for $R > 1$ and as an ordinary differential equation for $R < \sqrt{c^2 + 2c_0}$.

Theorem 3.3. *There exist two smooth functions*

$$c = c(R), \qquad c_0 = c_0(R) \qquad R \geq 1, \tag{3.14}$$

satisfying

$$c(1) = -1, \qquad c_0(1) = 0,$$

such that the family of periodic solutions of ordinary differential equation (3.11) with $(R, c(R), c_0(R))$ coincides with the family of time periodic solution of partial differential equation (3.6) satisfying (3.7).

Since the ordinary differential equation (3.11) has three parameters R, c and c_0, and has a family of periodic solutions for $R < c + \sqrt{c^2 + 2c_0}$ for fixed c and c_0, we can apply the implicit function theorem to get the correspondence between two bifurcations of periodic solutions. The detailed proof will be published in Htay Aung Win [12].

Remark If R increases further for the system (3.6) or we continue to trace the Hopf bifurcation branch above, we can obtain periodic doubling, torus and more complicated phenomena by the numerical computations. See also [11]. We will make reports on these elsewhere.

References

[1] J. T. Beale. The initial value problem for the Navier-Stokes equations with a free surface. *Comm. Pure Appl. Math.*, 34:359–392, 1980.

[2] J. T. Beale. Large time regularity of viscous surface waves. *Arch. Rat. Mech. Anal.*, 84:307–352, 1984.

[3] J. T. Beale and T. Nishida. Large time behavior of viscous surface waves. *Recent Topics in Nonlinear PDE II ed. by K. Masuda, M. Mimura. Math. Studies*, 128:1–14, 1985.

[4] D. J. Benney. Long waves on liquid film. *J. Math. Phys.*, 45:150–155, 1966.

[5] J. Carr. Centre manifold theory. *Springer-Verlag*, 1981.

[6] M. G. Crandall and P. H. Rabinowitz. Hopf bifurcation theorem for infinite dimensions. *J. Funct. Anal.*, 67:1978, 53–72.

[7] A. Matsumura and T. Nishida. Initial boundary value problem for the equation of motion of general fluid, *in Computing methods in Appl. Sci. and Engin. (ed. by R. Glowinsky and J. L. Lions)*, V:389–406, 1982.

[8] D. L. G. Sylvester. Large time existence of small viscous surface waves without surface tension. *Comm. Part. Diff. Eq.*, 15:823–903, 1990.

[9] Y. Teramoto. The initial value problem for viscous incompressible flow down an inclined plane. *Hiroshima Math. J.*, 15:619–643, 1985.

[10] Y. Teramoto. On the Navier-Stokes flow down an inclined plane. *J. Math. Kyoto University*, to appear.

[11] J. Topper and T. Kawahara Approximate equations for long nonlinear waves on a viscous fluid. *J. Phys. Soc. Japan*, 44:663–666, 1978.

[12] Htay Aung Win. Model equation of surface waves of viscous fluid down an inclined plane. *J. Math. Kyoto University*, to appear.

A REMARK ON THE CAUCHY PROBLEM OF NON-STRICT HYPERBOLICITY *

Yujiro Ohya

Department of Applied Mathematics and Physics
Faculty of Engineering, Kyoto University, Kyoto 606, Japan

1. To state precisely the situation of this talk, we first of all quote the paper by W. Craig [1], which considers the Cauchy problem for

$$\frac{\partial^2}{\partial t^2}u + F(t,\, x,\, u,\, \frac{\partial}{\partial t}u,\, \frac{\partial}{\partial x}Hu) = 0 \tag{1}$$

$$(t,\, x) \in [0,\infty) \times I\!R$$

where

$$Hu(t,\, x) = \frac{1}{\pi}\,\mathrm{vp}\int \frac{u(y)}{x-y}\,dy$$

(H is the Hilbert transform) under the hypothesis on F such that

$$\frac{\partial}{\partial\left(\frac{\partial}{\partial x}Hu\right)}F \geq {}^{\exists}\,\gamma > 0\ . \tag{2}$$

Equation (1) is closely related to the equations of fluids (Yoshihara [12], Nishida [7]).

Noting that $\frac{\partial u}{\partial t} = p$, $\frac{\partial}{\partial x}Hu = q$, equation (1) is written as

$$\partial_t^2 u + F(t,\, x,\, u,\, p,\, q) = 0 \tag{3}$$

If we take the derivative of (3) with respect to x, we morally have

$$\partial_t^2 q + F_x + F_u q + F_p \partial_t q + F_q \partial_x q = 0\ . \tag{4}$$

Now, we are reduced to study the linear equation

$$\partial_t^2 u + a(t,\, x)\,\frac{\partial}{\partial x}Hu + b(t,\, x)\,\partial_t u + c(t,\, x)\,u = f(t,\, x) \tag{5}$$

under the hypothesis $a(t,\, x) \geq \gamma > 0$, taking account of (2).

Remark 1. we recall the well known property of the Hilbert transform. Denoting the Fourier transform of $u(x) \in L^2(I\!R)$ by $\mathcal{F}[u(x)](\xi)$, we obtain

$$\begin{aligned}\mathcal{F}[Hu(x)](\xi) &= -i\,\mathrm{sgn}(\xi)\,\mathcal{F}[u(x)](\xi) \\ &= \begin{cases} -i\mathcal{F}[u(x)](\xi) & \xi > 0 \\ i\mathcal{F}[u(x)](\xi) & \xi < 0\ ; \end{cases}\end{aligned}$$

*work supported by KAWATETSU (Kawasaki Steel Corporation)

then we have

$$\begin{aligned}\mathcal{F}[\frac{\partial}{\partial x} Hu(x)](\xi) &= (i\xi)\mathcal{F}[Hu(x)](\xi) \\ &= (i\xi)\begin{cases} -i\mathcal{F}[u(x)](\xi) & \xi > 0 \\ i\mathcal{F}[u(x)](\xi) & \xi < 0 \end{cases} \\ &= |\xi|\,\mathcal{F}[u(x)](\xi)\,.\end{aligned}$$

Now, (5) is written again

$$(5)' \qquad \partial_t^2 u + a(t, x)\,|D|\,u(x) + b(t, x)\,\partial_t u + c(t, x)\,u = f(t, x)$$

Evidently (5)′ is a non-strict hyperbolic equation with constant (double) multiplicity.

Remark 2. Since 1964 (Ohya [8], Leray-Ohya [4]), the unique existence is known of solutions in the Gevrey class of C^∞ functions under suitable hypotheses on the coefficients and on the second member.

Remark 3. Since 1968 (Mizohata-Ohya [6]), more sensitively since E.E. Levi [13], it is already known ; we have to suppose that $a(t, x) \equiv 0$ for the Cauchy problem being well posed in the class of C^∞ functions.

2. If we suppose $a(t, x) \geq \gamma > 0$, then the classical energy inequality holds. Indeed, we define.

$$(6) \qquad E(t)^2 = \int \left[(\partial_t u)^2 + (\sqrt{a}\,|D|^{1/2}\,u)^2\right] dx \qquad \text{according to } (5)'.$$

Easily, we obtain

$$\begin{aligned}E(t)\,\frac{d}{dt}\,E(t) = \int [&\partial_t u\,\partial_{tt} u \\ &+ \sqrt{a}\,|D|^{1/2}\,u\,(\sqrt{a}\,|D|^{1/2}\,u)'_t\,]\,dx\end{aligned}$$

taking account of
<u>Lemma</u>
1)

$$\begin{aligned}&\int f(x)\cdot|D|^{1/2}\,g(x)\,dx \\ &\qquad = \int |D|^{1/2}\,f(x)\cdot g(x)\,dx\end{aligned}$$

for $f, g \in H^{1/2}(\mathbb{R})$,
2) $[a(x), |D|^{1/2}]$ is a bounded operator of $L^2(\mathbb{R})$, we shall obtain

$$\begin{aligned}&\int \sqrt{a}\,|D|^{1/2}\,u\,\left\{(\sqrt{a})'_t\,|D|^{1/2}\,u + \sqrt{a}\,|D|^{1/2}\,\partial_t u\right\} dx \\ &= \int \sqrt{a}\,(\sqrt{a})'_t\,(|D|^{1/2}\,u)^2\,dx \\ &+ \int a\,|D|\,u\cdot\partial_t u\,dx \\ &+ \int [\,|D|^{1/2}, a\,]\,|D|^{1/2}\,u\cdot\partial_t u\,dx\ ;\end{aligned}$$

from which we conclude

$$(7) \qquad |E'(t)| \leq const\,(E(t) + \|\,f(t)\,\|)$$

Remark 4. By the usual reasoning, we obtain the existence of solutions having the following regularity; given $f(t, x) \in C^0([0, \infty), H^n(\mathbb{R}))$ for any positive integer n, we recognize that

$$u(t, x) \in C^1([0, \infty), H^n) \cap C^0([0, \infty), H^{n+1/2})$$

3. We propose the following problem; the Cauchy problem for

$$L = \partial_t^2 + a(t, x)\,|D| + b(t, x)\,|D|^{1/2} \tag{8}$$

$$L\,[u] = f \tag{9}$$

with initial data at $t = 0$ where

$$a(t, x) \geq 0 \tag{10}$$

Remark 5. First, the energy inequality suggests that it is interesting to increase the term $|D|^{1/2}$ as that of lower order. Moreover, remember the hypothesis $a(t, x) \geq 0$ is much more general than $a(t, x) \geq \gamma > 0$.

We are interested in searching the necessary and sufficient condition for the Cauchy problem being well posed in the Sobolev spaces. We do not consider the local uniqueness theorem. Such a problem is very difficult, even though there are many papers concerning the Cauchy problem of non-strict hyperbolicity. We limit ourselves in treating it in a very special case.

4. In the following, we suppose that

$$a(t, x) = t^{2k}\ , \ b(t, x) = \alpha t^l \tag{11}$$

where k and l are non-negative integers, α being a real number. If we transform (9) by using the Fourier transform

$$\hat{u}(t, \xi) = \frac{1}{\sqrt{2\pi}} \int e^{-ix\xi}\, u(t, x)\, dx\ ,$$

then, we shall have

$$\begin{cases} \partial_t^2\, \hat{u} + t^{2k}\,|\xi|\,\hat{u} + \alpha t^l\,|\xi|^{1/2}\,\hat{u} = \hat{f}(t, \xi) \\ \hat{u}(0, \xi) = \hat{\varphi}(\xi)\ , \ \partial_t\,\hat{u}(0, \xi) = \hat{\psi}(\xi)\ , \end{cases} \tag{12}$$

that is an ordinary dofferential equation depending on a parameter ξ.

To show the existence of a solution in the Sobolev spaces, we refer to the following theorem.

<u>**Theorem**</u>

The Cauchy problem for

$$(\partial t^2 + a(t)\,|D| + b(t)\,|D|^{1/2})\,u = 0$$

is H^∞−well posed if and only if, for all $t \in [0, T]$, positive constants p and $C(T)$ exist such that

$$\begin{aligned} &|\,\hat{u}(t, \xi)\,|^2 + |\,\partial_t \hat{u}(t, \xi)\,|^2 \\ &\qquad \leq C(T)\,(1 + |\xi|)^p\,(\,|\hat{u}(0, \xi)\,|^2 + |\,\partial_t \hat{u}(0, \xi)\,|^2) \end{aligned} \tag{13}$$

Remark 6. It is apparent that inequality (13) is not important but for $|\xi| \geq 1$. We shall obtain the

<u>**Theorem 1.**</u>

If α is not negative, then the Cauchy problem (12) where $\hat{f}(t, \xi) \equiv 0$ is H^∞− well posed.

<u>**Theorem 2.**</u>

If α is negative, the necessary and sufficient condition for the Cauchy probrem (12), where $\hat{f} \equiv 0$, being H^∞− well posed, is $l + 1 \geq k$.

5. To prove these Theorems, we difine

$$\begin{aligned} E(t)^2 &= |\partial_t \hat{u}|^2 + |\xi|\,|\hat{u}|^2 \\ G(t)^2 &= |\partial_t \hat{u}|^2 + (t^{2k}|\xi| + \alpha t^l |\xi|^{1/2})\,|\hat{u}|^2 \\ H(t)^2 &= |\partial_t \hat{u}|^2 + t^{2k}|\xi|\,|\hat{u}|^2 \ ; \end{aligned}$$

we easily prove the

Property

For $t \in [\,|\xi|^{-1/2}, T]$, we have

(i) $$|\xi|^{-k}E(t) \le G(t) \le C(T)E(t)$$
if $\alpha \ge 0$.

(ii) $$|\xi|^{-k}E(t) \le H(t)$$
and $H(\,|\xi|^{-1/2}\,) \le E(\,|\xi|^{-1/2}\,)$.

Proof of Theorem 1.

It is sufficient to estimate $E(t)$.

i)
$$0 \le t \le |\xi|.^{-1/2}$$

$$\begin{aligned} \frac{d}{dt}E(t)^2 &= 2\,\partial_t\hat{u}\,\partial_t^2\hat{u} + 2\,|\xi|^{1/2}\hat{u}\,\partial_t\hat{u} \\ &\le (\,|\xi|^{1/2} + |\alpha|\,)2\,\partial_t\hat{u}\,|\xi|^{1/2}\hat{u} \\ &\le (\,|\xi|^{1/2} + |\alpha|\,)E(t)^2\ , \end{aligned}$$

using $\partial_t^2\hat{u} = -t^{2k}|\xi|\hat{u} - \alpha t^l|\xi|^{1/2}\hat{u}$; from which we deduce

$$E(t) \le const\, E(0).$$

ii)
$$|\xi|^{-1/2} \le t \le T.$$

$$\begin{aligned} \frac{d}{dt}G(t)^2 &= 2\,\partial_t\hat{u}\,\partial_t^2\hat{u} \\ &+ (2kt^{2k-1}|\xi| + \alpha l t^{l-1}|\xi|^{1/2})\hat{u}^2 \\ &+ (t^{2k}|\xi| + \alpha t^l|\xi|^{1/2})2\,\hat{u}\,\partial_t\hat{u} \\ &\le \frac{c}{t}\,G(t)^2\ ; \end{aligned}$$

from which we obtain
$$G(t) \le t^c|\xi|^{c/2}G(|\xi|^{-1/2}).$$

Taking in to account of the property (i), we obtain the inequality (13).

Proof of Theorem 2.

Sufficiency

We prove the inequality only for $|\xi|^{-1/2} \le t \le T$, since the evaluation for $0 \le t \le |\xi|^{-1/2}$ is the same of that in Theorem 1. Hence, we recognize

$$\begin{aligned} \frac{d}{dt}H(t)^2 &= 2\,\partial_t\hat{u}\,\partial_t^2\hat{u} + 2kt^{2k-1}|\xi|\,\hat{u}^2 \\ &+ t^{2k}|\xi|\,2\hat{u}\,\partial_t\hat{u} \end{aligned}$$

$$\leq \; const \, \frac{H(t)^2}{t} \,,$$

taking account of $l-(k-1)\geq 0$; from this we obtain

$$t^{-c}H(t)\leq (|\xi|^{-1/2})^{-c}H(|\xi|^{-1/2}).$$

<u>**Necessity**</u>

We show that, if we suppose $\beta=-\alpha>0$, $l<k-1$, then the inequality (13) never holds. Let $S(t)=\hat{u}\,\partial_t\hat{u}$, we obtain

$$\begin{aligned}\frac{d}{dt}S(t) &= (\partial_t\hat{u})^2+\hat{u}\,\partial_t^2\hat{u}\\ &= (\partial_t\hat{u})^2+(\beta-t^{2k-l}|\xi|^{1/2})t^l|\xi|^{1/2}\hat{u}^2.\end{aligned}$$

choosing t^* such that $t^*=\left(\frac{\beta}{2}|\xi|^{-1/2}\right)^{\frac{1}{2k-l}}$, we shall have

$$\begin{aligned}\frac{d}{dt}S(t) &\geq (\partial_t\hat{u})^2+\frac{\beta}{2}t^l|\xi|^{1/2}\hat{u}^2\\ &\geq 2\sqrt{(\partial_t\hat{u})^2\,\frac{\beta}{2}\,t^l|\xi|^{1/2}\hat{u}^2}\\ &= \sqrt{2\beta}\,t^{\frac{l}{2}}|\xi|^{1/4}|\partial_t\hat{u}|\,|\hat{u}|\end{aligned}$$

for $0\leq t\leq t^*$. From this we get

(14) $$S(t)\geq e^{\sqrt{2\beta}\frac{t^{l/2+1}}{l/2+1}|\xi|^{1/4}}S(0)$$

for $0\leq t\leq t^*$.

Let us fix $t=|\xi|^{-\sigma}$. First, $|\xi|^{-\sigma}$ has to belong to the interval $[0,\,t^*]$; thus requiring

$$|\xi|^{-\sigma}\leq\left(\frac{\beta}{2}\right)^{\frac{1}{2k-l}}|\xi|^{-\frac{1}{2(2k-l)}}\,,$$

that is

(15) $$\sigma>\frac{1}{2(2k-l)}$$

Then, we recognize that inequality (14) is in contradiction to (13) ; it is necessary that the power of $|\xi|$ is positive,

$$-\sigma\left(\frac{l}{2}+1\right)+\frac{1}{4}>0\,;$$

that is

(16) $$\frac{1}{2(l+2)}>\sigma\,.$$

For (15) and (16) being compatible, it must be

$$\frac{1}{2(l+2)}>\frac{1}{2(2k-l)}\,;$$

that is

$$k>l+1\,,$$

thus completing the proof of necessity in Theorem 2.

Remark 7. If we come back to statements of Theorems 1 and 2, then we shall recall similar results by Oleinik [10] for the operator $\partial t^2 - t^{2k}\partial_x^2 + \alpha t^l \partial_x$ (see also Ohya [9]).

6. Finally, we give an extension of the above results in the follwing form ; the Cauchy problem for

$$(\partial_t^2 + t^{2k}|D|^m + dt^l|D|^n)\, u(t,\, x) = 0$$

with data $u(0,\, x)$, $\partial_t u(0,\, x)$ where $2 \geq m > n > 0$ and $\alpha \in \mathbb{R}$. we state the
Theorem (Y. Shiozaki [11])

I) Case $m/2 \geq n$,
 i) If $\alpha \geq 0$, then the Cauchy problem is H^∞–well posed.
 ii) If $\alpha < 0$, then the necessary and sufficient condition for the Cauchy problem being H^∞–well posed is $m/2n \geq k + 1/l + 2$.
II) Case where $m/2 < n$, the Cauchy problem is H^∞– well posed, if and only if we have

$$m/2n \geq \frac{k+1}{l+2} \qquad \text{for all} \qquad \alpha \in \mathbb{R}.$$

References

[1] W. Craig, Nonstrictly hyperbolic nonlinear systems, Math. Ann. 277 (1987), 213-230.

[2] Y. Hattori-Y. Ohya, On a differential operator appearing in the analysis of water waves, Math. Jap. 36 (1991), 591-601.

[3] L. Hörmander, Pseudo-differential operators and hypoelliptic equations, "Singular integrals," Proc. Symp. Pure Math. 10(1976), 138-183.

[4] J. Leray-Y. Ohya, Systèmes linéaires, hyperboliques non stricts, CBRM, 1964, 105-144.

[5] S. Mizohata, On the Cauchy problem, Science press and Academic Press, Inc., 1985.

[6] S. Mizohata-Y. Ohya, Sur la condition de E. E. Levi concernant des équations hyperboliques, Publ. RIMS., Kyoto Univ. Ser. A4 (1968), 511-526.

[7] T. Nishida, Analysis of equations of fluid (in Japanese), Sûgaku 37 (1985), 289-304.

[8] Y. Ohya, Le problème de Cauchy pour les équations hyperboliques à caractéristique multiple, J. Math. Soc. Japan, 16 (1964), 268-286.

[9] Y. Ohya, Le problème de Cauchy à caractéristiques multiples, Ann. Scu. Norm. Sup., 4 (1977), 757-805.

[10] O. A. Oleinik, On the Cauchy problem for weakly hyperbolic equations, CPAM, 23 (1970), 569-586.

[11] Y. Shiozaki, On H^∞–well posed Cauchy problems for some weakly hyperbolic pseudo-differential equations, à paraître.

[12] H. Yoshihara, Gravity waves on the free surface of an incompressible perfect fluid of finite depth, Publ. RIMS. 18 (1982), 49-96.

[13] E. E. Levi, Caratteristiche multiple e problema di Cauchy, Ann. Mat. Pura. Appl., 16 (1909), 109-127.

translated by the editors
from French

FLOWS BETWEEN ROTATING PLATES

K.R. Rajagopal

Department of Mechanical Engineering
University of Pittsburgh
Pittsburgh, PA 15261

INTRODUCTION

In this article, I shall discuss some recent results regarding the flow of fluids due to a rotating plate or between rotating plates. Such flows have relevance to important problems in astrophysics, geophysics and engineering. The pioneering study of Karman [1] of the steady axially symmetric flow due to a rotating plate has been followed by literally hundreds of investigations that span the gamut from rigorous mathematical analyses to precise experimental studies. Such sedulous studies notwithstanding, the problem has continued to open new horizons hitherto unexplored. We shall discuss some of these new developments here. A more detailed presentation can be found in the recent review article by Rajagopal [2].

For the sake of completeness and continuity, we shall present a brief discussion of the classical results. Karman [1] used the similarity transformation (cf. Figure 1)

$$v_r = rF'(z), \quad v_\theta = rG(z) \quad and \quad v_z = -2F(z), \tag{1}$$

where v_r, v_θ, and v_z are the components of the velocity **v** in the r, θ and z directions. The velocity field (1) automatically satisfies the constraint of incompressibility

$$div\ v = 0\,. \tag{2}$$

Substituting (1) into the Navier-Stokes equations, we obtain

$$\left(\frac{\mu}{\rho}\right)F^{iv} + 2FF''' + 2GG' = 0, \tag{3}$$

$$\left(\frac{\mu}{\rho}\right)G'' + 2FG' - 2F'G = 0\,. \tag{4}$$

Developments in Partial Differential Equations and Applications to Mathematical Physics, Edited by G. Buttazzo *et al.*, Plenum Press, New York, 1992

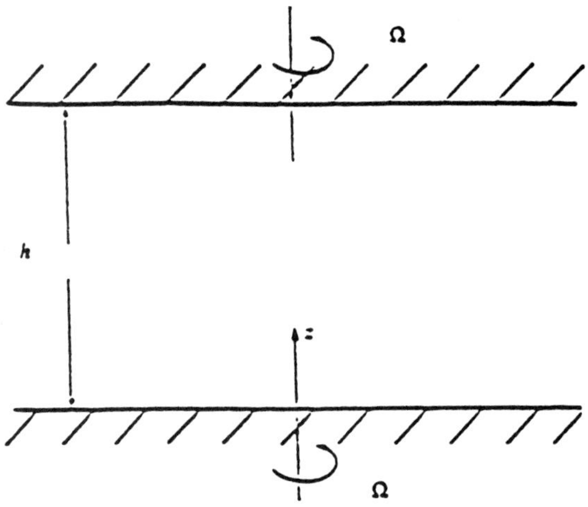

Figure 1. **Flow domain -- disks rotating about a common axis.**

Karman was interested in the flow due to a rotating plate above which a Newtonian fluid rests. Nearly thirty years later, Batchelor [3] used the same similarity transformation (1) to study the flow due to two impervious plates rotating about a common axis. The appropriate boundary conditions are

$$F(0) = F(h) = 0 \qquad (no\ penetration), \tag{5}$$

$$G(0) = \Omega_0, \quad G(h) = \Omega_h \qquad (no\ slip), \tag{6}$$

$$F'(0) = F'(h) = 0 \qquad (no\ slip), \tag{7}$$

where Ω_0 and Ω_h are the angular speeds of the plates at z = 0 and z = h. Equations (3), (4) subject to (5), (6) and (7) have been studied in great detail analytically and numerically. It is well known that at large enough Reynolds numbers there is a multiplicity of solutions. Rigorous equations regarding existence of solutions have also been considered and a thorough discussion of these various issues can be found in the review articles by Parter [4], Zandbergen and Dijkstra [5] and Rajagopal [2]. We would, however, like to mention another landmark paper concerning flows between rotating plates which has been instrumental in the subsequent intense research, namely the paper of Stewartson [6] in which he suggested solutions to the problem different from those of Batchelor [3].

A plethora of papers have been written on the problem of the flow due to rotating plates that are porous, when the flow is taking place in the presence of an axial magnetic field, and when the fluid under consideration is non-Newtonian. Most, if not all, of these studies are concerned with flow between rotating plates that are infinite. In fact, there have been few studies that have considered the flow between finite rotating plates. Here, we shall only be concerned with flow between infinite rotating plates.

There have been several experimental investigations of flows between rotating plates, which we shall not discuss here.

RECENT RESULTS

Solutions That Are Not Axially Symmetric

The similarity transformation (1) implies that the flow is axially symmetric. When $\Omega_0 = \Omega_h$, then equations (3)-(7) imply that the solution is the rigid body rotation. In 1979, Berker [7] found an ingenious solution to the flow of a Navier-Stokes fluid between two disks rotating about a common axis with the same angular speeds. He assumed a velocity field of the form

$$v_x = -\Omega[y - g(z)], \tag{8}$$

$$v_y = \Omega[x - f(z)], \tag{9}$$

$$v_z = 0, \tag{10}$$

where v_x, v_y and v_z denote the components of the velocity in the x, y and z directions. We notice that the flow defined by (8)-(10) automatically satisfies the constraint of incompressibility. The above motion corresponds to streamlines that in each z- constant plane are concentric circles, the location of the centers of the circles being defined by the curve $x = f(z)$, $y = g(z)$. Also, no flow takes place across z-constant planes. Such flows are subclasses of pseudo-planar flows that have been studied in great detail by Berker. The important fact to notice about the velocity field (8)-(10) is that if $f \neq 0$, and $g \neq 0$, then such flows are not axially symmetric. Of course, the case $f \equiv 0$, $g \equiv 0$ corresponds to rigid body rotation. We first observe that (8)-(10) into the Navier-Stokes equation leads to

$$\mu f''' + \rho\Omega g' = 0, \tag{11}$$

$$\mu g''' - \rho\Omega f' = 0. \tag{12}$$

Notice that we have eliminated the pressure by taking the curl of the Navier-Stokes equation and hence raised the order of the governing equation. The appropriate boundary conditions are

$$f(0) = 0 = f(h), \qquad (13)_1$$

$$g(0) = 0 = g(h). \qquad (13)_2$$

Since we have increased the order of the equations, the above boundary conditions need to be augmented. Let the locus of the centers of rotation intersect the mid-plane at $(\ell, 0, h/2)$. We can always ensure that such is the case by rotating the co-ordinate system appropriately about the z-axis. Then

$$f(h/2) = \ell, \quad g(h/2) = 0. \tag{14}$$

Berker [7] showed that (11) and (12) subject to (13)-(15) have a one parameter family of solutions, the parameter in this case being ℓ:

$$f(z)=\frac{\ell}{\Delta}\{(\cosh mh\ \cos mh-\cosh mz\ \cos mz)(\cosh mh\ \cos mh-1)$$
$$+(\sinh mh\ \sin mh-\sinh mz\ \sin mz)(\sinh mh\ \sin mh)\}, \tag{15$_1$}$$
$$g(z)=\frac{\ell}{\Delta}\{(\sinh mh\ \sin mh-\sin mz\ \sinh mz)(\cosh mh\ \cos mh-1)$$
$$-(\cosh mh\ \cos mh-\cosh mz\ \cos mz)(\sinh mh\ \sin mh)\}, \tag{15$_2$}$$

where

$$\Delta=(\cosh mh\ \cos mh-1)^2+(\sinh mh\ \sin mh)^2. \tag{15$_3$}$$

All solutions except for $\ell = 0$ are not axially symmetric. Berker [7] did not investigate the implications of his striking result when $\Omega_0 \neq \Omega_h$. This result, which has bearing on solutions that are not axially symmetric for the problem studied by Batchelor [3], was carried out by Parter and Rajagopal [8] who sought a similarity transformation of the form

$$v_x=\frac{x}{2}H'(z)-\frac{y}{2}G(z)+g(z), \tag{16}$$

$$v_y=\frac{y}{2}H'(z)+\frac{x}{2}G(z)-f(z), \tag{17}$$

$$v_z=-H(z). \tag{18}$$

When $f \equiv 0$, $g \equiv 0$, velocity field reduces to that studied by Karman [1] and Batchelor [3], while if $H \equiv 0$, $G(z) \equiv 2\Omega$, the velocity field reduces to that studied by Berker [7]. Substituting (16)-(18) into the Navier-Stokes equation leads to

$$\left(\frac{\mu}{\rho}\right)H^{iv}+HH'''+GG'=0, \tag{19}$$

$$\left(\frac{\mu}{\rho}\right)G''+HG'-HG'=0, \tag{20}$$

$$\left(\frac{\mu}{\rho}\right)f'''+Hf''+\frac{1}{2}H'f'-\frac{1}{2}H''f+\frac{1}{2}(Gg)'=0, \tag{21}$$

$$\left(\frac{\mu}{\rho}\right)g''+Hg''+\frac{1}{2}H'g'-\frac{1}{2}H''g-\frac{1}{2}(Gf)'=0, \tag{22}$$

and the appropriate boundary conditions are

$$H(0)=H(h)=0, \tag{23}$$

$$H'(0)=H'(h)=0, \tag{24}$$

$$G(0)=2\Omega_0,\quad G(h)=2\Omega_h, \tag{25}$$

$$f(0)=f(h)=0, \tag{26}$$

$$g(0) = g(h) = 0, \tag{27}$$

$$f\left(\tfrac{h}{2}\right) = \ell, \quad g\left(\tfrac{h}{2}\right) = 0. \tag{28}$$

We notice that (19) and (20) subject to (23)-(25) is precisely the problem governing the axially symmetric problem. We can thus ask, given an axially symmetric solution, when can we find a solution to the additional system (21) and (22) subject to (26)-(28)? Parter and Rajagopal [8] show that an axially symmetric solution is always imbedded in a one parameter family of solutions to the system (19)-(28).

A minor modification to the system (19)-(28) also governs the flow of a Navier-Stokes fluid due to the rotation of two infinite parallel plates about distinct axes with different angular speeds. In this case, (27) and (28) need to be changed to

$$g(0) = -\frac{a}{2}, \quad g(h) = \frac{a}{2}, \tag{29}$$

$$f\left(\tfrac{h}{2}\right) = \ell_1, \quad g\left(\tfrac{h}{2}\right) = \ell_2, \tag{30}$$

where 'a' is the distance between the two axes. Parter and Rajagopal [8] proved that in this case there is a two parameter family of solutions to the problem.

Flows of Non-Newtonian Fluids

The flow of fluids between rotating plates has relevance to the problem of the flow occurring in an orthogonal rheometer (cf. Figure 2) which is an instrument used to characterize the material moduli of non-Newtonian fluids (cf. Maxwell and Chartoff [9]). The instrument consists of two parallel plates rotating with the same angular speed about distinct axes. By measuring the normal forces and the torque that are exerted by the fluid on the plates, the material moduli of the fluid can be

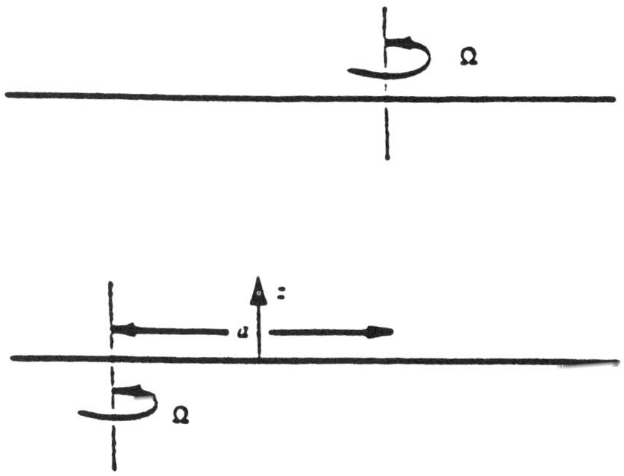

Figure 2. Flow domain -- disks rotating about distinct axes.

characterized. Due to its practical utility, the flow has been studied in some detail. However, most of these studies have been restricted to non-inertial flows for which a very simple kinematical assumption can be made. However, since the constitutive equations for the fluids are highly non-linear, there seems little sense in retaining the material non-linearities while ignoring the non-linearities in the equations due to inertia.

Rajagopal [10] showed that the flow field (8)-(10) is appropriate to model the flow taking place in an orthogonal rheometer, and even when inertial effects are included, showed that such a motion is one of constant principal relative stretch history (cf. Noll [11], Truesdell [12]). In fact Rajagopal [10] showed that it is a special motion with constant principal relative stretch history in that the flow of any simple fluid reduces to a second order partial differential equation.

Let $\mathbf{X}$ = (X, Y, Z) denote a particle in the reference configuration which at time τ occupies the position $\boldsymbol{\xi} = (\xi, \eta, \zeta)$ and at time t occupies the position $\mathbf{x}$ = (x, y, z). By the motion of a particle we mean the mapping $\mathbf{x}$ defined through

$$x = \chi(X,t). \tag{31}$$

We shall assume the motion is invertible so that

$$\xi = \chi(X,\tau) = x\left(\chi^{-1}(x,t),\tau\right) := \chi_t(x,\tau). \tag{32}$$

The relative deformation gradient is defined through

$$F_t(\tau) = \frac{\partial \chi_t}{\partial x}. \tag{33}$$

An incompressible simple fluid is a fluid in which the Cauchy stress $\mathbf{T}$ has the form

$$T = -p\mathbf{1} + \mathop{\mathcal{F}}_{s=0}^{\infty}\left[F_t(t-s)\right], \tag{34}$$

where $-p\mathbf{1}$ is the indeterminate part of the stress due to the constraint of incompressibility, and $\mathcal{F}$ is a functional.

It turns out that for the motion under consideration, the stress $\mathbf{T}$ is given by

$$T = -p\mathbf{1} + f(A_1, A_2), \tag{35}$$

where

$$A_1 = (grad\, v) + (grad\, v)^T \tag{36}$$

$$A_2 = \frac{dA_1}{dt} + A_1(grad\, v) + (grad\, v)^T A_1, \tag{37}$$

where d/dt is the usual material time derivative. Usually, the term $d\mathbf{A}_1/dt$ is one order higher spatially than $\mathbf{A}_1$. However, due to the special nature of the flow, we find that $d\mathbf{A}_1/dt \equiv 0$, and hence the balance of linear momentum of the fluid is the same order as the Navier-Stokes equation and the usual no-slip boundary condition is sufficient for determinacy.

In general, the balance of linear momentum has the form (cf. Rajagopal [10]):

$$-\frac{1}{\rho}\frac{\partial p}{\partial x}=-\frac{\partial \varphi}{\partial x}+\Omega^2[x-f]+\frac{1}{\rho}h_1(f',g',f'',g''), \tag{38}$$

$$-\frac{1}{\rho}\frac{\partial p}{\partial y}=-\frac{\partial \varphi}{\partial y}+\Omega^2[y-g]+\frac{1}{\rho}h_2(f',g',f'',g''), \tag{39}$$

$$-\frac{1}{\rho}\frac{\partial p}{\partial z}=-\frac{\partial \varphi}{\partial z}+\frac{1}{\rho}h_3(f',g',f'',g''), \tag{40}$$

where h_1, h_2 and h_3 depend on the specific constitutive equation for the non-Newtonian fluid.

It is interesting to note that even in the case of a non-Newtonian fluid model of the integral type, the equations have to reduce to equations of the form (38)-(40). To illustrate this, let us consider the flow of a K-BKZ fluid (cf. Kaye [13], Bernstein Kearsley and Zapas [14])

$$-p\mathbf{1}+2\int_{-\infty}^{t}\left\{\frac{\partial U}{\partial I}\mathbf{C}_t^{-1}(\tau)-\frac{\partial U}{\partial II}\mathbf{C}_t(\tau)\right\}d\tau, \tag{41}$$

where

$$\mathbf{C}_t(\tau)=\mathbf{F}_t^T(\tau)\mathbf{F}_t(\tau), \tag{42}$$

and

$$U=U(I,II,t-\tau) \tag{43}$$

is the stored energy for the material, with

$$I_1=tr\,\mathbf{C}_t^{-1}(\tau), \tag{44}$$

$$I_2=tr\,\mathbf{C}_t(\tau). \tag{45}$$

Rajagopal and Wineman [15] show that for the flow (8)-(10)

$$\mathbf{C}_t(\tau)=\mathbf{1}-\frac{s}{\Omega}\mathbf{A}_1+\frac{(1-c)}{\Omega}\mathbf{A}_2, \tag{46}$$

$$\mathbf{C}_t^{-1}(\tau)=\mathbf{1}+\frac{s}{\Omega}\left[1+2(1-c)(f'^2+g'^2)\right]\mathbf{A}_1-\frac{(1-c)}{\Omega^2}\left[1+2(1-c)(f'^2+g'^2)\right]\mathbf{A}_2$$
$$+\frac{s^2}{\Omega^2}\mathbf{A}_1^2+\frac{(1-c)}{\Omega^4}\mathbf{A}_2^2+\frac{s(1-c)}{\Omega^3}\left[\mathbf{A}_1\mathbf{A}_2+\mathbf{A}_2\mathbf{A}_1\right], \tag{47}$$

where

$$s=\sin\Omega(t-\tau)\ \ and\ \ c=\cos\Omega(t-\tau). \tag{48}$$

It immediately follows that (38)-(40) reduce to

$$\frac{d}{dz}\{f'B(\kappa)+g'A(\kappa)\}=\rho\Omega^2 f, \tag{49}$$

$$\frac{d}{dz}\{-f'A(\kappa)+g'B(\kappa)\}=\rho\Omega^2 g, \tag{50}$$

where

$$\kappa^2 := f'^2+g'^2, \tag{51}$$

$$A(\kappa)=2\int_0^\infty \tilde{U}\left[3+2(1-\cos\Omega\alpha)\kappa^2,\alpha\right]\sin\Omega\alpha\, d\alpha, \tag{52}$$

$$B(\kappa)=2\int_0^\infty \tilde{U}\left[3+2(1-\cos\Omega\alpha)\kappa^2,\alpha\right](1-\cos\Omega\alpha)\,d\alpha, \tag{53}$$

with

$$\tilde{U}:=\frac{\partial U}{\partial I}+\frac{\partial U}{\partial II}. \tag{54}$$

Here, κ is a generalized shear rate.

In the case of flow of rotating plates with the same angular speed about a common axis, Rajagopal and Wineman [16] are able to establish explicit exact solutions. Consider the subclass K-BKZ fluids such that

$$\frac{\partial U}{\partial I_1}=U_1(t-\tau),\quad \frac{\partial U}{\partial I_2}=U_2(t-\tau). \tag{55}$$

It immediately follows that

$$A(\kappa)=G_2(\Omega)\quad and\quad B(\kappa)=G_1(\Omega) \tag{56}$$

where $G_1(\Omega)$ and $G_2(\Omega)$ denote the real and imaginary parts of the complex shear modulus of linear viscoelasticity. In this case, the equations of motion reduce to

$$G_1(\Omega)f''+G_2(\Omega)g''-\rho\Omega^2 f=q_1, \tag{57}$$

$$G_2(\Omega)f''+G_1(\Omega)g''-\rho\Omega^2 g=q_2, \tag{58}$$

where q_1 and q_2 are constants. When q_1 and q_2 are zero,

$$\begin{aligned} f(z)=(2a/\Delta)\{&\sin\beta h\cosh\alpha h[\cos\beta z\sinh\alpha h\\ &+\cos\beta(z-h)\sinh\alpha(z-h)]-\cos\beta h\sinh\alpha h[\sin\beta z\cos\alpha z\\ &+\sin\beta(z-h)\cosh\alpha(z-h)]\}, \end{aligned} \tag{59}$$

$$g(z) = (2a/\Delta)\{cos\beta\, h\, sinh\,\alpha h[cos\beta\, z\, sinh\,\alpha z$$
$$+cos\beta\,(z-h) sinh\alpha\,(z-h)] - sin\beta\, h\, cosh\,\alpha h[sin\beta\, z\, cos\alpha z$$
$$+sin\beta\,(z-h) cosh\alpha\,(z-h)]\}, \tag{60}$$

also

$$\Delta = 4\left[sinh^2 \alpha h + sin^2 \beta\, h\right], \tag{61}$$

and

$$\alpha^2 = \frac{\rho\Omega^2}{2\left[\tilde{G}(\Omega)\right]^2}\left[\tilde{G}(\Omega) + G_1(\Omega)\right], \tag{62}$$

$$\beta^2 = \frac{\rho\Omega^2}{2\left[\tilde{G}(\Omega)\right]^2}\left[\tilde{G}(\Omega) + G_1(\Omega)\right], \tag{63}$$

where

$$\tilde{G}^2 = G_1^2 + G_2^2\,. \tag{64}$$

In the case of flow due to plates rotating about distinct axes with the same angular speed, the problem has been studied by Abbot and Walters [17] for the classical linearly viscous fluid, Rajagopal [18] for an incompressible fluid of second grade, Rajagopal, Renardy, Renardy, and Wineman [19] for a Currie fluid, and Bower, Wineman and Rajagopal [20] for a Wagner fluid. As we mentioned earlier, there have been various studies in which inertial effects have been neglected and we shall not discuss them here. In fact, it is the inertial effect that is interesting to study. The studies by Rajagopal and his co-workers clearly indicate the presence of sharp boundary layers (cf. Figure 3).

The tractions t_x and t_y acting on the plates can be expressed as

$$t_x(h) = B(\kappa) f'(h) + A(\kappa) g'(h), \tag{65}$$

$$t_y(h) = -A(\kappa) f'(h) + A(\kappa) g'(h)\,. \tag{66}$$

By measuring the forces and moments which act on the plate, we can characterize the material moduli of the fluid that is being tested. We refer the reader to Rajagopal [2] where this issue is discussed in greater detail.

The flow of non-Newtonian fluids due to plates rotating with different velocities is a much more complicated problem. We shall first discuss some recent results on axially symmetric solutions. The flow of a fluid of second grade has been studied by Erdogan [21]. However, there are problems associated with the prescription of boundary conditions (cf. Rajagopal [22], Kaloni [23]). Axially symmetric solution for a Maxwell fluid have been studied in detail by Walsh [24]. This has been followed by a study of an Oldroyd-B fluid between rotating plates by Ji, Rajagopal and Szeri [25] and the recent exhaustive memoir by Crewther, Huilgol and Josza [26] who also study unsteady flows. Ji, Rajagopal and Szeri [25] used an analytic

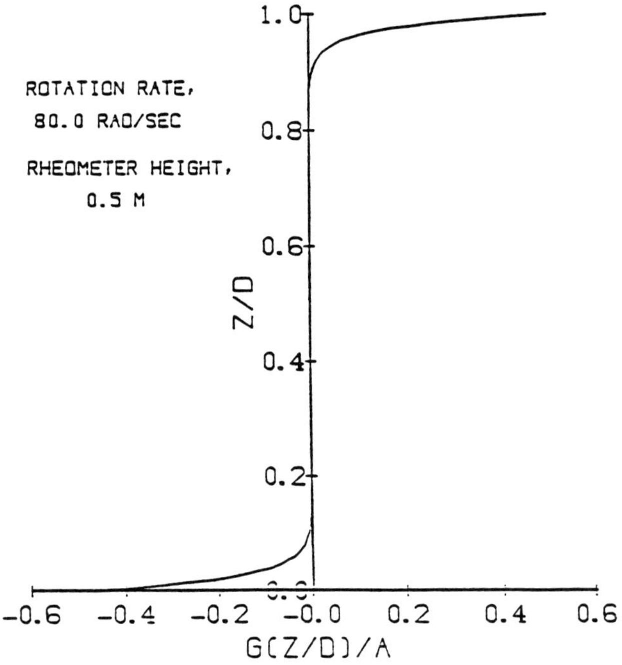
ROTATION RATE,
80.0 RAD/SEC
RHEOMETER HEIGHT,
0.5 M
1.0
0.8
0.6
0.4
0.2
Z/D
-0.6
-0.4
-0.2
-0.0
0.2
0.4
0.6
G[Z/D]/A

Figure 3

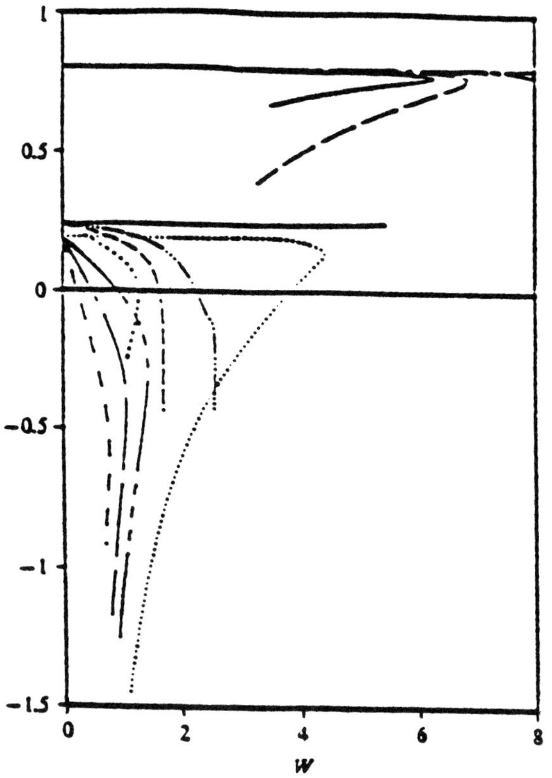

Figure 4. Variation of the Weissenberg number with k, for various values of β. ______, β = 0, s = 0.8; – – –, β = 0.5, s = 0.8; ___ ---___, β = 0, s = 0; ------, β = 0.5, s = 0; ___ ...___ , β = 0.75, s = 0; ___.___.___, β = 1.0, s = 0; ---, β = 0, s = –1; _____ __, β = 0.75, s = –1; ...., β = 1, s = –1.

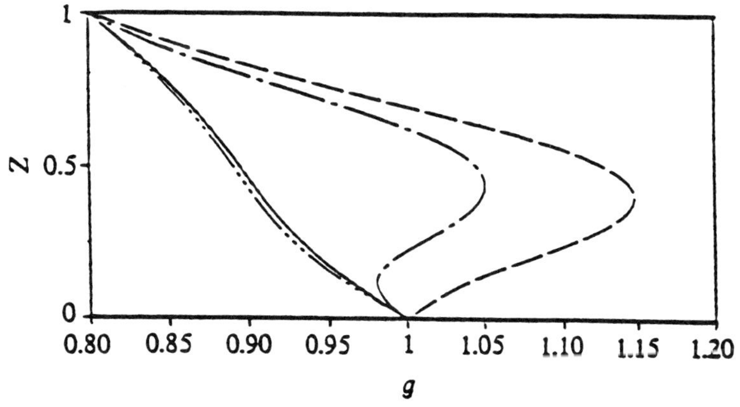

Figure 5. Non-dimensional azimuthal velocity profiles, ________, Branch I, β = 0.5; ___ ---- ____, Branch I, β = 0.0; – – – , Branch II, β = 0.5; ___ . ___, Branch II, β = 0.0.

continuation method that is suited for locating turning points and bifurcating points of non-linear equations. We shall not give the equations here as they are very lengthy and cumbersome but refer the reader to [25] for the same. There are four non-dimensional numbers which play important roles as far as the solutions are concerned: the Ekmann number E, the ratios of the speed λ, the Weissenberg number W (which is the ratio of the relaxation time to a characteristic time for the flow) and the non-dimensional number β that is the ratio of the relaxation time to the retardation time. They find that the solutions exhibit turning points as the Weissenberg number is increased for var'ous values of β (cf. Figure 4). The multiple solutions have clearly different structure in that one of them exhibits flow reversal while the other does not (cf. Figure 5).

Very little work has been done with regard to the flow of non-Newtonian fluids between plates that lack axial symmetry. Huilgol and Rajagopal [26] showed that the flow field used by Parter and Rajagopal [8] is also possible in the case of an Oldroyd-B fluid. Crewther, Huilgol and Josza [26] used this formulation to carry out an extensive numerical study of such solutions. An interesting feature of their study is the presence of shear layers.

There have been no studies of solutions that lack symmetry in the case of integral models. The problem is rather daunting, but it is one of the interesting and important open problems in the field of non-Newtonian fluid mechanics.

REFERENCES

[1] T. von Karman, Uber laminare und turbulente Reibung, *Z. Angew. Math. Mech.*, 1:232 (1921).

[2] K.R. Rajagopal, Flow of viscoelastic fluid between rotating disks, *Theoretical and Computational Fluid Dynamics*, 3:185 (1992).

[3] G.K. Batchelor, Note on a class of solutions of the Navier-Stokes equations representing steady rotationally-symmetric flow, *Quart. J. Mech. Appl. Math.*, 4:29 (1951).

[4] S.V. Parter, On the swirling flow between rotating coaxial disks: a survey, *Theory and Applications of Singular Perturbations*, Proc. of a Conf., Oberwolfach, 1981 (ed. W. Eckhaus and E.M. Dejager). Lecture Notes in Mathematics, Springer-Verlag-Berlin (1982).

[5] P.J. Zandbergen and D. Dijkstra, Von Karman swirling flows, *Annual Rev. Fluid Mech.*, 19:465 (1987).

[6] K. Stewartson, On the flow between two rotating coaxial disks, *Proc. Cambridge Philos. Soc.*, 49:333 (1953).

[7] R. Berker, A new solution of the Navier-Stokes equation, the vortex with curvilinear axis, *Internat. J. Engrg. Sci.*, 20:217 (1982).

[8] S.V. Parter and K.R. Rajagopal, Swirling flow between rotating plates, *Arch. Rational Mech. Anal.*, 86:305 (1984).

[9] B. Maxwell and R.P. Chartoff, Studies of a polymer melt in an orthogonal rheometer, *Trans. Soc. Rheol.*, 9:51 (1965).

[10] K.R. Rajagopal, On the flow of a simple fluid in an orthogonal rheometer, *Arch. Rational Mech. Anal.*, 79"{29 (1982).

[11] W. Noll, Motions with constant stretch history, *Arch Rational Mech. Anal.*, 11:97 (1962).

[12] C. Truesdell, *A First Course in Rational Continuum Mechanics*, Academic Press, New York (1977).

[13] A. Kaye, Note No. 134, College of Aeronautics, Cranfield Institute of Technology (1962).
[14] B. Bernstein, E.A. Kearsley and L.J. Zapas, A study of stress relaxation with finite strain, *Trans Soc. Rheol.*, 7:391 (1963).
[15] K.R. Rajagopal and A.S. Wineman, Flow of a BKZ fluid in an orthogonal rheometer, *J. Rheol.*, 27:509 (1983).
[16] K.R. Rajagopal and A.S. Wineman, A class of exact solutions for the flow of a viscoelastic fluid, *Arch. Mech. Stos.*, 34:747 (1983).
[17] T.N.G. Abbot and K. Walters, Rheometrical flow systems, Part 2. Theory for the orthogonal rheometer, including an exact solution of the Navier-Stokes equations, *J. Fluid Mech.*, 40:205 (1970).
[18] K.R. Rajagopal, The flow of a second order fluid between rotating parallel plates, *J. Non-Newtonian Fluid Mech.*, 9:185 (1981).
[19] K.R. Rajagopal, M. Renardy, Y. Renardy and A.S. Wineman, Flow of viscoelastic fluids between plates rotating about distinct axes, *Rheol. Acta* 25:459 (1986).
[20] M.V. Bower, K.R. Rajagopal and A.S. Wineman, Flow of K-BKZ fluids between parallel plates rotating about distinct axes: shear thinning and inertial effects, *J. Non-Newtonian Fluid Mech.*, 22:289 (1987).
[21] M.E. Erdogan, Non-Newtonain flow due to non-coaxially rotations of a disk and a fluid at infinity, *Z. Angew. Math. Mech.*, 56:141 (1967).
[22] K.R. Rajagopal, On the creeping flow of the second order fluid, *J. Non-Newtonian Fluid Mech.*, 15:239 (1984).
[23] P.N. Kaloni, Several comments on the paper "Some remarks on useful theorems for second order fluid," *J. Non-Newtonian Fluid Mech.*, 36:70 (1990).
[24] W.P. Walsh, On the flow of a non-Newtonian fluid between rotating, coaxial disks, *Z. Angew Math. Phys.*, 38:495 (1987).
[25] Z.H. Ji, K.R. Rajagopal and A.Z. Szeri, Multiplicity of solutions in von Karman flows of viscoelastic fluids, *J. Non-Newtonian Fluid Mech.*, 36:1 (1990).
[26] I. Crewther, R.R. Huilgol and R. Josza, Axisymmetric and non-axisymmetric flows of a non-Newtonian fluid between coaxial rotating disks, Preprint (personal communication).

H^2-CONVERGENT LINEARIZATIONS TO THE NAVIER-STOKES INITIAL VALUE PROBLEM

Reimund Rautmann

Fachbereich Mathematik-Informatik der Universität
Warburger Str. 100, D 4790 Paderborn

Summary

Let u denote a sufficiently smooth solution of the Navier-Stokes initial value problem in a smoothly bounded 3-dimensional domain Ω on a compact time interval $[0, T]$. Then u can be approximated in $H^2(\Omega)$ by functions u^ε, each of which is the unique solution of a sequence of K linear initial value problems on the subsequent intervals $(k\varepsilon, (k+1)\varepsilon]$ *in* $[0, T]$ with $k = 0, \ldots, K-1$, $\varepsilon = \frac{T}{K}$.

The convergence rate for $K \to \infty$ will be given by explicit estimates. The approximation scheme established below is H^2-stable in so far, as any sufficiently small error at the initial time $t = 0$ can be controlled by error estimates in $H^2(\Omega)$ on the whole interval $[0, T]$. In addition in [26] our linearization method leads to new estimates of the convergence rate of the product formula [2], [20], which recently has been proved to be useful for flow computations at higher Reynolds numbers.

0.1 Introduction

The efficient numerical treatment of a nonlinear problem normally requires a suitable linearization, which of course should preserve characteristic features of the original problem at least approximately. In case of the Navier-Stokes initial value problem, the numerical approximation of the solution's gradient ∇u (or, equivalently, of the vorticity curl u of the flow) including its boundary values, and the control of the approximations by error estimates would be highly desirable because of the decisive rôle, which curl u plays in flow problems [3], [17].

Since the existence of a unique global solution of the Navier-Stokes initial value problem in 3-dimensional domains remained an open question until now, we restrict our investigation as follows: We assume a smooth solution u of the Navier-Stokes initial value problem in a bounded 3-dimensional domain to be given on a compact time interval $[0, T]$. We will consider a sequence of partitions of $[0, T]$ in subintervals of length ε tending to zero. Then we will show that a time shift inside of each subinterval combined with a suitable mollification in the nonlinearity of the equation leads to a sequence of linear problems, the unique solutions of which are converging to u together with their spatial derivatives up to the second order (i.e. in the H^2-sense). The scheme presented below is H^2-stable in so far, as any sufficiently small error at the initial time $t = 0$ can be controlled by error estimates in H^2 on the whole time interval $[0, T]$. In addition in [26] our linearization method leads to new estimates of the convergence rate of the product formula [2],[20], which recently has been proved to be useful for flow computations at higher Reynolds numbers [4], [10].

The main emphasis in this paper will be laid on the convergence of the approximation scheme in H^2. The possible strengthening of its convergence rate will be studied elsewhere.

0.2 Notations. Formulation of the problem

We consider a viscous incompressible flow at times $t \geq 0$ in a bounded open set Ω of the Euclidean (x_1, x_2, x_3)-space $I\!R^3$, the boundary $\partial\Omega$ being a compact 2-dimensional C^3-submanifold of $I\!R^3$. The velocity vector $u(t,x) = (u_1, u_2, u_3)$ and the kinematic pressure $p(t,x) \geq 0$ of the flow fulfil the Navier-Stokes equations

$$\begin{aligned} \frac{\partial}{\partial t}u - \Delta u + u \cdot \nabla u + \nabla p &= F, \;\; \nabla \cdot u = 0 \quad for\ t > 0, x \in \Omega \\ u_{|\partial\Omega} &= 0, \;\; u(o, \cdot) = u_o \end{aligned} \tag{0.1}$$

with the prescibed density $F(t,x)$ of the outer forces, if we assume the constant mass density and the kinematic viscosity constant to equal 1. Δ denotes the Laplacean, ∇ the gradient operator in $I\!R^3$.

The unsolved difficulty of finding global unique solutions to the problem (0.1) stems from its nonlinearity and the fact that the stationary case of (0.1) is elliptic in the more general sense of Douglis-Nirenberg only. Cancelling the term ∇p and the condition $\nabla \cdot u = 0$ leads to a weakly coupled parabolic system, for which in case $F \doteq 0$ the maximum principle holds and which therefore has a unique global solution.

We will investigate solutions $u(t)$ of (0.1) with values in one of the Hilbert spaces H^m of measurable real vector functions, which are square integrable on Ω together with their spatial derivatives $\partial_x^\alpha u$ up to the order $m \geq |\alpha| = \alpha_1 + \ldots + \alpha_n$, $\alpha_j = 0, 1, \ldots, m$. The incompressibility condition $\nabla \cdot u = 0$ and the condition of adherence $u_{|\partial\Omega} = 0$ at the boundary $\partial\Omega$ are defined in the usual generalized sense in the closure H or V with respect to the space $H^0 = L^2(\Omega)$ or H^1, respectively, of the space D of real test functions on Ω. The latter functions are divergence free, have spatial derivatives of any order and compact support in Ω. Using H.Weyl's orthogonal projection $P : H^0 \to H$, which sends into zero exactly the generalized gradients $\nabla q \in H^0$, we get from (0.1) the initial value problem of the evolution Navier-Stokes equation

$$\begin{aligned} \partial_t u + Au + Pu \cdot \nabla u &= PF, \qquad t > 0, \\ u(0) &= u_0 \end{aligned} \tag{0.2}$$

for the function $u : [0, T] \to D_A$ with its values in the domain of definition $D_A = V \cap H^2$ of the Stokes operator $A = -P\Delta$.

The spaces $\tilde{D}, \tilde{H}^m, \tilde{H}, \tilde{V}, \tilde{D}_A$of complex valued vector functions on Ω are defined in an analogous way, e.g. the complex Hilbert space $\tilde{H}^0 = \tilde{L}^2(\Omega)$ being equipped with the Hermitean sesquilinearform

$$< f, g >= \int_{\Omega} f(x) \cdot \overline{g}(x)\, dx \quad for\ f, g \in \tilde{H}^0, \tag{0.3}$$

$\overline{g}(x)$ denoting the conjugate complex vector of $g(x)$. We will always write

$$\|v\| =< v, v >^{\frac{1}{2}} \quad or\ \|\nabla v\| =< \nabla v, \nabla v >^{\frac{1}{2}}$$

for the $L^2(\Omega)$-norm of the (real or complex valued) vector function v or ∇v, respectively. Let J be a real interval, B a Banach space. Then by $C^0(J, B)$ we denote the set of all uniformly bounded and continuous functions f defined on J with values in B. By $c, c_0, \ldots$ we will denote positive constants which may have different values at different places below.

Let us assume now that the function $u \in C^0([0,T], D_A)$ is a solution of (0.2) with $PF = 0$. In order to approximate u by the solutions of a sequence of linear initial value problems, we divide the interval $(0,T]$ in K parts $J_k = (t_k, t_{k+1}] = (k\varepsilon, (k+1)\varepsilon]$ of length $\varepsilon = \frac{T}{K}$, $k = 0, \ldots, K-1$. On each J_k we consider the linear initial value problem

$$\begin{aligned} \partial_t u^\varepsilon + Au^\varepsilon + Pu_k^{\varepsilon *} \cdot \nabla u^\varepsilon &= 0, \quad t_k < t \leq t_{k+1} \\ u^\varepsilon(t_k) &= u_k^\varepsilon, \end{aligned} \tag{0.4}$$

where

$$u_k^{\varepsilon *} = (1 + rA)^{-1} u_k^\varepsilon \tag{0.5}$$

denotes the Yosida-approximation to the initial value u_k^ε, $r > 0$, and compute $u_{k+1}^\varepsilon = u^\varepsilon(t_{k+1}), k = 0, \ldots, K-1$ one after the other from the given initial value u_0^ε.

Because of the linearity of problem (0.4) its solution will be represented by the semigroup generated by the operator

$$C_k = A + Pu_k^{\varepsilon *} \cdot \nabla. \tag{0.6}$$

We will prove

Theorem I [25]:

Assume $v \in D_A$. Then the operator $C = A + Pv \cdot \nabla$ with domain $D_C = D_A$ is m-sectorial and generates the contractive semigroup e^{-Ct}, which is holomorphic in the sector $\sum_\omega = \{t \in \mathbb{C} \mid |arg\ t| < \omega\}$ of the complex plane $\mathbb{C}$ where $\omega = \frac{\pi}{2} - arc\ tg\, 2c\|v\|_{L^3(\Omega)}$, the constant $c > 0$ being independent of v. On the Hilbert space D_A, the operators A and C define equivalent norms, i.e. the inequalities $c_1\|Aw\| \leq \|Cw\| \leq c_2\|Aw\|$ hold with constants $c_1, c_2 > 0$ depending on $\|A^\vartheta v\|$ and $\vartheta \in (\frac{1}{4}, 1)$ only.

Let $u_0^\varepsilon \in D_A$ be given. In virtue of Theorem I we have $u_1^\varepsilon = e^{-\varepsilon C_0} u_0^\varepsilon$. More generally having already computed $u_k^\varepsilon \in D_A$ stepwise we get

$$u^\varepsilon(t) = e^{-(t-t_k)C_k} u_k^\varepsilon \quad on\ \overline{J}_k,$$

from which $u^\varepsilon \in C^0(\overline{J}_k, D_A)$ follows, $k = 0, \ldots, K-1$. therefore we have

$$u^\varepsilon(t) = e^{-(t-t_k)C_k} \cdot e^{-\varepsilon C_{k-1}} \cdot \ldots \cdot e^{-\varepsilon C_0} u_0^\varepsilon, \quad t \subset \overline{J}_k. \tag{0.7}$$

Evidently

$$u^\varepsilon \in C^0([0,T], D_A) \tag{0.8}$$

results from the stepwise construction of u^ε. We will prove

Theorem II:

Assume $u \in C^0([0,T], D_A)$ denotes a solution of the Navier-Stokes initial value problem (0.2) with right hand side $PF = 0$, and the initial values $u_0, u_0^\varepsilon \in D_A$ fulfil

$$\|u_0^\varepsilon - u_0\| \leq c_0\varepsilon \text{ and } \|A(u_0^\varepsilon - u_0)\| \leq c\varepsilon^{\frac{1}{2}}$$

with constants c_0, $c \geq 0$. Then the solutions u^ε of (0.4) converge to u in D_{A^α} with $\varepsilon \to 0$, $0 < r \leq \varepsilon$ for all $\alpha \in [0,1]$. For $\varepsilon \in (0, \varepsilon_0], \varepsilon_0$ being sufficiently small, the error estimate

$$\|A^\alpha(u^\varepsilon(t) - u(t))\| \leq a_\alpha \varepsilon^{1-\frac{\alpha}{2}} \tag{0.9}$$

holds on $[0,T]$ with the constant a_α depending on α, T, c_0, c, and $\sup_{t\in[0,T]} \|Au(t)\|$.

Our assumption $PF = 0$ implies that the density F of the outer force in the Navier-Stokes equation is the gradient of a scalar function. We use this unessential restriction only in order to simplify the notations below. In the next sections 1-4 we provide basic tools from functional analysis and semigroup theory and some preliminary results, which we need to prove the main theorems. The proof of the first part of Theorem I follows by a straightforward application of the Lax-Milgram Lemma and an additional estimate for the numerical range of C. In order to prove Theorem II in section 6 we use a bootstrap argument: By energy methods we prove error estimates in H^0 and H^1 and then strenghten the results by estimates in H^2, which use the semigroup methods provided in section 3.

1 Lemmata on the Stokes operator A

By its definition the Stokes operator A is the closure in H of the operator $-P\Delta$, which is symmetric and positive definite on the dense subspace $D \subset H$. Therefore A is a selfadjoint, positive definite operator.

Because of the compact imbedding $H^2 \cap H \subset H$, the special conclusion

$$\|u\|_{H^2} \leq c\|Au\| \tag{1.1}$$

from Cattabriga and Solonnikov's a priori estimate [5], [27] includes that A has a compact inverse. Therefore the eigenfunctions (e_k) related to the eigenvalues (λ_k), $0 < \lambda_1 \leq \lambda_2 \leq \ldots \leq \lambda_k \to \infty$ with $k \to \infty$ form a fundamental system in H, which we assume to be orthonormalized. In addition the fractional powers A^α with domain of definition $D_{A^\alpha} \subset H$ are defined for any real α by means of the spectral representation of A, [7, p. 281; 29, p. 10, p. 44, Theorem 2.3.2]. For $\alpha < \beta$ the imbedding $D_{A^\beta} \subset D_{A^\alpha}$ is compact and D_{A^β} is dense in D_{A^α}.

LEMMA 1.1 : *(a) The equality*

$$\|A^{\frac{1}{2}}f\| = \|\nabla f\| \tag{1.2}$$

and Poincaré's inequality

$$\|f\| \leq c_0 \cdot \|\nabla f\| \tag{1.3}$$

hold for any $f \in V$ with a constant c_0 greater than zero. We have $D_{A^{\frac{1}{2}}} = V$ and

$$< A^{\frac{1}{2}}f, A^{\frac{1}{2}}g >=< Af, g > \quad \textit{for any} f \in D_A,\ g \in V. \tag{1.4}$$

(b) Assume $\gamma > \frac{3}{4}$. *Then there exists a constant* $c = c(\gamma)$, *such that*

$$\|f\|_{L^\infty} \leq c\|A^\gamma f\| \tag{1.5}$$

holds for any $f \in D_{A^\gamma}$.

The proof is given in [7, p. 270, 277–278]. Equation (1.2) is an immmediate consequence of the definition of A.

LEMMA 1.2 : *For each real number* $\delta \in (\frac{1}{4}, \frac{1}{2}]$ *the domain of definition* D_{A^δ} *of* A^δ *is continuously imbedded in* $L^r(\Omega)$ *for any* $r \in [2, \frac{6}{3-4\delta}]$.

PROOF:

Let $\overset{\circ}{H}{}^1$ denote the closure in H^1 of the space C_c^∞ of C^∞-vector functions having compact support in Ω. Then the closure B in H^0 of the Laplacean $-\Delta$ on Ω has the domain of definition $D_B = H^2 \cap \overset{\circ}{H}{}^1$. By Fujita and Morimoto's result [8] we have

$$D_{A^\delta} = D_{B^\delta} \cap H$$

for any $\delta \in [0,1]$, $D_{B^\delta} \subset H^0$ denoting the domain of definition of the fractional power B^δ of the selfadjoint, positive definite operator B. By $\overset{\circ}{H}{}^s$ we denote the closure of C_c^∞ in the norm

$$\|\varphi\|_{H^s} = \{\|\varphi\|^2_{H^{[s]}} + \sum_{|\alpha|=[s]} \int_{\Omega\times\Omega} \frac{|\partial_x^\alpha \varphi(x) - \partial_y^\alpha \varphi(y)|^2}{|x-y|^{3+2(s-[s])}}\, dx\, dy\}^{\frac{1}{2}}$$

of the fractional order Sobolev space $H^s(\Omega)$, where $[s]$ denotes the greatest integer not exceeding s, i.e. $[s] \leq s < [s]+1$. Then for any $s \in (0,1)$, $s \neq \frac{1}{2}$, $\overset{\circ}{H}{}^s$ equals the interpolation space $D_{B^{\frac{s}{2}}} = [\overset{\circ}{H}{}^1, H^0]_{1-s}$, [18, p. 64, Theorem 11.6] (The proof of this theorem extends to our special case with $\partial\Omega$ being a C^3-manifold only.) Therefore $D_{A^\delta} = \overset{\circ}{H}{}^{2\delta} \cap H$ is continuously imbedded in $L^r(\Omega)$ for any $\delta \in [0, \frac{1}{2}], \delta \neq \frac{1}{4}$ and any $r \in [2, \frac{6}{3-4\delta}]$, [1, p. 217, Theorem 7.57].

LEMMA 1.3 : *For any* $f \in H$, $r > 0$ *let*

$$f^* = (1 + rA)^{-1} f \tag{1.6}$$

denote the Yosida approximation of f. *Then* $f^* \in D_A$, *and the inequalities*

$$\|A^\alpha f^*\| \leq c r^{\beta-\alpha} \|A^\beta f\| \tag{1.7}$$

for any $0 \leq \beta \leq \alpha \leq 1$, *and*

$$\|A^\gamma (f^* - f)\| \leq c r^{\beta-\gamma} \|A^\beta f\| \tag{1.8}$$

for any $0 \leq \gamma \leq \beta \leq 1$ *hold with constants* $c = c(\alpha, \beta)$ *or* $c = c(\beta, \gamma)$, *respectively, provided that* $f \in D_{A^\beta}$.

PROOF:

Since the Stokes operator A is maximal accretive,

$$0 \leq < Au, u > \tag{1.9}$$

holds for any $u \in D_A$, and for $Re\lambda > 0$ the inverse $(\lambda + A)^{-1}$ exists which maps H on D_A and fulfils

$$\|(1 + rA)^{-1}\| \leq 1 \quad for \ r > 0, \tag{1.10}$$

[13, p. 279]. Then for any $f \in H$ and $\lambda > 0$ the equation $(\lambda + A)u = f$ is equivalent to

$$A(\lambda + A)^{-1} f = f - \lambda u = Au. \tag{1.11}$$

Taking the inner product of this equation with Au and using Cauchy's inequality, in virtue of (1.9) we see $\|Au\|^2 \leq \|f\|^2$, which together with (1.9) gives

$$\|rA(1 + rA)^{-1}\| \leq 1 \quad for \ any \ r = \frac{1}{\lambda} > 0. \tag{1.12}$$

Now assume $f \in D_{A^\beta}$, $0 \leq \beta \leq \alpha \leq 1$. From the identity

$$A^\alpha (1 + rA)^{-1} f = r^{\beta - \alpha}(rA)^{\alpha - \beta}(1 + rA)^{-1} A^\beta f$$

(1.7) follows because of (1.10), (1.12) and the moment inequality

$$\|(rA)^{\alpha-\beta}(A^\beta f)^*\| \leq c\|rA(A^\beta f)^*\|^{\alpha-\beta}\|(A^\beta f)^*\|^{1-(\alpha-\beta)}.$$

In the second case $0 \leq \gamma \leq \beta \leq 1$, from $A^\gamma(f^* - f) = -r^{\beta-\gamma}(rA)^{1+\gamma-\beta}(1 + rA)^{-1}A^\beta f$ using the moment inequality

$$\|(rA)^{1+\gamma-\beta} f^*\| \leq c\|(rA)f^*\|^{1+\gamma-\beta} \cdot \|f^*\|^{\beta-\gamma}$$

and (1.10), (1.12) again we get (1.8).

2 Lemmata on the convection term $Pu \cdot \nabla v$

LEMMA 2.1 : *Assume $u, v, w \in V$. Then the trilinear form*

$$b(u, v, w) = \int_\Omega (Pu \cdot \nabla v) w \, dx$$

is continuous in (u, v, w) and skew-symmetric in v and w:

$$b(u, v, w) = -b(u, w, v). \tag{2.1}$$

For the proof see [30, p. 162].

LEMMA 2.2 [9]: *Assume $0 \leq \delta < \frac{5}{4}$. Then*

$$\|A^{-\delta} Pu \nabla v\| \leq c\|A^\Theta u\| \, \|A^\rho v\| \tag{2.2}$$

holds for any $u \in D_{A^\Theta}$, $v \in D_{A^\rho}$ with some constant $c = c(\delta, \Theta, \rho)$, provided that $\frac{5}{4} \leq \delta + \Theta + \rho$, $0 < \Theta$, $0 < \rho, \frac{1}{2} < \delta + \rho$.

This lemma is a special case of Giga and Miyakawa's Lemma 2.2 in [9, p. 270–271], [14].

COROLLARY 2.1 : *Let δ, Θ, ρ and $u_j \in D_{A^\Theta}$, $v_j \in D_{A^\rho}$ satisfy the assumptions of Lemma 2.2, $j = 1, 2$. Then*

$$\|A^{-\delta}P(u_2\cdot\nabla v_2 - u_1\cdot\nabla v_1)\| \leq c\|A^\Theta(u_2-u_1)\|\cdot\|A^\rho v_2\| + c\|A^\Theta u_1\|\cdot\|A^\rho(v_2-v_1)\|. \quad (2.3)$$

(2.3) is an immediate consequence of (2.2).

COROLLARY 2.2 : *Assume (a) $u \in H$, $v \in D_{A^{\frac{3}{4}+\xi}}$ or (b) $u \in D_{A^{\frac{3}{4}+\xi}}$, $v \in H$ with some $\xi > 0$. Then*

$$\|A^{-\frac{1}{2}}Pu\,\nabla v\| \leq c\|u\|\,\|A^{\frac{3}{4}+\xi}v\| \text{ or} \quad (2.4)$$

$$\|A^{-\frac{1}{2}}Pu\,\nabla v\| \leq c\|A^{\frac{3}{4}+\xi}u\|\,\|v\| \quad (2.5)$$

holds, respectively, with a constant $c = c(\xi)$.

PROOF:

For any $\varphi, \Psi \in D$ we see by the symmetry of $A^{-\delta}$ and (2.1)

$$I = \int_\Omega (A^{-\delta}P\varphi\,\nabla v)\Psi\,dx = \int_\Omega (P\varphi\,\nabla v)A^{-\delta}\Psi\,dx = -\int_\Omega (P\varphi\,\nabla A^{-\delta}\Psi)v\,dx, \quad (2.6)$$

with $\delta = \frac{1}{2}$ therefore

$$|I| \leq c\|\varphi\|\cdot\|\Psi\|\cdot\|A^{\frac{3}{4}+\xi}v\| \quad (2.7)$$

holds in case (a) because of Lemma 1.1. Taking the limit $\varphi \to u$, $\Psi \to w$ in H for any $w \in H$ we get (2.4) from (2.7) by duality. In case (b) with $v \in H$ we define $A^{-\frac{1}{2}}Pu\,\nabla v$ by means of (2.6) for each $\Psi \in H$ and conclude analogously.

COROLLARY 2.3 : *Assume $\delta \in (\frac{1}{4}, \frac{1}{2}]$. Then*

$$\|A^{-\delta}Pu\,\nabla v)\| \leq c\|u\|\,\|Av\| \quad (2.8)$$

holds for any $u \in H$, $v \in D_A$ with a constant $c = c(\delta)$.

PROOF:

For the first equation (2.6) using (7.2), (7.7) and Lemma 1.2 we find $|I| \leq \|\varphi\|\,\|\nabla v\|_{L^6}\,\|A^{-\delta}\Psi\|_{L^3} \leq c\|\varphi\|\cdot\|Av\|\cdot\|\Psi\|$, which in the limit $\varphi \to u$, $\Psi \to w$ in H for any $w \in H$ proves (2.8) by duality. In a more general framework of Besov spaces, Kobayashi and Muramatu have proved limit cases of Giga and Miyakawa's inequality [9] recently in [14].

3 Lemmata on the Stokes semigroup e^{-tA}

The selfadjoint, positive definite Stokes operator A is the generator of the contractive holomorphic semigroup e^{-tA}. On D_{A^α} the fractional powers A^α commute with e^{-tA}.

LEMMA 3.1 : *For $t > 0$ the estimates*

$$\|A^{\alpha+\beta}e^{-tA}u\| \leq t^{-\alpha}\cdot\|A^\beta u\| \text{ for any } u \in D_{A^\beta} \text{ and } 0 < \alpha \leq e = 2,718\ldots, \quad (3.1)$$

and

$$\|(e^{-tA}-1)f\| \leq \frac{1}{\alpha}t^\alpha\|A^\alpha f\| \text{ for any } 0 < \alpha \leq 1,\ f \in D_{A^\alpha} \quad (3.2)$$

hold.

The proof can be found in [7, p. 281–282]. Let f denote a function defined on a real interval J with values in H. For any $\alpha \in (0,1]$, by

$$[f]_{\alpha,J} = \sup_{\substack{s,t \in J \\ s \neq t}} \| \frac{f(t) - f(s)}{|t - s|^{\alpha}} \| \tag{3.3}$$

we define the Hölder quotient of f with exponent α. In case $[f]_{\alpha,J} < \infty$ f is called uniformly Hölder continuous on J.

LEMMA 3.2 : *Let (t_k), $k = 0, \ldots, K$ with $t_0 = 0, t_K = T, t_k < t_{k+1}$ denote a partition of the interval $J = [0,T]$. Assume the function $f : (0,T] \to H$ restricted to (t_k, t_{k+1}) has continuous extensions to $(0, t_1]$ as well as to each $[t_k, t_{k+1}], k = 1, \ldots, K-1$ and satisfies*

$$\sup_{0 < s \leq t} s^{\lambda} \|f(s)\| \leq M(t) < \infty \tag{3.4}$$

for a constant $\lambda \in [0,1)$ and a real valued function M. Consider

$$u(t) = \int_0^t e^{-(t-s)A} f(s)\, ds, \ t \in [0,T]. \tag{3.5}$$

Then for any $\alpha \in [0,1)$,

$$A^{\alpha} u(t) = \int_0^t A^{\alpha} e^{-(t-s)A} f(s)\, ds \tag{3.6}$$

exists for each $t \in (0,T]$ and fulfils the inequalities

$$\|A^{\alpha}\, u(t)\| \leq t^{1-\alpha-\lambda} M(t)\, B(1-\alpha, 1-\lambda), \tag{3.7}$$

$$[A^{\alpha} u]_{\vartheta,J} \leq c\, M(T) \tag{3.8}$$

for any $\vartheta \in (0, 1-\alpha-\lambda)$ with a constant $c = c(\alpha, \lambda, \vartheta, T)$ and Euler's betafunction $B(\cdot,\cdot)$.

The proof in [7, p. 282] holds verbally on each of the intervals $[t_k, t_{k+1}]$, $k = 0, \ldots, K-1)$ and therefore in their union J, too.

LEMMA 3.3 [7]: *Let α, ϑ and μ be real numbers such that $0 \leq \alpha < 1$ and $0 < \mu < \vartheta - \alpha$. Assume $[f]_{\vartheta,J} < \infty$ and consider*

$$v(t) = \int_0^t e^{-(t-s)A} \{f(s) - f(t)\}\, ds, \ t \in [0,T]. \tag{3.9}$$

Then

$$A^{1+\alpha} v(t) = \int_0^t A^{1+\alpha} e^{-(t-s)A} \{f(s) - f(t)\}\, ds \tag{3.10}$$

exists and fulfils

$$\|A^{1+\alpha} v\| \leq c_0 [f]_{\vartheta,J} \text{ and } [A^{1+\alpha} v]_{\mu,J} \leq c[f]_{\vartheta,J} \tag{3.11}$$

with constants $c_0 = c_0(\alpha, \vartheta, T)$ and $c = c(\alpha, \vartheta, \mu, T)$.

The proof is given in [7, p. 283]

COROLLARY 3.1 : *For any $0 \leq r \leq t$, the equations*

$$\int_r^t e^{-sA} ds = A^{-1}(e^{-rA} - e^{-tA}) \tag{3.12}$$

and

$$\int_r^t e^{-(t-s)A} ds = A^{-1}(1 - e^{-(t-r)A}) \tag{3.13}$$

hold.

For the proof see [13, p. 488–489].

4 Special bounds for smooth Navier-Stokes solutions

LEMMA 4.1 : *Let $u \in C^0([0,T], D_A)$ denote a strong solution of (0.2) with $PF = 0$. Then*

$$\partial_t u \in C^0([0,T], H) \text{ and } \tag{4.1}$$

$$\int_0^t \|\nabla \partial_t u\|^2 \, dt \le c \tag{4.2}$$

hold with a constant $c = c(T, \sup_{t\in[0,T]} \|Au(t)\|)$.

For the proof first of all we note that under our assumption (4.1) follows from the equation (0.2), since $Pu\nabla u$ is uniformly bounded and strongly continuous on $[0,T]$ because of Lemma 2.2 and its Corollary 2.1 with $\delta = 0$, $\Theta = \rho = 1$. Next we consider the Galerkin approximations

$$u_k(t,x) = \sum_{j=1}^{k} a_{kj}(t) \cdot e_j(x) \tag{4.3}$$

on the basis of the orthonormal eigenfunctions e_j of A. By P_k we denote the orthogonal projection of H onto the k-dimensional subspace spanned by $e_1, \ldots, e_k$ in H.

For the k unknown coefficients

$$a_{ki}(t) = \int_\Omega u_k(t,x) \cdot e_i \, dx,$$

the inner products with e_i in L^2, $i = 1, \ldots, k$ of the approximate equations

$$\begin{aligned} \partial_t u_k - P\Delta u_k &= -P_k(u_k \cdot \nabla u_k) \quad for\ t > 0, \\ u_k &= P_k u_0 \qquad\qquad\quad for\ t = 0 \end{aligned} \tag{4.4}$$

constitute a system of k ordinary differential equations and initial conditions. The differential equations are quadratic in the unknown functions a_{ki}, thus locally Lipschitz-continuous. According to a wellknown theorem on ordinary differential equations, the global existence and uniqueness of the $a_{ki} \in C^\infty[0,\infty)$ results from the energy equation

$$\partial_t \|u_k(t,\cdot)\|^2 + 2\|\nabla u_k\|^2 = 0 \tag{4.5}$$

which follows by taking the inner product in L^2 of (4.4) with $a_{ki}e_i$ and summing up about $i = 1, \ldots, k$. The nonlinear term drops out because of (2.1).

Therefore we can differentiate (4.4) with respect to t. Then taking the inner product with $\partial_t u_k$ in H and using (1.2), (7.2), (7.5) and (7.6) leads to

$$\begin{aligned} \frac{d}{dt}\|\partial_t u_k\|^2 + 2\|\nabla \partial_t u_k\|^2 &= -2 < (\partial_t u_k) \cdot \nabla u_k, \partial_t u_k > \\ &\le 2\|\partial_t u_k\|_{L^3} \|\nabla u_k\| \, \|\partial_t u_k\|_{L^6} \\ &\le c\|\partial_t u_k\|^{\frac{1}{2}} \|\nabla u_k\| \, \|\nabla \partial_t u_k\|^{\frac{3}{2}}, \end{aligned}$$

which gives

$$\frac{d}{dt}\|\partial_t u_k\|^2 + 2(1-\mu)\|\nabla \partial_t u_k\|^2 \le c_\mu \|\nabla u_k\|^4 \|\partial_t u_k\|^2, \quad 0 < \mu < 1$$

by (7.4) and thus

$$2(1-\mu)\int_\delta^t \|\nabla\partial_t u_k\|^2 dt \le \|\partial_t u_k(\delta)\|^2 - \|\partial_t u_k(t)\|^2 + c_\mu \int_\delta^t \|\nabla u_k\|^4 \cdot \|\partial_t u_k\|^2\, dt = c_{k,\delta} \tag{4.6}$$

by integration for any $\delta \in (0,t)$. Now we use the result from [11, p. 661–666; 21,p. 432–433] that under our assumption on u, the approximations u_k and $\partial_t u_k$ converge in H^1 to u or $\partial_t u$, respectively, uniformly on $[\delta, T]$ with $k \to \infty$. Since the norm values on the right hand side in (4.6) with u in place of u_k are continuous on $[0,T]$, we can take the limit $\delta \to 0$. Therefore the analogous limit of the left hand side exists and fulfils (4.2) with $c = lim_{\delta\to 0}(lim_{k\to\infty} c_{k,\delta}) \cdot \frac{1}{2(1-\mu)}$ for $0 < \mu < 1$.

REMARK 4.1 : *Even in the case of the homogeneous Stokes initial value problem*

$$\begin{aligned} \partial_t v + Av &= 0, \\ v(0) &= v_0 \in D_A \end{aligned}$$

the general estimate (3.1) gives the result

$$\|\nabla\partial_t v(t)\|^2 = \|A^{\frac{1}{2}}\partial_t v(t)\|^2 \le \frac{1}{t}\|A\, v_0\|^2$$

only, which is essentially weaker than (4.2).

COROLLARY 4.1 : *Under the assumption of Lemma 4.1, the estimates*

$$\|u(s) - u(t)\| \le |s-t| \sup_t \|\partial_t u(t)\|, \tag{4.7}$$

$$\|A^{\frac{1}{4}}(u(s) - u(t))\| \le c|s-t|^{\frac{3}{4}}(\sup_t \|\partial_t u(t)\|)^{\frac{1}{2}} \cdot |\int_t^s \|\nabla\partial_t u\|^2\, d\tau|^{\frac{1}{4}}, \tag{4.8}$$

$$\|A^{\frac{1}{2}}(u(s) - u(t))\| \le c|s-t|^{\frac{1}{2}}\, |\int_t^s \|\nabla\partial_t u\|^2\, d\tau\, |^{\frac{1}{2}} \tag{4.9}$$

hold for any $s,t \in [0,T]$.

For the proof, using (7.8) we get $\|A^{\frac{1}{4}}(u(s)-u(t))\| \le c\|u(s)-u(t)\|^{\frac{1}{2}} \cdot$ $\cdot \|A^{\frac{1}{2}}(u(s)-u(t))\|^{\frac{1}{2}}$. Expressing the differences on the left hand sides by $\partial_t u$, we find (4.7), (4.8) and (4.9) by application of the Cauchy-Schwarz inequality.

COROLLARY 4.2 : *Let $u \in C^0([0,T], D_A)$ denote a solution of (0.2) with $PF = 0$. Then*

$$u(t) = e^{-tA}u_0 - \int_0^t e^{-(t-s)A}\,(Pu\cdot\nabla u)(s)\, ds,\ 0 \le t \le T \tag{4.10}$$

holds. For $\alpha \in [0,1)$, $\gamma \in (0, 1-\alpha)$ the function $A^\alpha u(t)$ has the bounds

$$\|A^\alpha u(t)\| \le \|A^\alpha u_0\| + c\,M \tag{4.11}$$

and

$$[A^\alpha u]_{\gamma,J} \le c_1\|A^{\alpha+\gamma}u_0\| + c_2\, M \tag{4.12}$$

on $J = [0,T]$ with $M = \sup_{t\in J}\|Au(t)\|^2$. The constants c, c_1, c_2 depend on α, T or α, γ, T, respectively.

The proof of (4.10) follows from [29, p. 64, Theorem 3.2.1] because of (4.1). From this, the inequalities (4.11), (4.12) result by (3.2) and Lemma 3.2.

5 Proof of Theorem I. Some properties of e^{-tC}.

For any $u \in \tilde{D}_A$, $v \in D_A$, the estimate

$$\|Pv\,\nabla u\| \leq c\|Av\|\,\|A^{\frac{1}{2}}u\| \tag{5.1}$$

follows from (1.5) and (1.2), therefore the operator $Pv \cdot \nabla$ is defined on $\tilde{D}_{A^{\frac{1}{2}}} \supset \tilde{D}_A$, and the sum $C = A + Pv \cdot \nabla$ has the domain of definition $\tilde{D}_C = \tilde{D}_A$ (or D_A in the real case). Using (1.4), (7.2) and (7.5) we see that for any complex number λ

$$< u, w >_{C_\lambda} = < (\lambda + C)u, w > = \int_\Omega (\lambda u + v \cdot \nabla u)\overline{w}\,dx + \int_\Omega (A^{\frac{1}{2}}u)(A^{\frac{1}{2}}\overline{w})\,dx \tag{5.2}$$

defines a bounded sesquilinearform on $\tilde{V}$, which on account of (1.2), (1.3) fulfils

$$| < v, w >_{C_\lambda} | \leq (|\lambda| + 1 + \|Av\|)\|\nabla u\|\,\|\nabla w\|. \tag{5.3}$$

From the representation

$$u = f + ig,\ \overline{u} = f - ig \tag{5.4}$$

of the complex valued vector function $u \in \tilde{D}_A$ by real valued vector functions $f, g \in D_A$, i denoting the imaginary unit, we find

$$< Pv \cdot \nabla u, u > = -2i \int_\Omega \{v \cdot \nabla f\} g\,dx \tag{5.5}$$

by (2.1), thus

$$Re < Pv \cdot \nabla u, u > = 0 \tag{5.6}$$

From this and (1.4) we get

$$\begin{aligned} Re < \lambda u + Cu, u > &= Re\lambda\|u\|^2 + \|\nabla u\|^2 \\ &\leq \|(\lambda + C)u\| \cdot \|u\| \\ &\leq |\lambda|\,\|u\|^2 + \|Cu\|\,\|u\|, \end{aligned} \tag{5.7}$$

which in the special case $\lambda = 0$ together with Poincaré's inequality (1.3) gives

$$\|u\| \leq c_0^2\|Cu\|,\ \|\nabla u\| \leq c_0\|Cu\|. \tag{5.8}$$

The first equation in (5.7) shows that $< u, u >_{C_\lambda}$ is positive definite on $\tilde{V}$ for $Re\lambda \geq 0$. Since any $f \in \tilde{H}$ defines the bounded semilinearform $< f, w >$ on $\tilde{V}$, in virtue of the Lax-Milgram Lemma [6, p. 41] there exists a unique function $u \in \tilde{V}$ which satisfies

$$< u, w >_{C_\lambda} = < f, w > \tag{5.9}$$

for all $w \in \tilde{V}$. The variational problem (5.9) is in case

$$\tilde{f} = f - Pv \cdot \nabla u - \lambda u \in \tilde{H}$$

equivalent to the weak form of the linear Stokes system for $u \in V$ with the right hand side $\tilde{f}$, [30, p. 22]. Since $u \in \tilde{V}$, $v \in D_A$ implies $\tilde{f} \in H$ by Lemma 2.2, applying Cattabriga's theorem[5] we see $u \in \tilde{D}_A$. Then (5.9) reads

$$\lambda u + Cu = f$$

or, equivalently

$$u = (\lambda + C)^{-1} f \tag{5.10}$$

with the inverse operator $(\lambda + C)^{-1} : \tilde{H} \to \tilde{D}_A$ being defined on $\tilde{H}$. Putting u from (5.10) in (5.7) we see

$$\|\lambda + C)^{-1}\| \leq (Re\lambda)^{-1}$$

for $Re\lambda > 0$. Therefore C is maximal accretive in $\tilde{H}$, [13, p. 279].

In order to show that C is even m-sectorial, we consider the elements of the numerical range

$$\Theta_C = \{z =< Cu, u > |u \in \tilde{D}_A,\ \|u\| = 1\}.$$

Θ_C is contained in the right halfplane $Re\ z > 0$ because of (5.7). From (5.4), (5.5) and (1.2), (1.4) we see

$$z = \|\nabla u\|^2 - 2i \int_\Omega \{v \cdot \nabla f\} \cdot g\, dx.$$

The absolute value of the latter integral has the bound $c\|v\|_{L^3}\|\nabla f\|\,\|\nabla g\| \leq c\|v\|_{L^3}\|\nabla u\|^2$ because of (7.2) and (7.5). Thus we conclude that z is contained in the sector $\sum_{\omega'} = \{\xi \in \mathbb{C} \mid |arg\ \xi| \leq \omega'\}$ where $\omega' = arc\ tg\ 2c\|v\|_{L^3}$. Consequently in virtue of a well known theorem [13, p. 492, Theorem 1.24], C is the generator of the contractive semigroup e^{-tC} on $\tilde{H}$, which is holomorphic in t for

$$|arg\ t| < \omega = \frac{\pi}{2} - \omega'. \tag{5.11}$$

In order to show the equivalence of the norms $\|Au\|$ and $\|Cu\|$ for $u \in D_A$, from Lemma 2.2 we get

$$\|Cu\| \leq \|Au\|(1 + c\|A^\vartheta v\|), \vartheta \geq \frac{1}{4}. \tag{5.12}$$

In addition the inequality

$$\|Pv \cdot \nabla u\| \leq c\|A^\vartheta v\|\,\|A^\rho u\| \leq c_1\|A^\vartheta v\|\,\|Au\|^\rho\|u\|^{1-\rho},\ \frac{1}{2} < \rho < 1,\ \vartheta + \rho \geq \frac{5}{4}$$

results from Lemma 2.2 and (7.8). Therefore using the Cauchy-Young inequality (7.4) we get

$$\begin{aligned} 2\|Au\|\,\|Pv \cdot \nabla u\| &\leq 2c_1\|A^\vartheta v\|\,\|Au\|^{1+\rho}\|u\|^{1-\rho} \\ &\leq \mu\|Au\|^2 + c_\mu\|A^\vartheta v\|^{\frac{2}{1-\rho}}\|u\|^2,\ \mu > 0. \end{aligned}$$

With this estimate and (5.8) the explicit expression for $\|Cu\|^2$ leads to

$$(1-\mu)\|Au\|^2 + \|Pv \cdot \nabla u\|^2 \leq \|Cu\|^2(1 + c_0^4 c_\mu\|A^\vartheta v\|^{\frac{2}{1-\rho}})\ for\ any\ \mu \in (0,1),\ \vartheta > \frac{1}{4}. \tag{5.13}$$

The inequalities (5.12) and (5.13) together prove the equivalence of the norms induced by C and A on D_A.

COROLLARY 5.1 : *Under the assumption of Theorem I we have*
(a) C and e^{-tC} commute on $\tilde{D}_A$.
(b) For any $u_0 \in D_A$ the function $u(t) = e^{-tC}u_0$ is the unique solution of the initial value problem

$$\begin{aligned} \partial_t u + Cu &= 0\ , t \geq 0, \\ u(0) &= u_0, \end{aligned} \tag{5.14}$$

(c) $u \in C^0([0,T], D_A)$ *and*
(d) $\partial_t u \in C^0([0,T], H)$ *hold.*

For the proof of (a), (b) c.p. [13, p. 483]. For real t the semigroup e^{-tC} operates on the real space D_A since in case of a real u_0 (5.14) has a unique real solution. The statement (c) follows from (a) and the continuity of $e^{-tC}\, Cu_0$ on $[0,T]$ in case $u_0 \in D_A$, since the norms $\|Cu\|$ and $\|Au\|$ are equivalent on D_A. Then (d) follows from (5.14).

6 Proof of Theorem II

In (0.7) we have written the solution u^ε of (0.4) in terms of the semigroups e^{-tC_k}. Since the proof of Theorem II requires a comparison of u^ε with the solution u of (0.2), which has the representation (4.10) by means of e^{-tA}, first of all we will express u^ε in terms of e^{-tA}, too.

COROLLARY 6.1 : *Assume* $u \in D_A$. *Then the solutions* u^ε *of (0.4) satisfies the integral equation*

$$u^\varepsilon(t) = e^{-tA}u_0^\varepsilon + \int_0^t e^{-(t-s)A} f^\varepsilon(s)\, ds, \quad 0 \le t \le T, \tag{6.1}$$

where the function $f^\varepsilon : [0,T] \to H$ *is defined by*

$$\begin{aligned} f^\varepsilon(s) &= -Pu_k^{\varepsilon *} \cdot \nabla u^\varepsilon(s),\ t_k < s \le t_{k+1}, \quad k = 0,\dots,K-1, \\ f^\varepsilon(0) &= -Pu_0^{\varepsilon *} \nabla u_0^\varepsilon. \end{aligned} \tag{6.2}$$

PROOF:

Since $u^\varepsilon \in C^0([0,T], D_A)$ holds in virtue of Corollary 5.1, we see from Lemma 1.3 and Lemma 2.2 that the function f^ε takes its values in H. The restriction of f^ε to $J_k = (t_k, t_{k+1}]$ has the continuous extension $f_k^\varepsilon = -Pu_k^{\varepsilon *} \cdot \nabla u^\varepsilon$ to $\overline{J}_k$. Therefore $u^\varepsilon(t) = \tilde{u}_k^\varepsilon(t)$, $k = 0,\dots,K-1$, is the unique solution of the initial value problem

$$\begin{aligned} \partial_t \tilde{u}_k^\varepsilon + A\tilde{u}_k^\varepsilon &= f_k^\varepsilon \quad , t \in \overline{J}_k, \\ \tilde{u}_k^\varepsilon(t_k) &= u^\varepsilon(t_k) \end{aligned} \tag{6.3}$$

and $\partial_t \tilde{u}_k^\varepsilon \in C^0(\overline{J}_k, H)$ holds. Thus due to a wellknown theorem [29, p. 64, Theorem 3.2.1] u^ε has the representation

$$\tilde{u}_k^\varepsilon(t) = e^{-(t-t_k)A} u^\varepsilon(t_k) + \int_{t_k}^t e^{-(t-s)A} f^\varepsilon(s)\, ds, \ t \in \overline{J}_k \tag{6.4}$$

where under the integral sign we have written $f^\varepsilon(s)$ instead of $f_k^\varepsilon(s) = f^\varepsilon(s)$ a.e. on $\overline{J}_k$. From this by induction with respect to $k = 0,\dots,K-1$, using the boundedness of e^{tA} and the semigroup relation $e^{-t_1 A} \cdot e^{-t_2 A} = e^{-(t_1+t_2)A}$ we get (6.1).

We assume the solution $u \in C^0([0,T], D_A)$ of (0.2) with $PF = 0$ is given, and denote by u^ε the solution of (0.4) for $0 < \varepsilon = \frac{T}{K}$, $0 < r$, $K = 1,2,\dots$ In the following the real number M stands for an upper bound of the functions $\|Au(t)\|^2$, $\|u(t)\|_{L^\infty}^2$, $\|\partial_t u(t)\|^2$, $\int_0^t \|\nabla \partial_t u\|^2 dt$, $t \in [0,T]$. We have seen in Lemma 1.1 and Lemma 4.1 that

M can be represented by a suitable function of T and $\sup_{t\in[0,T]} \|Au(t)\|^2 < \infty$.

On the interval $J_k = (t_k, t_{t+1}]$ the difference

$$w = u^\varepsilon - u$$

solves the initial value problem

$$\begin{aligned} \partial_t w + Aw &= f \ , t \in J_k, \\ w(t_k) &= w_k = u^\varepsilon(t_k) - u(t_k), \end{aligned} \tag{6.5}$$

where

$$f(t) = -P\{w_k^* \cdot \nabla w + u^*(t_k) \cdot \nabla w + w_k^* \cdot \nabla u + (u^*(t_k) - u(t_k)) \cdot \nabla u + (u(t_k) - u) \cdot \nabla u\}. \tag{6.6}$$

<u>**Bounds in $\mathbf{H}^0$**</u>

Each term in (6.5) belongs to H. Taking the inner product with w in H we find

$$\frac{1}{2}\frac{d}{dt}\|w\|^2 + \|\nabla w\|^2 = g, \quad \|w(t_k)\|^2 = \|w_k\|^2 \tag{6.7}$$

where

$$g = < w_k^* \cdot \nabla w, u > + < (u^*(t_k) - u(t_k)) \cdot \nabla w, u > + < (u(t_k) - u) \cdot \nabla w, u >$$

due to the symmetry (1.4) of A and the skew symmetry (2.1) of $< v \cdot \nabla u, w >$. Using the Cauchy-Schwarz inequality and then applying Lemma 1.3 we get

$$\begin{aligned} |g| &\le \{\|w_k^*\| + \|u^*(t_k) - u(t_k)\| + \|u(t_k) - u\|\}\|\nabla w\| \, \|u\|_{L^\infty} \\ &\le \{c\|w_k\| + crM^{\frac{1}{2}} + |t_k - t|M^{\frac{1}{2}}\}\|\nabla w\| \, cM^{\frac{1}{2}}, \end{aligned} \tag{6.8}$$

since $\|u\|_{L^\infty} \le c\|Au\|$ by (1.5). Because of the Cauchy-Young inequality (7.4) the last right hand side in (6.8) has the bound $3\mu\|\nabla w\|^2 + c_\mu M\{\|w_k\|^2 + (r^2 + |t_k - t|^2)M\}$. Thus from (6.7) we get

$$\frac{d}{dt}\|w\|^2 + 2(1 - 3\mu)\|\nabla w\|^2 \le 2c_\mu(M\,\|w_k\|^2 + M^2(r^2 + |t_k - t|^2)) \tag{6.9}$$

for any $\mu \in (0, \frac{1}{3}]$, the constant $c_\mu > 0$ depending on μ, c, and therefore

$$\|w(t)\|^2 + 2(1 - 3\mu)\int_{t_k}^{t} \|\nabla w\|^2 dt \le \|w_k\|^2 (1 + c_1[t - t_k]) + c_2\varepsilon^3 \tag{6.10}$$

with constants $c_1 = 2c_\mu M$, $c_2 = 4c_\mu M^2$.

With the help of (6.10) we estimate the initial values $\Phi_k = \|w_k\|^2$ of $\|w(t)\|^2$ on $\bar{J}_k$, $k = 0, 1, \ldots, K-1$. By induction we find

$$\Phi_k \le q^k\|w_0\|^2 + (q^k - 1)2M \cdot \varepsilon^2 \le (\frac{3}{2})^{3c_1T}\|w_0\|^2 + [(\frac{3}{2})^{3c_1T} - 1]2M\varepsilon^2 \tag{6.11}$$

with $q = 1 + c_1\varepsilon$. The second inequality in(6.11) results with $y = \frac{K}{c_1T}$ from $q^k = (1 + \frac{1}{y})^k \le (1 + \frac{1}{y})^{yc_1T}$, since the function $(1 + \frac{1}{[y]})^{[y]+1}$ of the natural number $[y] \le y < [y]+1$ is monotonously decreasing for $[y] \ge 2$. In the following we assume

$$\|w_0\|^2 \le c_0^2\varepsilon^2. \tag{6.12}$$

Then the inequality (6.11) shows the desired estimate (0.9) in case $\alpha = 0$, with the constant

$$a_0 = \{(\frac{3}{2})^{3c_1T}(c_0^2 + 2M)\}^{\frac{1}{2}}. \tag{6.13}$$

Bounds in $\mathbf{H}^1$

Taking the inner product of (6.5) by Aw in H gives

$$\begin{aligned}\frac{1}{2}\frac{d}{dt}\|\nabla w\|^2 + \|Aw\|^2 = & - < w_k^* \cdot \nabla w, Aw > - < u^*(t_k) \cdot \nabla w, Aw > \qquad (6.14)\\ & - < w_k^* \cdot \nabla u, Aw > - < [u^*(t_k) - u(t_k)] \cdot \nabla u, Aw > \\ & - < [u(t_k) - u] \cdot \nabla u, Aw > \\ = & - \sum_{j=1}^{5} T_i\end{aligned}$$

With the help of (7.2) or (7.1), (7.5),(7.7) and the Cauchy-Young inequality (7.4) we find

$$|T_1| \leq \|\nabla w_k^*\| \, \|\nabla w\|^{\frac{1}{2}} \, \|Aw\|^{\frac{3}{2}} \leq \mu \, \|Aw\|^2 + c_\mu \|\nabla w_k\|^4 \, \|\nabla w\|^2$$

because of (1.2) and Lemma 1.3,

$$|T_2| \leq \|u^*(t_k)\|_{L^\infty} \|\nabla w\| \, \|Aw\| \leq \mu \|Aw\|^2 + c_\mu \|Au(t_k)\|^2 \|\nabla w\|^2$$

with (1.5), Lemma 1.1 and Lemma 1.3. In addition, Lemma 2.2, Lemma 1.3, the moment inequality (7.8), (7.1) and (7.4) lead to

$$|T_3| \leq c\|A^{\frac{1}{4}} w_k^*\| \, \|Au\| \, \|Aw\| \leq \mu \|Aw\|^2 + c_\mu \|w_k\| \, \|\nabla w_k\| \, \|Au\|^2,$$

$$|T_4| \leq c\|A^{\frac{1}{4}}[u^*(t_k) - u(t_k)]\| \, \|Au\| \, \|Aw\| \leq \mu \|Aw\|^2 + c_\mu r^{\frac{3}{2}} \|Au(t_k)\|^2 \, \|Au\|^2,$$

$$\begin{aligned}|T_5| & \leq c\|A^{\frac{1}{4}}[u(t_k) - u]\| \, \|Au\| \, \|Aw\| \\ & \leq \mu \|Aw\|^2 + c_\mu |t - t_k|^{\frac{3}{2}} \sup_t \|\partial_t u\| \Big(\int_{t_k}^{t} \|\partial_t \nabla u\|^2 \, dt\Big)^{\frac{1}{2}} \|Au\|^2\end{aligned}$$

because of Lemma 1.3, and Corollary 4.1 for the last two inequalities.

Using these bounds in (6.14) together with the bound M defined above and a_0 from (6.13) we get

$$\frac{d}{dt}\|\nabla w\|^2 + 2(1 - 5\mu)\|Aw\|^2 \leq c_\mu \{\|\nabla w\|^2 (\|\nabla w_k\|^4 + M) + a_0 \varepsilon \|\nabla w_k\| M + 2\varepsilon^{\frac{3}{2}} M^2\}. \qquad (6.15)$$

For a fixed k we assume

$$\|\nabla w(t_k)\|^2 = \|\nabla w_k\|^2 \leq 1 \; if \; \varepsilon \in (0, \varepsilon_0], \; \varepsilon_0 \; sufficiently \; small. \qquad (6.16)$$

We write $M_0 = M(a_0 + 2\varepsilon^{\frac{1}{2}} M)$ and $M_1 = c_\mu(1 + M)$. Then integrating (6.15) we find

$$\|\nabla w_{k+1}\|^2 \leq e^{M_1 \varepsilon}\{\|\nabla w_k\|^2 + c_\mu \varepsilon^2 M_0\} \; for \; 0 < \mu \leq \frac{1}{5}. \qquad (6.17)$$

If we suppose (6.16) for a fixed $m < K$ and all $k = 0, \ldots, m$, from (6.17) we see by induction that the real numbers

$$\begin{aligned}\varphi_0 & = \|\nabla w_0\|^2 \qquad (6.18)\\ \varphi_k & = e^{k\varepsilon M_1} \cdot \varphi_0 + \sum_{j=1}^{k} e^{j\varepsilon M_1} \cdot c_\mu \varepsilon^2 M_0, \; k = 1, \ldots, K,\end{aligned}$$

which evidently satisfy

$$\varphi_k \le e^{TM_1} \cdot \varphi_0 + \varepsilon \frac{M_0}{1+M}(e^{M_1 T} - 1)e^{\varepsilon M_1}, \tag{6.19}$$

fulfil

$$\|\nabla w_k\|^2 \le \varphi_k \; for \; k = 0, \ldots, m. \tag{6.20}$$

In addition inequality (6.16) results from (6.19) for all these values of k, if we take

$$0 < \varepsilon \le \varepsilon_0 = \min\{\frac{1}{2}, \frac{1+M}{2M_0}e^{-M_1(T+1)}\} \tag{6.21}$$

and

$$0 \le \varphi_0 \le \frac{1}{2}e^{-M_1 T}.$$

Then (6.17) leads to

$$\|\nabla w_{m+1}\|^2 \le \varphi_{m+1}$$

by definition of φ_{m+1}, and (6.16) holds with $k = m+1$ because of (6.19). This proves (6.20) and thus (6.16) for all $k = 0, \ldots, K$ inductively. Therefore if we assume

$$\|\nabla w_0\|^2 \le \frac{\varepsilon}{2}e^{-M_1 T}, \tag{6.22}$$

the inequality

$$\|A^{\frac{1}{2}}(u^\varepsilon(t) - u(t))\| \le a_{\frac{1}{2}}\varepsilon^{\frac{1}{2}} \; with \; a_{\frac{1}{2}} = \{\frac{1}{2} + \frac{M_0}{1+M}e^{M_1\varepsilon}(e^{M_1 T} - 1)\}^{\frac{1}{2}} \tag{6.23}$$

holds.

Summarizing what we have proved above concerning the error w in H^0 and H^1 we state

LEMMA 6.1 : *Assume $u \in C^0([0,T], D_A)$ denotes a solution of the Navier-Stokes initial value problem (0.2) with $PF = 0$. Then with $\varepsilon \to 0$, $0 < r \le \varepsilon$, the solutions u^ε of (0.4) converge in H^1 to u. For the difference $u^\varepsilon - u$ the error estimate*

$$\|u^\varepsilon(t) - u(t)\| \le a_0\varepsilon$$

holds for all $t \in [0,T]$, if we suppose (6.12) $\|u^\varepsilon(0) - u(0)\| \le c_0\varepsilon$ at the initial time $t = 0$. In addition, the estimate

$$\|\nabla_x(u^\varepsilon(t) - u(t))\| \le a_{\frac{1}{2}}\varepsilon^{\frac{1}{2}}$$

holds for all $t \in [0,T]$ and $\varepsilon \in (0, \varepsilon_0]$, ε_0 being sufficiently small, if we suppose (6.22) $\|\nabla(u^\varepsilon(0) - u(0))\|^2 \le \frac{\varepsilon}{2}e^{-M_1 T}$. The constants $a_0, a_{\frac{1}{2}}$ are from (6.13) or (6.23), respectively.

In the next section by means of semigroup methods we will obtain a stronger result.

Bounds in $\mathbf{H}^2$

From the H^1-bound established in Lemma 6.1 we will get error bounds in H^2 by means of the semigroup representation of the solution $w(t)$ of (6.5). Subtracting the semigroup representation (4.10) of u from (6.1) for u^ε we get

$$w(t) = e^{-tA} w_0 + \int_0^t e^{-(t-s)A} f(s)\, ds, \tag{6.24}$$

where f again stands for the right hand side (6.6) of (6.5). In order to estimate f first of all we apply $A^{-\frac{1}{2}}$ to the 5 summands of f. We find the inequalities

$$\begin{aligned}
\|A^{-\frac{1}{2}} P w_k^* \cdot \nabla w\| &\leq c\|A^{\frac{3}{8}} w_k^*\|\,\|A^{\frac{3}{8}} w\| \leq c_1 a_0^{\frac{1}{2}} a_{\frac{1}{2}}^{\frac{3}{2}} \cdot \varepsilon, \\
\|A^{-\frac{1}{2}} P u^*(t_k) \cdot \nabla w\| &\leq c\|A u^*(t_k)\|\,\|w\| \leq c_1 M^{\frac{1}{2}} a_0 \varepsilon, \\
\|A^{-\frac{1}{2}} P w_k^* \cdot \nabla u\| &\leq c\|A u\| \cdot \|w_k^*\| \leq c_1 M^{\frac{1}{2}} \varepsilon, \\
\|A^{-\frac{1}{2}} P(u^*(t_k) - u(t_k)) \cdot \nabla u\| &\leq c\|A u\| \cdot \|u^*(t_k) - u(t_k)\| \leq c_1 M r \\
\|A^{-\frac{1}{2}} P(u(t_k) - u) \cdot \nabla u\| &\leq c\|A u\|\,\|u(t_k) - u\| \leq c_1 M \varepsilon
\end{aligned}$$

with the help of Lemma 2.2, its Corollary 2.2, (7.8), Lemma 1.3 and Corollary 4.1. Consequently the inequality

$$\|A^{-\frac{1}{2}} f(t)\| \leq c_2 \varepsilon, \; t \in [0, T] \tag{6.25}$$

holds with a constant c_2 calculated from the above inequalities. Therefore if we write (6.24) in the equivalent form

$$w(t) = e^{-tA} w_0 + \int_0^t A^{\frac{1}{2}} e^{-(t-s)A} A^{-\frac{1}{2}} f(s)\, ds, \tag{6.26}$$

the estimate

$$\|A^\beta w\| \leq \|A^\beta w_0\| + c_3 \varepsilon \quad \mathit{for}\ \beta \in [0, \frac{1}{2}) \tag{6.27}$$

results from Lemma 3.1 and Lemma 3.2. In the following we will assume

$$\|A w_0\| \leq c \varepsilon^{\frac{1}{2}}. \tag{6.28}$$

From this and our assumption $\|w_o\| \leq c_0 \varepsilon$ above using the moment inequality (7.8) we see

$$\|A^\beta w_0\| \leq c_1 \varepsilon^{1 - \frac{\beta}{2}} \tag{6.29}$$

with a constant c_1 depending on $\beta \in [0, 1]$. Consequently on account of (6.27) the estimate

$$\|A^\beta w\| \leq c_4 \varepsilon^{1 - \frac{\beta}{2}} \quad \mathit{for}\ \beta \in [0, \frac{1}{2}) \tag{6.30}$$

holds. The latter inequality (6.30) again leads to sharper estimates. Applying $A^{-\delta}$, $\delta \in (\frac{1}{4}, \frac{1}{2})$ to the right hand side in (6.6) we get the inequalities

$$\begin{aligned}
\|A^{-\delta} P w_k^* \cdot \nabla w\| &\leq c\|\nabla w_k^*\|\,\|\nabla w\| \leq c_1 a_{\frac{1}{2}}^2 \varepsilon, \\
\|A^{-\delta} P u^*(t_k) \cdot \nabla w\| &\leq c\|A u^*(t_k)\|\,\|A^{\frac{1}{4}} w\| \leq c_1 M^{\frac{1}{2}} \varepsilon^{\frac{7}{8}}, \\
\|A^{-\delta} P w_k^* \cdot \nabla u\| &< c\|A u\|\,\|A^{\frac{1}{4}} w_k^*\| \leq c_1 M^{\frac{1}{2}} c^{\frac{7}{8}}, \\
\|A^{-\delta} P(u^*(t_k) - u(t_k)) \cdot \nabla u\| &\leq c\|u^*(t_k) - u(t_k)\|\,\|A u\| \leq c_1 r M, \\
\|A^{-\delta} P(u(t_k) - u) \cdot \nabla u\| &\leq c\|u(t_k) - u\|\,\|A u\| \leq c_1 M \varepsilon
\end{aligned}$$

by means of Lemma 2.2 and its Corollary 2.3, (1.2), Lemma 1.3 and Corollary 4.1. Therefore under the assumptions (6.12), (6.28) the inequality

$$\|A^{-\delta} f\| \leq c_3 \varepsilon^{\frac{7}{8}} \tag{6.31}$$

holds with a constant $c_3 > 0$ fixed by the above inequalities. If we rewrite the integral equation (6.24) in the equivalent form

$$w(t) = e^{-tA} w_0 + \int_0^t A^{\delta} e^{-(t-s)A} A^{-\delta} f(s)\, ds$$

and apply Lemma 3.1 and Lemma 3.2 again, using (6.29) we get the estimates

$$\begin{aligned} \|A^{\beta} w\| \;&\leq\; c\varepsilon^{1-\frac{\beta}{2}} \; for \; \beta \in [\frac{1}{4}, 1-\delta) \\ and & \\ [A^{\beta} w]_{\gamma,[0,T]} \;&\leq\; c\varepsilon^{1-\frac{\beta}{2}} \; for \; \beta \in [\frac{1}{4}, 1-\delta), \; \gamma \in (0, 1-\beta-\delta). \end{aligned} \tag{6.32}$$

The constant c depends on β, γ and δ and c_1 from (6.29). The latter two inequalities will give us bounds for $\|f\|$ and $[f]_{\gamma,J_k}$: In the same way as above we obtain

$$\begin{aligned} \|Pw_k^* \cdot \nabla w\| \;&\leq\; c\|A^{\frac{5}{8}} w_k^*\| \, \|A^{\frac{5}{8}} w\| \leq c_1 \varepsilon, \\ \|Pu^*(t_k) \cdot \nabla w\| \;&\leq\; c\|Au^*(t_k)\| \cdot \|\nabla w\| \leq c_1 \varepsilon^{\frac{3}{4}} M^{\frac{1}{2}} \\ \|Pw_k^* \cdot \nabla u\| \;&\leq\; c\|A^{\frac{1}{4}} w_k^*\| \, \|Au\| \leq c_1 \varepsilon^{\frac{7}{8}} M^{\frac{1}{2}} \\ \|P(u^*(t_k) - u(t_k)) \cdot \nabla u\| \;&\leq\; c\|A^{\frac{1}{4}}(u^*(t_k) - u(t_k))\| \, \|Au\| \leq c_1 r^{\frac{3}{4}} M, \\ \|P(u(t_k) - u) \cdot \nabla u\| \;&\leq\; c\|A^{\frac{1}{4}}(u(t_k) - u)\| \, \|Au\| \leq c_1 \varepsilon^{\frac{3}{4}} M \end{aligned}$$

by Lemma 2.2, Lemma 1.3 and Corollary 4.1. Under the assumption $0 < r \leq \varepsilon \leq 1$, this gives

$$\|f\| \leq c\varepsilon^{\frac{3}{4}}. \tag{6.33}$$

The constant c can be calculated from the above inequalities. Similarly we find the Hölder-bounds

$$[Pw_k^* \cdot \nabla w]_{\gamma,J_k} \leq c\|A^{\frac{5}{8}} w_k^*\| \cdot [A^{\frac{5}{8}} w]_{\gamma,J_k} \leq c_1 \varepsilon, \quad 0 < \gamma < \frac{1}{8},$$

$$[Pu^*(t_k) \cdot \nabla w]_{\gamma,J_k} \leq c\|A^{\frac{5}{8}} u^*(t_k)\| \, [A^{\frac{5}{8}} w]_{\gamma,J_k} \leq c_1 \varepsilon^{\frac{5}{8}}, \quad 0 < \gamma < \frac{1}{8},$$

$$[Pw_k^* \cdot \nabla u]_{\gamma,J_k} \leq c\|A^{\frac{1}{2}} w_k^*\| \, [A^{\frac{3}{4}} u]_{\gamma,J_k} \leq c_1 \varepsilon^{\frac{3}{4}}, \quad 0 < \gamma < \frac{1}{4},$$

$$[P(u^*(t_k) - u(t_k)) \cdot \nabla u]_{\gamma,J_k} \leq c\|A^{\frac{1}{2}}(u^*(t_k) - u(t_k))\| \cdot [A^{\frac{3}{4}} u]_{\gamma,J_k} \leq c_1 r^{\frac{1}{2}}, \; 0 < \gamma < \frac{1}{4},$$

$$\begin{aligned} [P(u(t_k) - u) \cdot \nabla u]_{\gamma,J_k} \;&\leq\; \sup_{\substack{s \neq t \\ s,t \in J_k}} \|\frac{u(t) - u(s)}{|t-s|^{\gamma}} \cdot \nabla u(t)\| + \sup_{\substack{s \neq t \\ s,t \in J_k}} \|u(s) \cdot \frac{\nabla u(t) - \nabla u(s)}{|t-s|^{\gamma}}\| \\ &\leq\; c \cdot \sup_{\substack{s \neq t \\ s,t \in J_k}} \frac{\|A^{\frac{1}{4}}(u(t) - u(s))\|}{|t-s|^{\gamma}} \|Au(t)\| \\ &+\; c \cdot \sup_{\substack{s \neq t \\ s,t \in J_k}} \|u(s)\|_{L^{\infty}} \frac{\|\nabla u(t) - \nabla u(s)\|}{|t-s|^{\gamma}}. \end{aligned}$$

Under our assumption $0 < r \le \varepsilon$, and because of Corollary 4.1 this leads to

$$[f]_{\gamma,J_k} \le c_\gamma \varepsilon^{\frac{1}{2}-\gamma} \ \ if\ \varepsilon \le 1,\ \ 0 < \gamma < \frac{1}{8}. \tag{6.34}$$

The constant c_γ can be calculated from the above inequalities.

Thus on each interval J_k the function f is uniformly Hölder continuous. We write (6.24) in the form

$$\begin{aligned} w(t) &= e^{-tA}w_0 + \int_0^{t_{k-1}} e^{-(t-s)A} f(s)\,ds + \int_{t_{k-1}}^{t_k} e^{-(t-s)A} f(s)\,ds + \\ &+ \int_{t_k}^{t} e^{-(t-s)A} f(s)\,ds \qquad for\ t \in J_k \end{aligned} \tag{6.35}$$

Using the uniform boundedness of $\|Ae^{-(t-s)A}\|$ $for\ t - s \ge \varepsilon$, from (6.35) with Corollary 3.1 and Lemma 3.3 we get

$$\begin{aligned} Aw(t) &= e^{-tA}Aw_0 + \int_0^{t_{k-1}} Ae^{-(t-s)A} f(s)\,ds + \\ &+ (e^{-(t-t_k)A} - e^{-(t-t_{k-1})A}) f(t_k) + \int_{t_{k-1}}^{t_k} Ae^{-(t-s)A}(f(s) - f(t_k))\,ds + \\ &+ (1 - e^{-(t-t_k)A}) f(t) + \int_{t_k}^{t} Ae^{-(t-s)A}(f(s) - f(t))\,ds. \end{aligned} \tag{6.36}$$

On account of (6.33), (6.34), (6.35) using Lemma 3.1 we find

$$\begin{aligned} \|Aw(t)\| &\le c\varepsilon^{\frac{1}{2}} + c_1\varepsilon^{\frac{3}{4}} \int_0^{t_{k-1}} \frac{ds}{t-s} + 4c_1\varepsilon^{\frac{3}{4}} + 2\int_0^{\varepsilon} \frac{ds}{s^{1-\gamma}} c_1 \varepsilon^{\frac{1}{2}-\gamma} \\ &\le c\varepsilon^{\frac{1}{2}} + c_1\varepsilon^{\frac{3}{4}}\{\ln T - \ln\varepsilon + 4\} + \frac{2c_1}{\gamma}\varepsilon^{\frac{1}{2}} \\ &\le a_1\varepsilon^{\frac{1}{2}} \end{aligned} \tag{6.37}$$

which proves (0.9) in case $\alpha = 1$. The constant $a_1 > 0$ can be calculated from the above inequalities. Finally, by interpolation we see

$$\|A^\alpha w(t)\| \le c_\alpha \|Aw(t)\|^\alpha \|w(t)\|^{1-\alpha} \le c_\alpha a_1^\alpha a_0^{1-\alpha} \varepsilon^{\frac{\alpha}{2}} \varepsilon^{(1-\alpha)} \le \tilde{a}_\alpha \varepsilon^{1-\frac{\alpha}{2}}. \tag{6.38}$$

Therefore (0.9) holds with $a_\alpha = \tilde{a}_\alpha$.

7 Appendix

The following basic inequalities apply in the proofs above several times. They are quoted below in the special form in which we use them.

Hölder's inequalities

$$\int_\Omega |f|\,|g|\,dx \le \|f\|_{L^q} \cdot \|g\|_{L^{q'}}, \qquad \frac{1}{q} + \frac{1}{q'} = 1,\ q \ge 1, \tag{7.1}$$

$$\int_\Omega |f|\,|g|\,|h|\,dx \le \|f\|_{L^2} \cdot \|g\|_{L^3} \cdot \|h\|_{L^6}, \tag{7.2}$$

$$\|f \cdot g\|_{L^2} \leq \|f\|_{L^q} \cdot \|g\|_{L^{q*}}, \quad \frac{1}{q} + \frac{1}{q*} = \frac{1}{2}, \; q \geq 2. \tag{7.3}$$

Cauchy-Young's inequality

$$|a| \cdot |b| \leq \mu |a|^q + c_\mu |b|^{q'} \tag{7.4}$$

with $\mu = \frac{\varepsilon^q}{q}$, $c_\mu = \frac{\varepsilon^{-q'}}{q'}$, $\varepsilon > 0$, $\frac{1}{q} + \frac{1}{q'} = 1$, $q \geq 1$.

Sobolev's inequality

$$\|f\|_{L^6} \leq c\|\nabla f\|, \quad f \in V. \tag{7.5}$$

Multiplicative inequalities

$$\|u\|_{L^q} \leq c\|\nabla u\|^\alpha_{L^p} \cdot \|u\|^{1-\alpha}_{L^r}, \qquad if\ u \in \overset{\circ}{W}{}^{1,p}, \tag{7.6}$$

(or $u \in W^{1,p}$ and $\int_\Omega u\, u(x)\, dx = 0$).

$$\|\nabla u\|_{L^q} \leq c\|\nabla^2 u\|^\alpha_{L^p} \cdot \|\nabla u\|^{1-\alpha}_{L^r}, \tag{7.7}$$

$$with \quad \alpha = (\frac{1}{r} - \frac{1}{q})(\frac{1}{3} - \frac{1}{p} + \frac{1}{r})^{-1} \in [0,1],$$

if $\nabla u \in W^{1,p}$ and $\int_\Omega \nabla u\, dx = 0$, $1 < p, r$, $\Omega \subset \mathbb{R}^3$ open bounded, [16, p. 62–63, Theorem 2.2; 6, p. 27, Theorem 10.1].

Moment inequality

for an m-sectorial operator A

$$\|A^\beta f\| \leq c\|A^\gamma f\|^{\frac{\beta-\alpha}{\gamma-\alpha}} \cdot \|A^\alpha f\|^{\frac{\gamma-\beta}{\gamma-\alpha}}, \alpha \leq \beta \leq \gamma, \; \alpha < \gamma, \; f \in D_{A^\gamma}, \tag{7.8}$$

c. p. [6, p. 159, Theorem 14.1].

Cattabriga-Solonnikov's a priori estimate

$$\|\nabla^2 u\| \leq c\|Au\|, \tag{7.9}$$

[5], [27] for the Stokes operator A.

Acknowledgement

The author likes to thank Professor Kyûya Masuda and all Japanese colleagues who had invited him for their friendly guidance and warm hospitality as well as for many fruitful discussions during his stay in Japan. He gratefully appreciates the support from the Japan Society for the Promotion of Science.

References

[1] Adams, R.A.: Sobolev spaces, Academic Press, New York 1975.

[2] Beale, J.T.: The approximation of the Navier-Stokes equations by fractional time steps, Lecture on the conference: The Navier-Stokes equations, theory and numerical methods, Oberwolfach 18. -24.8.91.

[3] Beale, J.T., Kato, T., Majda, A.: Remarks on the breakdown of smooth solutions of the 3-D Euler equations, Comm. Math. Phys; 94 (1984) 61–66.

[4] Borchers, W.: A splitting algorithm for incompressible Navier-Stokes equations, in :H. Niki, M. Kawahara (eds.): Int. Conf. on computional methods in flow analysis, Okayama, Japan (1988) 454–461.

[5] Cattabriga, L.: Su un problema al contorno relativo al sistema di equazioni di Stokes, Rend. Mat. Sem. Univ. Padova 31 (1961) 308–340.

[6] Friedman, A.: Partial differential equations, Holt, Rinehart and Winston, New York 1964.

[7] Fujita, H., Kato, T.: On the Navier-Stokes initial value problem, I., Arch. Rational Mech. Anal. 16 (1964) 269–315.

[8] Fujita, H., Morimoto, H.: On fractional powers of the Stokes operator, Proc. Japan. Acad. 46 (1970) 1141–1143.

[9] Giga, Y., Miyakawa, T.: Solutions in L_r of the Navier-Stokes initial value problem, Arch. Rational Mech. Anal. 89 (1985) 267–281.

[10] Hebeker, F. K.: On a new boundary element spectral method, in: P. Deuflhard, B. Enquist (Ed.) Large Scale Scientific Computations, Boston (1987) 180–193.

[11] Heywood, J.G.: The Navier-Stokes equations: On the existence, regularity and decay of solutions, Indiana Univ. Math. J. 29 (1980) 639–681.

[12] Kato,T., Fujita, H.: On the nonstationary Navier-Stokes system, Rend. Sem. Math. Univ. Padova 32 (1962) 243–260.

[13] Kato, T.: Perturbation theory for linear operators, Second ed. Springer Berlin 1976.

[14] Kobayashi, T., Muramatu, T.: Abstract Besov space approach to the nonstationary Navier-Stokes equations, Inst. Math. Univ. of Tsukuba, Preprint 1991.

[15] Ladyženskaja, O.A.: The mathematical theory of viscous incompressible flow, Second ed., New York 1969.

[16] Ladyženskaja, O.A., Solonnikov, V.A., Ural'ceva, N.N., Linear and quasilinear equations of parabolic type, AMS Providence, Rhode Island 1968.

[17] Lighthill, M.J.: Introduction, Boundary layer theory, Rosenhead, L. (ed.), Laminar boundary layer, Oxford 1963, 46–113.

[18] Lions, J.L., Magenes, E.: Non-homogeneous boundary value problem and applications I, Springer Berlin 1972.

[19] Masuda, K.: On the stability of incompressible viscous fluid motion past objects, J. Math. Soc. Japan 27 (1975) 294–327.

[20] Masuda, K., Rautmann, R.: Convergence rates for product formula approximations to Navier-Stokes problems. To appear.

[21] Rautmann, R.: On the convergence-rate of nonstationary Navier-Stokes approximations, Proc. IUTAM Symp. Paderborn 1979, Springer Lecture Notes in Math. 771 (1980) 425–449.

[22] Rautmann, R.: On the Navier-Stokes initial value problem (G.I. Taylor memorial lecture), in Kaul, C.N., Maiti, M. (eds), Proc. Intern. Symposium on Nonlinear Continuum Mechanics, Kharagpur 1980, 1–16.

[23] Rautmann, R.: A semigroup approach to error estimates for nonstationary Navier- Stokes approximations, Proc. Conference Oberwolfach 1982, Methoden Verfahren Math. Physik 27 (1983) 63–77.

[24] Rautmann, R.: On optimum regularity of Navier-Stokes solutions at time $t = 0$, Math. Z. 184 (1983) 141–149.

[25] Rautmann, R.: A convergent product formula approach to threee dimensional flow computations, Finite Approximations in Fluid Mechanics II (1989) 322–325.

[26] Rautmann, R., Masuda, K.: Error estimates for approximation schemes to Navier-Stokes problems. To appear.

[27] Solonnikov, V.A.: On differential properties of the solutions of the first boundary-value problem for nonstationary systems of Navier-Stokes equations, Trudy Mat. Inst. Steklov 73 (1964) 221–291, Transl.: British Library Lending Div., RTS 5211.

[28] Solonnikov, V.A.: Estimates for the solutions of nonstationary Navier-Stokes equations, Zap. Nauchn. Sem. Leningrad Mat. Inst. Steklova 38 (1973) 153–231, J. Sov. Math. 8 (1977) 467–529.

[29] Tanabe, H.: Equations of evolution, Pitman London 1979.

[30] Temam, R.: Navier-Stokes equations, North-Holland, Amsterdam 1977, rev. ed. 1979.

[31] Varnhorn, W.: Zur Numerik der Gleichungen von Navier-Stokes, Dissertation Paderborn 1985.

[32] Wahl, W. von: The equations of Navier-Stokes and abstract parabolic equations, Vieweg Braunschweig 1985.

ON THE PROBLEM OF THE HYPOELLIPTICITY OF THE LINEAR PARTIAL DIFFERENTIAL EQUATIONS

Luigi Rodino

Dipartimento di Matematica
Università di Torino - Via Carlo Alberto, 10
10123 Torino, Italy

Introduction

According to Schwartz [25], a linear partial differential operator P with coefficients in $C^\infty(\Omega)$, Ω open subset of $\mathbf{R}^n$, is said to be *hypoelliptic* if $Pu \in C^\infty(\Omega')$ implies $u \in C^\infty(\Omega')$ for any $\Omega' \subset \Omega$ and all distribution $u \in \mathscr{D}'(\Omega')$.

At present, one usually prefers to base on the notion of wave front set WFu of Hörmander [10], and say that P is *micro-hypoelliptic* when $WF(Pu) = WFu$ for all $u \in \mathscr{D}'(\Omega)$. Elliptic operators are micro-hypoelliptic, and more generally we have

$$WF(Pu) \subset WFu \subset WF(Pu) \cup \Sigma \qquad \textit{for all} \qquad u \in \mathscr{D}'(\Omega)\,, \tag{0.1}$$

where Σ is the characteristic manifold of P. Micro-hypoellipticity implies hypoellipticity, since the projection of WFu on Ω is the singular support of u.

In turn, if P is hypoelliptic then it is *homogeneous hypoelliptic*, in the sense that every distribution u solution of $Pu = 0$ is C^∞ in its domain of existence.

For several relevant classes of operators (operators with constant coefficients, operators of principal type, etc.) all the preceding definitions of hypoellipticity are actually equivalent, being satisfied, or not satisfied, at the same time; however there exist in the case of the multiple characteristics examples of hypoelliptic operators which are not micro-hypoelliptic (Parenti-Rodino [16], Morimoto-Morioka [14]) and also examples of homogeneous hypoelliptic operators which are not hypoelliptic (Rodino [21]; cf. Baouendi-Trèves [1], Rodino [20], Okaji [15], Cattabriga-Rodino-Zanghirati [4], Gramchev [6]).

Main object of this paper is to discuss similar counter-examples in the Gevrey category. We recall that $u \in C^\infty(\Omega)$ is in the Gevrey class $G^s(\Omega)$, $s \geqslant 1$, if for all $K \subset\subset \Omega$

$$\sup_{x \in K} |\partial^\alpha f(x)| \leqslant C^{|\alpha|+1} (\alpha!)^s\,, \tag{0.2}$$

for a suitable constant C independent of α; when $s = 1$ we recapture the analytic class $A(\Omega)$.

Arguing on linear partial differential operators with analytic coefficients and replacing C^∞-regularity with G^s-regularity, one may introduce the notion of s-hypoellipticity, s-micro-hypoellipticity and homogeneous s-hypoellipticity.

We shall prove in particular the following: *for any fixed* $s = 1 + \epsilon$, $\epsilon > 0$ *as small as* *we want, there exist examples of s-hypoelliptic operators which are C^∞-micro-hypoelliptic but not s-micro-hypoelliptic.* The analytic case $s=1$ will remain, unfortunately, outside our arguments. Other counter-examples will concern *homogeneous s-hypoelliptic operators which are neither s-micro-hypoelliptic, nor C^∞-hypoelliptic.*

The next Section 1 contains basic notations and definitions on Gevrey ultradistributions and wave front sets. Section 2 provides some preliminary results; it can be also red with independent interest, we hope, as a short survey on the problem of the hypoellipticity. Section 3 contains the above-mentioned examples; proofs are modelled on Parenti-Rodino [16], Rodino [21].

1. Distributions, ultradistributions and wave front sets

We shall use the notations which are standard in the theory of the linear partial differential operators; in particular for a multi-index $\alpha=(\alpha_1, \ldots, \alpha_n) \in \mathbf{Z}_+^n$ we shall write

$$D^\alpha = D_1^{\alpha_1} \ldots D_n^{\alpha_n}$$

where $D_j = -i\,\partial/\partial x_j$. We expect the reader is familiar with the spaces $C_0^\infty(\Omega)$, $C^\infty(\Omega)$, $\mathscr{S}(\mathbf{R}^n)$, and their duals in Schwartz [25] $\mathscr{D}'(\Omega)$, $\mathscr{E}'(\Omega)$, $\mathscr{S}'(\mathbf{R}^n)$; we shall also base on the related theory of kernels, and in particular on the notion of *regular, very regular, regularizing* and *properly supported kernel.*

For sake of completeness, let us recall here the definition of wave front set of Hörmander [10]; as usual $\hat{f}$ will denote the Fourier transform of a distribution f in $\mathscr{E}'(\Omega)$ or $\mathscr{S}'(\mathbf{R}^n)$.

Definition 1.1. *Let Ω be an open subset of $\mathbf{R}^n$ and let u be in $\mathscr{D}'(\Omega)$. For fixed $x_0 \in \Omega$ and $\xi_0 \neq 0$, we say that u is micro-regular (in the C^∞ sense) at (x_0, ξ_0) if there exists $\varphi \in C_0^\infty(\Omega)$, $\varphi(x)=1$ in a neighborhood of x_0, such that for all $m \in \mathbf{Z}_+$ and suitable $c_m > 0$*

$$|(\varphi u)^\wedge(\xi)| \leqslant c_m (1+|\xi|)^{-m} \tag{1.1}$$

for ξ in a conic neighborhood Γ of ξ_0 independent of m. The wave front set of u, denoted by WFu, is then defined as the complement in $\Omega \times (\mathbf{R}^n \backslash 0)$ of the set of all (x_0, ξ_0) where u is micro-regular.

WFu is a closed subset of $\Omega \times (\mathbf{R}^n \backslash 0)$, conic with respect to the dual variables; we have

$$\text{projection}_\Omega\, WFu = \text{sing supp}\, u\ . \tag{1.2}$$

Let us now turn attention to the Gevrey classes $G^s(\Omega)$, $1 \leqslant s < \infty$, defined in the Introduction. For $s > 1$, write $G_0^s(\Omega)$ for $G^s(\Omega) \cap C_0^\infty(\Omega)$. Giving a natural topology to $G_0^s(\Omega)$ and taking the dual, one obtains $\mathscr{D}_s'(\Omega)$, space of the Gevrey ultradistributions of order s; $\mathscr{D}_s'(\Omega)$ contains as a subspace $\mathscr{D}'(\Omega)$, distributions of Schwartz. Let us write $\mathscr{E}_s'(\Omega)$ for the class of the ultradistributions in $\mathscr{D}_s'(\Omega)$ with compact support.

One may develop in $\mathscr{D}_s'(\Omega)$ a theory of kernels similar to that of the distributions of Schwartz, associating to every kernel $K \in \mathscr{D}_s'(\Omega \times \Omega)$ a map K: $G_0^s(\Omega) \to \mathscr{D}_s'(\Omega)$.

We say that K is *s-regular* if it maps $G_0^s(\Omega)$ into $G^s(\Omega)$ and it can be extended as a continuous linear map from $\mathscr{E}_s'(\Omega)$ into $\mathscr{D}_s'(\Omega)$. We say *s-very regular kernel* for a kernel K

which is s-regular and which, moreover, is a G^s function in the complement of the diagonal in $\Omega\times\Omega$; the corresponding map K is s-pseudolocal, i.e.

$$s\text{-sing supp } Ku \subset s\text{-sing supp } u \qquad \textit{for all} \qquad u\in \mathscr{E}_s'(\Omega)\,. \tag{1.3}$$

Finally, K is called *s-regularizing kernel* if it is an element of $G^s(\Omega\times\Omega)$; this is equivalent to say that the corresponding map can be extended to a continuous linear map of $\mathscr{E}_s'(\Omega)$ into $G^s(\Omega)$.

Similarly we may define *analytic-regular*, *analytic very regular* and *analytic-regularizing* kernels in $\mathscr{D}'(\Omega\times\Omega)$, or $\mathscr{D}_s'(\Omega\times\Omega)$, $s>1$.

For a more detailed exposition on ultradistributions and ultradistribution-kernels see Komatsu [11] and the forthcoming book of Rodino [22].

Let us also recall the definition of Gevrey and analytic wave front sets.

Definition 1.2. *Let $s>1$. For fixed $x_0\in\Omega$ and $\xi_0\in\mathbf{R}^n$, $\xi_0\neq 0$, we say that $u\in\mathscr{D}_s'(\Omega)$ is s-micro-regular at (x_0,ξ_0) if there exists $\varphi\in G_0^s(\Omega)$, $\varphi(x)=1$ in a neighborhood of x_0, and a conic neighborhood Γ of ξ_0, such that for some C, $\epsilon>0$*

$$|(\varphi u)^\wedge(\xi)| \leqslant C\exp\left[-\epsilon|\xi|^{1/s}\right] \qquad \textit{for} \qquad \xi\in\Gamma\,. \tag{1.4}$$

The s-wave front set of u, $WF_s u$, is then defined as the complement in $\Omega\times(\mathbf{R}^n\backslash 0)$ of the set of all (x_0,ξ_0) where u is s-micro-regular.

We have for all $u\in\mathscr{D}_s'(\Omega)$

$$\text{projection}_\Omega\, WF_s u = s\text{-sing supp } u\,. \tag{1.5}$$

To define the analytic wave front set of $u\in\mathscr{D}_s'(\Omega)$, $WF_a u$, one will replace (1.4) by

$$|\hat{u}_N(\xi)| \leqslant C(CN/|\xi|)^N\,, \quad N=1,2,\ldots,\ \xi\in\Gamma\,, \tag{1.6}$$

where $u_N\in\mathscr{E}_s'(\Omega)$ is a bounded sequence with $u_N=u$ in a fixed neighborhood of x_0.

2. The problem of the hypoellipticity

Let Ω be an open subset of $\mathbf{R}^n$ and consider a linear partial differential operator with coefficients in Ω:

$$P=P(x,D)=\sum_{|\alpha|\leqslant m} c_\alpha(x)\,D^\alpha\,, \tag{2.1}$$

where we shall assume now and later on $c_\alpha(x)\in A(\Omega)$. We may consider in particular the case when the coefficients are constants $c_\alpha\in\mathbf{C}$:

$$P=P(D)=\sum_{|\alpha|\leqslant m} c_\alpha D^\alpha\,. \tag{2.2}$$

We shall regard P as a map

$$P: \mathscr{D}'(\Omega) \to \mathscr{D}'(\Omega) \tag{2.3}$$

or else for a fixed $s > 1$

$$P: \mathscr{D}'_s(\Omega) \to \mathscr{D}'_s(\Omega) . \tag{2.4}$$

Consider the equation $Pu = f$ and assume the datum f is smooth (C^∞, Gevrey or analytic); can we then exclude the existence of non-smooth solutions $u \in \mathscr{D}'(\Omega)$, or $u \in \mathscr{D}'_s(\Omega)$, of the equation?

This is the problem of the *hypoellipticity*, or *interior regularity of solutions*, which has a fundamental role in the modern theory of the linear partial differential operators. We shall specify in the following different properties of hypoellipticity; they are all satisfied by the elliptic operators but not only by these, that gives the reason for the term «hypoelliptic».

First let us argue in the C^∞-set-up. We say that P in (2.1) is *hypoelliptic* (in the C^∞-sense) in Ω if

$$\text{sing supp } Pu = \text{sing supp } u \qquad \textit{for all} \qquad u \in \mathscr{D}'(\Omega). \tag{2.5}$$

This is equivalent to say that

$$\textit{for any open set } \Omega' \subset \Omega: \tag{2.5$'$}$$
$$\text{sing supp } Pu = \text{sing supp } u \qquad \textit{for all} \qquad u \in \mathscr{D}'(\Omega') ;$$

or else

$$\textit{for any open set } \Omega' \subset \Omega, \textit{ for all } u \in \mathscr{D}'(\Omega') , \tag{2.5$''$}$$
$$Pu \in C^\infty(\Omega') \quad \textit{implies} \quad u \in C^\infty(\Omega') .$$

In fact (2.5)$'$ → (2.5)$''$ is obvious; (2.5) follows from (2.5)$''$ by considering a small neighborhood Ω' of any $x_0 \notin \text{sing supp } Pu$. Finally one proves (2.5) → (2.5)$'$ taking $\varphi \in C_0^\infty(\Omega')$, $\varphi(x) = 1$ in a neighborhood Ω'' of $x_0 \notin \text{sing supp } u$, for $u \in \mathscr{D}'(\Omega')$, and then observing that $(P(\varphi u))\big|_{\Omega''} = (Pu)\big|_{\Omega''}$.

The same arguments show that in (2.5) we may limit ourselves to consider distributions $u \in \mathscr{E}'(\Omega)$; this gives the following equivalent definition, which we way refer also to maps $K: \mathscr{E}'(\Omega) \to \mathscr{D}'(\Omega)$. Precisely, consider a very regular kernel $K \in \mathscr{D}'(\Omega \times \Omega)$; the corresponding map K is pseudo-local, i.e. $\text{sing supp } Ku \subset \text{sing supp } u$ for all $u \in \mathscr{E}'(\Omega)$. We say that K is hypoelliptic if

$$\text{sing supp } Ku = \text{sing supp } u \qquad \textit{for all} \qquad u \in \mathscr{E}'(\Omega) . \tag{2.5$'''$}$$

It is evident that hypoellipticity in Ω implies hypoellipticity in any open set $\Omega' \subset \Omega$. In the opposite direction, if $\{\Omega_J\}$ is an open covering of Ω with $\Omega_J \subset \Omega$, and for all J the restriction $K\big|_{\Omega_J}: u \in \mathscr{E}'(\Omega_J) \to (Ku)\big|_{\Omega_J} \in \mathscr{D}'(\Omega_J)$ is hypoelliptic in Ω_J, then K is hypoelliptic in Ω.

A powerful tool to prove hypoellipticity is given by the *method of the parametrix.*

Definition 2.1. *Let K be a very regular map. We say that the regular properly supported map K' is a right (left) parametrix of K if $KK' = I + R_1$ ($K'K = I + R_2$) where I is the identity and R_1 (respectively R_2) is regularizing. When K itself is properly supported, we shall not require K' properly supported.*

Observe that K' is a left (right) parametrix of K if and only if the transpose ${}^tK'$ is a right (left) parametrix of the transpose ${}^tK$. If in the previous definition $R_1 \equiv 0$ (or $R_2 \equiv 0$) we say that K' is a *regular right (or left) inverse of K.*

Proposition 2.2. *Assume there exists a very regular left parametrix K' of K; then K is hypoelliptic.*

In fact for all $u \in \mathscr{E}'(\Omega)$

$$\text{sing supp}\, u = \text{sing supp}\, (K'Ku - R_1 u) = \text{sing supp}\, K'Ku \subset \text{sing supp}\, Ku \ ,$$

since R_1 is regularizing and K' is pseudo-local.

Obviously the conclusions of Proposition 2.2 hold if parametrices are replaced by inverses. The kernel $e(x, y) \in \mathscr{D}'(\Omega \times \Omega)$ of a right inverse of the operator $P(x, D)$ in (2.1) is a *fundamental kernel* for $P(x, D)$, i.e.

$$\sum_{|\alpha| \leqslant m} c_\alpha(x)\, D_x^\alpha\, e(x, y) = \delta(x-y) \ , \tag{2.6}$$

where $\delta(x-y)$, Dirac measure on the diagonal of $\Omega \times \Omega$, is the kernel of the identity map; the transpose of the kernel $e(x, y)$ of a left inverse is a fundamental kernel for the transpose:

$$\sum_{|\alpha| \leqslant m} (-1)^{|\alpha|}\, D_y^\alpha\, (c_\alpha(y)\, e(x, y)) = \delta(x-y) \ . \tag{2.6$'$}$$

If $P(D)$ has constant coefficients as in (2.2) and $E \in \mathscr{D}'(\mathbf{R}^n)$ is a *fundamental solution* of $P(D)$, i.e.

$$P(D)\, E = \delta \ , \tag{2.7}$$

then the operator of convolution with kernel $e(x, y) = E(x-y)$, mapping $u \in \mathscr{E}'(\mathbf{R}^n)$ into $E * u \in \mathscr{D}'(\mathbf{R}^n)$, is a right and left regular inverse of $P(D)$; in fact for all $u \in \mathscr{E}'(\mathbf{R}^n)$

$$P(D)\,(E * u) = E * (P(D)\, u) = (P(D)\, E) * u = \delta * u = u \ . \tag{2.8}$$

Proposition 2.3. *If the operator $P(D)$ has a fundamental solution $E \in \mathscr{D}'(\mathbf{R}^n)$ with sing supp $E = \{0\}$, then $P(D)$ is hypoelliptic in $\mathbf{R}^n$.*

In fact the kernel $e(x, y)=E(x-y)$ is very regular and we may apply Proposition 2.2.

Let us finally return to the operator $P=P(x, D)$ in (2.1) and observe that from (2.5)″ it follows the *homogeneous hypoellipticity* of P:

$$\begin{aligned}&\textit{for any open set } \Omega' \subset \Omega, \quad \textit{for all} \quad u \in \mathscr{D}'(\Omega'),\\ &Pu=0 \quad \textit{in} \quad \Omega' \quad \textit{implies} \quad u \in C^\infty(\Omega').\end{aligned} \tag{2.9}$$

Does (2.9) imply hypoellipticity? The answer is positive for operators with constant coefficients, but negative in general, as we shall see in the sequel.

Consider now hypoellipticity in Gevrey classes. We fix $s > 1$ and we regard P as in (2.4). We say that P is *G^s-hypoelliptic* (*s-hypoelliptic* for short) in Ω if

$$s\text{-sing supp } Pu = s\text{-sing supp } u \qquad \textit{for all} \qquad u \in \mathscr{D}_s'(\Omega), \tag{2.10}$$

or equivalently

$$\begin{aligned}&\textit{for any open set } \Omega' \subset \Omega,\\ &s\text{-sing supp } Pu = s\text{-sing supp } u \qquad \textit{for all} \qquad u \in \mathscr{D}_s'(\Omega');\end{aligned} \tag{2.10$'$}$$

$$\begin{aligned}&\textit{for any open set } \Omega' \subset \Omega, \textit{ for all } u \in \mathscr{D}_s'(\Omega'),\\ &Pu \in G^s(\Omega') \textit{ implies } u \in G^s(\Omega').\end{aligned} \tag{2.10$''$}$$

Therefore if P is s-hypoelliptic in Ω then it is *homogeneous s-hypoelliptic*:

$$\begin{aligned}&\textit{for any open set } \Omega' \subset \Omega, \textit{ for all } u \in \mathscr{D}_s'(\Omega'),\\ &Pu=0 \textit{ in } \Omega' \textit{ implies } u \in G^s(\Omega'),\end{aligned} \tag{2.11}$$

but the converse is not true in general.

As before we may extend considerations to general operators. Precisely, let $K \in \mathscr{D}_s'(\Omega \times \Omega)$ be s-very regular; we say that the corresponding map K is s-hypoelliptic if

$$s\text{-sing supp } Ku = s\text{-sing supp } u \qquad \textit{for all} \quad u \in \mathscr{E}_s'(\Omega) \tag{2.10$'''$}$$

which is equivalent to (2.10) when K is a linear partial differential operator.

We then define a *properly supported s-regular map* K as in the C^∞-case, so that $K: G_0^s(\Omega) \to G_0^s(\Omega)$, $G^s(\Omega) \to G^s(\Omega)$, $\mathscr{E}_s'(\Omega) \to \mathscr{E}_s'(\Omega)$, $\mathscr{D}_s'(\Omega) \to \mathscr{D}_s'(\Omega)$ continuously.

A *right (left) parametrix* K' of a s-very regular map K is a s-regular map satisfying $KK' = I + R_1$ ($K'K = I + R_2$) where now R_1 (respectively R_2) is s-regularizing.

Arguing as in the proof of Proposition 2.2, we have that *the existence of a s-very regular left parametrix K' of K implies the s-hypoellipticity of K.*

Right and left inverses of $P(x, D)$ in (2.1) are characterized in the frame of the Gevrey classes by the previous identities (2.6), (2.6)′, where the fundamental kernel $e(x, y)$ is allowed to be in $\mathscr{D}_s'(\Omega\times\Omega)$. Moreover, admitting fundamental solutions $E\in\mathscr{D}_s'(\mathbf{R}^n)$ for the operator with constant coefficients $P(D)$ in (2.7), we have that s-sing supp $E=\{0\}$ *implies the s-hypoellipticity of* $P(D)$.

Let us come to hypoellipticity in the analytic case. Taking initially $\mathscr{D}'(\Omega)$ as universe set, we say that P in (2.1) is *analytic-hypoelliptic* in Ω if

$$a\text{-sing supp } Pu=a\text{-sing supp } u \qquad \textit{for all} \quad u\in\mathscr{D}'(\Omega)\ . \tag{2.12}$$

Equivalent definitions can be set as before; the definition of analytic hypoellipticity can be obviously referred to an analytic very regular map $K\colon \mathscr{E}'(\Omega)\to\mathscr{D}'(\Omega)$.

We have that *the existence of an analytic very regular left parametrix* $K'\in\mathscr{D}'(\Omega\times\Omega)$ *of* K *implies the analytic-hypoellipticity of* K. In particular *if the operator with constat coefficients* $P(D)$ *has a fundamental solution* $E\in\mathscr{D}'(\mathbf{R}^n)$ *with* a-sing supp $E=\{0\}$, *then* $P(D)$ *is analytic hypoelliptic in* $\mathbf{R}^n$

The preceding arguments can be restated in the frame of the ultradistributions. In particular, assume $K\in\mathscr{D}_s'(\Omega\times\Omega)$ is analytic-very regular, for some $s>1$; we say that K is *analytic-hypoelliptic on* $\mathscr{E}_s'(\Omega)$ if

$$a\text{-sing supp } Ku=a\text{-sing supp } u \qquad \textit{for all} \quad u\in\mathscr{E}_s'(\Omega)\ . \tag{2.13}$$

If there exists an analytic very regular left parametrix $K'\in\mathscr{D}_s'(\Omega\times\Omega)$ *of* K, *then* K *is analytic-hypoelliptic on* $\mathscr{E}_s'(\Omega)$; this applies in particular to the case when $K\in\mathscr{D}'(\Omega\times\Omega)$, and K is then also analytic-hypoelliptic according to (2.12).

We recall here some basic examples of linear partial differential operators whose hypoellipticity can be achieved by means of fundamental solutions, fundamental kernels or parametrices.

In the case of the operator with constant coefficients $P(D)$ in (2.2) hypoellipticity can be characterized in terms of the associated *characteristic polynomial*

$$p(\xi)=\sum_{|\alpha|\leqslant m} c_\alpha\,\xi^\alpha\ . \tag{2.14}$$

Precisely, consider the subset of $\mathbf{C}^n$

$$V=\{\zeta\in\mathbf{C}^n\ ,\ P(\zeta)=0\} \tag{2.15}$$

and define then for $\xi\in\mathbf{R}^n$

$$d(\xi)=\textit{distance}\ (\xi,V)\ . \tag{2.16}$$

Theorem 2.4. *The following properties are equivalent:*

(i) $P(D)$ *is hypoelliptic in* $\mathbf{R}^n$.

(ii) *For some open subset* $\Omega\subset\mathbf{R}^n$, *every* $u\in\mathscr{D}'(\Omega)$ *such that* $Pu=0$ *belongs to* $C^\infty(\Omega)$.

(iii) *The polynomial* $p(\xi)$ *in* (2.14) *is hypoelliptic, i.e. there exist* C, $\sigma>0$ *such that*

$$|\xi|^{\sigma} \leqslant C\, d(\xi) \qquad \textit{for} \quad |\xi| > C\,. \tag{2.17}$$

(iv) *There exists a very regular left parametrix of* $P(D)$.

Moreover for any fixed $s \geqslant 1$, *we have equivalence of the following properties:*

$(\mathrm{i})_s$ $P(D)$ *is s-hypoelliptic (analytic-hypoelliptic if* $s = 1$) *in* $\mathbf{R}^n$.

$(\mathrm{ii})_s$ *For some open subset* $\Omega \subset \mathbf{R}^n$, *every* $u \in \mathscr{D}'(\Omega)$ *such that* $Pu = 0$ *belongs to* $G^s(\Omega)$.

$(\mathrm{iii})_s$ *The estimate* (2.17) *is satisfied with* $\sigma = 1/s$.

$(\mathrm{iv})_s$ *There exists a s-very regular (analytic very regular if* $s = 1$) *left parametrix of* $P(D)$.

Since every operator with constant coefficients $P(D)$ admits a fundamental solution $E \in \mathscr{D}'(\mathbf{R}^n)$ (see for example Hörmander [10]), we may replace (iv) and $(\mathrm{iv})_s$ with the condition sing supp $E = \{0\}$, respectively s-sing supp $E = \{0\}$ for any such E.

It follows from Theorem 2.4 that $P(D)$ is analytic-hypoelliptic if and only if it is elliptic; in fact, the condition (2.17) with $\sigma = 1$

$$|\xi| \leqslant C\, d(\xi) \quad \textit{for} \quad |\xi| > C$$

is equivalent to each of the following two:

$$|\xi|^m \leqslant C\,|p(\xi)| \quad \textit{for} \quad |\xi| > C\,,$$

for a suitable $C > 0$, and

$$p_m(\xi) = \sum_{|\alpha| = m} c_\alpha \xi^\alpha \neq 0 \quad \textit{for} \quad |\xi| \neq 0\,.$$

The Laplace operator and the Cauchy-Riemann operator are examples of elliptic operators, whereas the characteristic polynomial of the heat operator

$$iD_n + \sum_{j=1}^{n-1} D_j^2 \tag{2.18}$$

satisfies (2.17) only for $\sigma \leqslant 1/2$.

Theorem 2.4 was first obtained by Hörmander (see [10]). The preceding arguments of this section give (iv) $\rightarrow$ (i) $\rightarrow$ (ii), $(\mathrm{iv})_s \rightarrow (\mathrm{i})_s \rightarrow (\mathrm{ii})_s$. Let us refer to [10] for a complete proof of the statement.

Turning attention to operators with variable coefficients, we consider now in $\mathbf{R}^2$ the operator of Mizohata [13]

$$P = \frac{\partial}{\partial x_1} + i x_1^h \frac{\partial}{\partial x_2} = i\,(D_1 + i x_1^h D_2) \tag{2.19}$$

for an even integer h.

A fundamental kernel for P and ${}^tP$ is given by

$$e\,(x, y) = \frac{1}{2\pi}\,(x_1^{h+1}/(h+1) + ix_2 - y_1^{h+1}/(h+1) - iy_2)^{-1}\,. \tag{2.20}$$

In fact we have to prove:

$$\left(\frac{\partial}{\partial x_1}+ix_1^h\frac{\partial}{\partial x_2}\right)e(x,y)=-\left(\frac{\partial}{\partial y_1}+iy_1^h\frac{\partial}{\partial y_2}\right)e(x,y)=\delta(x-y)\ . \tag{2.21}$$

Since (2.19) reduces to the Cauchy-Riemann operator if $x_1^{k+1}/(k+1)$ is introduced as a new variable instead of x_1, then (2.21) is satisfied for $x\neq y$. Moreover for $\varphi\in C_0^\infty(\mathbf{R}^2)$ the formal integral

$$-\int\varphi(y)\left(\frac{\partial}{\partial y_1}+iy_1^h\frac{\partial}{\partial y_2}\right)e(x,y)\,dy$$

can be defined at any $x\in\mathbf{R}^2$ as

$$\lim_{\epsilon\to 0}\int_{|x-y|>\epsilon}e(x,y)\left(\frac{\partial}{\partial y_1}+iy_1^h\frac{\partial}{\partial y_2}\right)\varphi(y)\,dy$$

which by Gauss formula is given by

$$\lim_{\epsilon\to 0}\int_{|x-y|=\epsilon}e(x,y)\,\varphi(y)\,(iy_1^h\,dy_1-dy_2)=\varphi(x)\ ,$$

since the argument variation of $y_1^{h+1}/(h+1)+iy_2-x_1^{h+1}/(h+1)-ix_2$ around $|x-y|=\epsilon$ in the positive sense is 2π.

This proves the second part of (2.21); the same argument gives the first part.

The kernel $e(x,y)\in\mathscr{D}'(\mathbf{R}^2\times\mathbf{R}^2)$ is analytic very regular; therefore P in (2.19) is analytic-hypoelliptic, s-hypoelliptic for all $s>1$ and C^∞-hypoelliptic.

Observe finally that when the integer h in (2.19) is odd, the operator of Mizohata P is not hypoelliptic, a solution of $Pu=0$ being given by $u(x)=(x_1^{h+1}/(h+1)+ix_2)^{-1}$.

Referring to the notion of wave front set, we may also introduce the following definitions of micro-hypoellipticity.

Again, we begin with the C^∞-case. Consider a linear partial differential operator as in (2.1) or, more generally, a kernel $K\in\mathscr{D}'(\Omega\times\Omega)$ with $WFK\subset$ Ⓗ , where

$$\text{Ⓗ}=\{(x,y,\xi,\eta);\ x=y,\ \xi=-\eta\}\ ; \tag{2.22}$$

then K is very regular and micro-local, i.e. $WF(Ku)\subset WFu$ for all $u\in\mathscr{E}'(\Omega)$. We say that K is *micro-hypoelliptic* (in the C^∞-sense) if

$$WF(Ku)=WFu\qquad \textit{for all}\qquad u\in\mathscr{E}'(\Omega)\ . \tag{2.23}$$

Similarly we say that $K\in\mathscr{D}_s'(\Omega\times\Omega)$, $s>1$, with $WF_sK\subset$ Ⓗ is *s-micro-hypoelliptic* if

$$WF_s(Ku)=WF_su\qquad \textit{for all}\qquad u\in\mathscr{E}_s'(\Omega)\ . \tag{2.24}$$

Analytic-micro-hypoellipticity is defined for $K \in \mathscr{D}'(\Omega \times \Omega)$ with $WF_a K \subset Ⓗ$ by

$$WF_a(Ku) = WF_a u \qquad \textit{for all} \quad u \in \mathscr{E}'(\Omega) . \tag{2.25}$$

If K is micro-hypoelliptic (s-micro-hypoelliptic with $s > 1$ or analytic-micro-hypoelliptic) then it is also hypoelliptic (respectively s-hypoelliptic or analytic-hypoelliptic), since the projection of the wave front set on Ω is the corresponding singular support.

The existence of a left parametrix K' *of* K *gives* (2.23), (2.24), (2.25), *provided we have* $WFK' \subset Ⓗ$, *respectively* $WF_s K' \subset Ⓗ$, $WF_a K' \subset Ⓗ$.

Using pseudo-differential operators, one may construct such parametrices for all the elliptic linear partial differential operators; that gives for them micro-hypoellipticity and hypoellipticity in the C^∞, Gevrey and analytic classes. See for example Hörmander [19].

More generally, P being defined as in (2.1) and Σ its characteristic manifold, $\Sigma = \{(x, \xi);$ $p_m(x, \xi) = 0,\ \xi \neq 0\}$ with $p_m(x, \xi) = \sum_{|\alpha| = m} c_\alpha(x)\, \xi^\alpha$, we may prove for any $s > 1$

$$WF_s u \subset WF_s(Pu) \cup \Sigma \qquad \textit{for all} \quad u \in \mathscr{D}_s'(\Omega), \tag{2.26}$$

$$WF_a u \subset WF_a(Pu) \cup \Sigma \qquad \textit{for all} \quad u \in \mathscr{D}_s'(\Omega), \tag{2.27}$$

which are the Gevrey-analytic version of (0.1); when $\Sigma = \phi$ one recaptures the micro-hypoellipticity of the elliptic equations.

Turning attention to the non-elliptic case, we consider first hypoelliptic partial differential operators with constant coefficients $P(D)$; *the existence of a fundamental solution* $E \in \mathscr{D}'(\mathbf{R}^n)$ *with* sing supp $E = \{0\}$ *implies the micro-hypoellipticity of* $P(D)$.

In fact the fundamental kernel $e(x, y) = E(x - y)$ satisfies $WFe \subset Ⓗ$. This can be proved directly, or else deduced from (0.1), since

$$(D_{x_j} + D_{y_j})\, e = 0 , \qquad j = 1, \dots, n ,$$

implies $\xi = -\eta$ for $(x, y, \xi, \eta) \in WFe$. *Similarly the existence of a fundamental solution* $E \in \mathscr{D}_s'(\mathbf{R}^n)$ *with* s-sing supp $E = \{0\}$ *implies the s-micro-hypoellipticity of* $P(D)$.

For example, the heat operator is s-micro-hypoelliptic for $s \geqslant 2$.

Consider then the operator of Mizohata in (2.19), with even h. We have already obtained a left inverse with kernel $e(x, y)$ given by (2.20). Let us prove that $WF_a e \subset Ⓗ$. We know already that a-sing supp e is the diagonal of $\mathbf{R}^2 \times \mathbf{R}^2$. Applying (2.27), for $(x, y, \xi, \eta) \notin Ⓗ = WF_a \delta(x - y)$ and $(x, y, \xi, \eta) \in WF_a e$ we would obtain from (2.21) that $\xi_1 = \eta_1 = 0$; on the other hand $\xi_2 = -\eta_2$ for $(x, y, \xi, \eta) \in WF_a e$, since $(D_{x_2} + D_{y_2})\, e = 0$.

We conclude $WF_a e = Ⓗ$; therefore the operator in (2.19) is analytic-micro-hypoelliptic, as well as s-micro-hypoelliptic for all $s > 1$ and micro-hypoelliptic in the C^∞-sense.

We end by quoting two general results of micro-hypoellipticity, which can be obtained by means of a technique of parametrices. The first, proved in Cattabriga-Rodino-Zanghirati [4], concerns operators which are, modulo lower order terms, m-th power of the Mizohata operator, $m \geqslant 2$.

Theorem 2.5. *Let P be an operator in* $\mathbf{R}^2$ *of the form*

$$P=(D_1+i x_1^h D_2)^m+\sum_{\alpha+\beta\leqslant m-1} c_{\alpha\beta}(x_1,x_2)\, D_1^\alpha D_2^\beta\,, \tag{2.28}$$

where h is even and the coefficients $c_{\alpha\beta}$ *are analytic; then P is analytic-micro-hypoelliptic and also s-micro-hypoelliptic for* $1<s<m/(m-1)$.

The result is independent of the lower order terms, which on the contrary may have a strong influence when arguing in C^∞ or in G^s for large s, as we shall see in the next Section.

The second result concerns the so-called (ϱ,δ)-operators; the following statement contains the theorem of Hörmander [8] concerning C^∞-hypoellipticity and its Gevrey variant in Liess-Rodino [12], Bolley-Camus-Métivier [2], Hashimoto-Matsuzawa-Morimoto [7] (see Zanghirati [27], Taniguchi [24] for further generalizations).

Theorem 2.6. *Let P be the operator in* (2.1), *where we first allow* $c_\alpha(x)\in C^\infty(\Omega)$, $\Omega\subset\mathbf{R}^n$, *and write* $p(x,\xi)$ *for its symbol*

$$p(x,\xi)=\sum_{|\alpha|\leqslant m} c_\alpha(x)\,\xi^\alpha\,. \tag{2.29}$$

Let ϱ,δ *be given with* $0\leqslant\delta<\varrho\leqslant 1$. *Assume for every* $K\subset\subset\Omega$ *there exist* $m'\in\mathbf{R}$ *and positive constants* ϵ, *C, B and* $c_{\alpha\beta}$ *such that*

$$|p(x,\xi)|\geqslant\epsilon|\xi|^{m'}\qquad \textit{for}\quad x\in K,\quad |\xi|>C\,, \tag{2.30}$$

$$|D_x^\alpha D_\xi^\beta p(x,\xi)|\leqslant c_{\alpha\beta}|p(x,\xi)|\,(1+|\xi|)^{-\varrho|\beta|+\delta|\alpha|}\qquad \textit{for}\quad x\in K,\ |\xi|>B\,. \tag{2.31}$$

Then P is micro-hypoelliptic in Ω.

Assume now the coefficients $c_\alpha(x)$ *are analytic and* (2.30), (2.31) *are satisfied, with*

$$c_{\alpha\beta}=c^{|\alpha|+|\beta|+1}\alpha!\,\beta! \tag{2.32}$$

in (2.31) *for some positive constant c. Then P is also s-micro-hypoelliptic for* $s\geqslant 1/(\varrho-\delta)$.

Applying Theorem 2.6 to the operators with constant coefficients we obtain (iii) → (i) and $(\text{iii})_s\to(\text{i})_s$ in Theorem 2.4; in fact if (2.17) is satisfied for some $\sigma>0$, then (2.31) is satisfied with $\varrho=\sigma$, $\delta=0$.

When (2.30), (2.31) are valid only for (x,ξ) in Λ, open subset of $\Omega\times(\mathbf{R}^n\backslash 0)$ conic with respect to ξ, one may replace the conclusions with

$$\Lambda\cap WF\,Pu=\Lambda\cap WF u\qquad \textit{for all}\quad u\in\mathscr{E}'(\Omega)\,, \tag{2.33}$$

and respectively

$$\Lambda\cap WF_s Pu=\Lambda\cap WF_s u\qquad \textit{for all}\quad u\in\mathscr{E}_s'(\Omega)\,. \tag{2.34}$$

For a survey on other results concerning analytic and Gevrey hypoellipticity, see Bolley-Camus-Rodino [2].

3. Counter-examples

In this Section we shall argue always in two dimensions; changing notations with respect to §§ 1, 2 we shall write now x, y for the variables in $\mathbf{R}^2$ and ξ, η for the respective dual variables.

Theorem 3.1. *The operator in* $\mathbf{R}^2$ *of order* $2m+2$, $m \geqslant 1$,

$$P = D_x^{2m} + x^{2h} D_y^{2m+2}, \tag{3.1}$$

is s-hypoelliptic if $1 + 1/m \leqslant s < +\infty$ *and also hypoelliptic for any integer* $h \geqslant 1$, *whereas P is neither s-hypoelliptic if* $1 < s < 1 + 1/m$ *nor analytic-hypoelliptic. The s-micro-hypoellipticity of P for* $1 + 1/m \leqslant s < +\infty$ *depends on h; precisely*:

(i) *if* $h = 1$, *P is neither micro-hypoelliptic nor s-micro-hypoelliptic, for any s.*

(ii) *if* $1 < h \leqslant m$, *P is micro-hypoelliptic, and s-micro-hypoelliptic if and only if*

$$1 + 1/(h-1) \leqslant s < +\infty\ . \tag{3.2}$$

(iii) *if* $h > m$, *P is micro-hypoelliptic and s-micro-hypoelliptic for all s with* $1 + 1/m \leqslant s < +\infty$.

We have in particular that for $1 < h \leqslant m$ the operator P in (3.1) is $(1+1/m)$-hypoelliptic and also C^∞-micro-hypoelliptic, but not $(1+1/m)$-micro-hypoelliptic.

Proof. The first part of the statement is a straightforward consequence of known results. In fact, for $x \neq 0$ the properties of hypoellipticity of P in (3.1) are the same of

$$D_x^{2m} + D_y^{2m+2}\ ; \tag{3.1$'$}$$

the two operators actually coincide after a change of variable, for $x \neq 0$, modulo lower order terms. Applying Theorem 2.4, for (3.1)′ we have hypoellipticity, and s-hypoellipticity if and only if $1 + 1/m \leqslant s < +\infty$. The same conclusion for P in (3.1), $x \neq 0$, can be obtained by strict arguments by using Theorem 2.6 in the preceding section, Theorem 13.4.4 in Hörmander [10] and Theorem 1.1 in Rodino-Zanghirati [23].

That hypoellipticity in (3.1) keeps valid for $x = 0$ follows from Parenti-Rodino [17]; this depends on the fact that the ordinary differential equation

$$D_x^{2m} \varphi + x^{2h} \varphi = 0$$

does not admit non-trivial solutions $\varphi \in \mathscr{S}(\mathbf{R})$.

Also s-hypoellipticity holds for $x = 0$, in view of the same fact and the results of Rodino [18], [19]. Precisely, we may use the obvious extension of Theorem 1.1 of [19] to the case of the weight $(\varrho m/(m+1), \varrho)$, with $0 < \varrho \leqslant 1$; the corresponding result of micro-hypoellipticity with respect to the anisotropic Gevrey wave front set used there implies s-hypoellipticity for $1 + 1/m \leqslant s < +\infty$.

Let us now prove (i), (ii), (iii) in the second part of the statement. If $h = 1$, P is not micro-hypoelliptic in view of the result of Chazarain [5]; in fact, writing in this case

$$P = (x D_y)^2 D_y^{2m} + D_x^{2m} ,$$

we recognize that at $x=0$, $\eta \neq 0$ we have double characteristics and the Levi condition is satisfied.

Suppose then $1 \leqslant h \leqslant m$ and take $1 < s < 1 + 1/(h-1)$; writing

$$P = (x D_y)^{2h} D_y^{2m-2h+2} + D_x^{2m}$$

we recognize that at $x=0$, $\eta \neq 0$, the principal symbol is characteristic with multiplicity $2h$ and the ϱ-Levi condition of Rodino-Zanghirati [23] is satisfied with $\varrho = 1 - 1/h$. It follows from Theorem 1.1 in [23] that P is not s-micro-hypoelliptic for

$$1 < s < 1/\varrho = 1 + 1/(h-1) ,$$

where we understand the right hand side is $+\infty$ when $h=1$.

Under the same assumptions on h and s, we obtain also from Theorem 1.1 in [23] a precise result of propagation of singularities in the fibers, which is worth the following independent statement.

Proposition 3.2. *Fix a point* $(x=0, y)$ *in* $\mathbf{R}^2$, *denote by* Γ_y *the fiber* $\mathbf{R}^2_{\xi,\eta}$ *over it, and split* $\Gamma_y = \Gamma_y^+ \cup \Gamma_y^- \cup \Gamma_y^\circ$, *with*

$$\Gamma_y^\pm = \Gamma_y \cap \{\eta \gtrless 0\} , \quad \Gamma_y^\circ = \Gamma_y \cap \{\eta = 0\} . \tag{3.3}$$

Consider now $z = (0, y; \xi, \eta)$ *with* $\eta \neq 0$ *in* Γ_y^+ *(or* Γ_y^-*). Let* u *be in* $\mathscr{D}_s'(\Omega)$ *and let* $V \subset \mathbf{R}^2 \times (\mathbf{R}^2 \backslash 0)$ *be a conic neighborhood of* z *such that* $WF_s(Pu) \cap V = \phi$, *and* $V \cap \Gamma_y$ *is connected. Under the assumption* $1 < s < 1 + 1/(h-1)$, *if* $z \in WF_s u$ *we have*

$$V \cap (\Gamma_y^+ \cup \Gamma_y^\circ) \subset WF_s u \qquad (\text{or } V \cap (\Gamma_y^- \cup \Gamma_y^\circ) \subset WF_s u) . \tag{3.4}$$

A similar result is valid for the analytic wave front set, for any value of $h \geqslant 1$.

We may also give a more direct proof of the non-s-micro-hypoellipticity in (i), (ii), modelling on Parenti-Rodino [16]. Let s be fixed, $1 < s < h/(h-1)$, $h \leqslant m$. Consider the operator

$$\hat{P} = \eta^{2m+2} (-D_\xi)^{2h} + \xi^{2m} \tag{3.5}$$

that we obtain from P conjugating by Fourier transform. If $U(t)$ is a solution of

$$(-D_t)^{2h} U + t^{2m} U = 0 \tag{3.6}$$

then

$$\omega(\xi, \eta) = U(\eta^{-(m+1)/(m+h)} \xi) \tag{3.7}$$

is a solution of $\hat{P}\omega=0$ for $\eta > 0$. It follows from the theory of the asymptotic integration (see for example Wasow [25]) that we may choose a solution $U(t)$ of (3.6) satisfying for targe $|t|$ the estimates

$$c \leqslant |U(t)| \leqslant C \exp\,[C|t|^{(m+h)/h}]\,, \tag{3.8}$$

with suitable positive constants C, c. With such $U(t)$ we obtain for $\omega(\xi,\eta)$ in (3.7) the double estimates

$$c \leqslant |\omega(\xi,\eta)| \leqslant C \exp\,[C|\xi|^{(m+h)/h}\,\eta^{-(m+1)/h}]\,. \tag{3.9}$$

Now take $\varphi(t) \in G_0^s(\mathbf{R})$, $\varphi(t)=1$ for $|t| \leqslant 1$, $\varphi(t)=0$ for $|t| \geqslant 2$, and $\psi(\xi,\eta) \in G^s(\mathbf{R}^2)$, $\psi(\xi,\eta)=0$ for $|\xi|^2+|\eta|^2 \leqslant 1/4$, $\psi(\xi,\eta)=1$ for $\xi^2+\eta^2 \geqslant 1$. Let us define $H(\xi,\eta) \in G^s(\mathbf{R}^2)$, $H(\xi,\eta)=0$ for $\eta \leqslant 0$, and for $\eta > 0$

$$H(\xi,\eta)=\psi(\xi,\eta)\,\varphi(\xi\eta^{-1-\epsilon})\,\omega(\xi,\eta)\exp\,[-B\eta^{1/\sigma}]\,, \tag{3.10}$$

where σ is fixed with $s < \sigma < h/(h-1)$, and $\epsilon > 0$, $B > 0$ will be defined later. Observe that $\hat{P}H(\xi,\eta)=0$ in the region

$$\{\eta > 0,\ \ |\xi| < \eta^{1+\epsilon},\ \ \xi^2+\eta^2 > 1\}\,, \tag{3.11}$$

and

$$\operatorname{supp} H \subset \{\eta > 0,\ \ |\xi| \leqslant 2\eta^{1+\epsilon}\}\,. \tag{3.12}$$

It follows from (3.9), (3.12) that for $\xi, \eta \in \operatorname{supp} H$ and large $|\xi|+|\eta|$

$$c \leqslant |\omega(\xi,\eta)| \leqslant C \exp\,[C\eta^{(h-1)/h+\epsilon(m+h)/h}]\,, \tag{3.13}$$

with a new constant C. We now define ϵ in (3.10) such that

$$(h-1)/h+\epsilon(m+h)/h=1/\sigma$$

and take B in (3.10) larger than C in (3.13); it follows then for a suitable $c' > 0$

$$|H(\xi,\eta)| \leqslant C \exp\,[-c'\eta^{1/\sigma}]\,, \tag{3.14}$$

whereas in the region (3.11) we have

$$c \exp\,[-B\eta^{1/\sigma}] \leqslant |H(\xi,\eta)|\,. \tag{3.15}$$

Let us define $u=\mathscr{F}^{-1}(H)$, where $\mathscr{F}^{-1}$ denotes the inverse Fourier transform; from (3.14) we obtain $u \in C^\infty(\mathbf{R}^2) \cap \mathscr{S}'(\mathbf{R}^2)$. It is then easy to evaluate the projections $\pi(WF_s u)$ and $\pi(WF_s Pu)$ on the dual space $\mathbf{R}^2_{\xi,\eta}$; precisely, since $s < \sigma$, from (3.15) in the region (3.11) we obtain

$$\pi(WF_s u) = \{(\xi, \eta) \neq (0,0), \quad \eta \geqslant 0\} , \tag{3.16}$$

whereas

$$\pi(WF_s Pu) \subset \{(\xi, \eta) \neq (0,0), \quad \eta = 0\}, \tag{3.17}$$

since $(Pu)^\wedge(\xi, \eta) = \hat{P}H(\xi, \eta) = 0$ in the region (3.11) and for $\eta < 0$. Comparing (3.16), (3.17), we conclude that P is not s-micro-hypoelliptic.

It remains to prove micro-hypoellipticity and s-micro-hypoellipticity according to (ii), (iii). We remark initially that for $x \neq 0$ the operator P is micro-hypoelliptic and s-micro-hypoelliptic if $1 + 1/m \leqslant s < +\infty$; this is easily obtained arguing as in the beginning of the proof. It will be then actually sufficient to prove that for every $y \in \mathbf{R}$, and for s as in (ii), (iii):

$$\Gamma_y \cap WF_s Pu = \Gamma_y \cap WF_s u \qquad \textit{for all} \quad u \in \mathscr{E}_s'(\mathbf{R}^2) , \tag{3.18}$$

where we use the notations of Proposition 3.2; define also

$$\Gamma_y^{\pm\pm} = \Gamma_y^{\pm} \cap \{\xi \gtrless 0\} \tag{3.19}$$

We shall prove:

Lemma 3.3. *Assume $h > 1$ and $1 + 1/(h-1) \leqslant s < +\infty$. Then*

$$\Gamma_y^{\pm\pm} \cap WF_s Pu = \Gamma_y^{\pm\pm} \cap WF_s u \qquad \textit{for all} \quad u \in \mathscr{E}_s'(\mathbf{R}^2) . \tag{3.20}$$

So (3.18) would be achieved, but for the rays through the points $(0, \pm 1)$, $(\pm 1, 0)$ in Γ_y. Taking then advantage from the s-hypoellipticity of P for $1 + 1/m \leqslant s < +\infty$, which is already proved, we shall conclude the proof by using the following lemma with s as in (ii), (iii), i.e.

$$\max(1 + 1/m,\ 1 + 1/(h-1)) \leqslant s < +\infty . \tag{3.21}$$

We return here to the n-dimensional notations.

Lemma 3.4. *Let P be a linear partial differential operator with analytic coefficients in $\Omega \subset \mathbf{R}^n$. Assume P is s-hypoelliptic in Ω, for some $s > 1$. Moreover, for every $x \in \Omega$ let a finite number of points $\xi_{x,1}, \ldots, \xi_{x,m} \in \mathbf{R}^n_\xi \backslash 0$ be fixed (m may depend on x) and write*

$$Ⓗ = \{(x, \xi) \in \Omega \times (\mathbf{R}^n \backslash 0), \quad (x, \xi) \neq (x, \xi_{x,j}) \quad \textit{for all} \quad j = 1, \ldots, m\} . \tag{3.22}$$

If

$$Ⓗ \cap WF_s Pu = Ⓗ \cap WF_s u \qquad \textit{for} \quad u \in \mathscr{E}_s'(\Omega) , \tag{3.23}$$

then P is s-micro-hypoelliptic in Ω .

Lemma 3.3 and Lemma 3.4 keep valid for the C^∞ wave front set; the same argument will then give the micro-hypoellipticity of P.

Proof of Lemma 3.3. Let us prove (3.20) for Γ_y^{++} (the proof in the other cases in similar). Consider the conic region

$$\Gamma_R = \{(x, y, \xi, \eta);\ \eta < R\xi,\ \xi < R\eta,\ \xi > 0,\ \eta > 0\} \tag{3.24}$$

with $R > 0$ as large as we want. We shall check that (2.30), (2.31), (2.32) in the preceding Section are satisfied in Γ_R by the symbol of P:

$$p = \xi^{2m} + x^{2h} \eta^{2m+2} , \tag{3.25}$$

with $\varrho = 0$, $\delta = 1/h$. Using Theorem 2.6, we shall apply (2.34) with $s \geqslant 1/(1-\delta) = 1 + 1/(1-h)$, $\Lambda = \Gamma_R$ and obtain then the conclusion, since

$$\Gamma_y \cap (\cup \Gamma_R) = \Gamma_y^{++} .$$

Observe first that in Γ_R

$$\sqrt{\xi^2 + \eta^2} \sim \xi \sim \eta , \tag{3.26}$$

$$p \sim (1 + x^{2h} \eta^2)\, \eta^{2m} . \tag{3.27}$$

The estimate (2.30) is therefore obvious, with $m' = 2m$. Fix then attention on (2.31) and observe that (2.32) will be automatically satisfied, since p is now a polynomial of order $2h + 2m + 2$ and its derivatives of hygher order vanish. In view of (3.26), (3.27) we have to prove for $(\xi, \eta) \in \Gamma_R$:

$$|D_x^\alpha D_\xi^\beta D_\eta^\gamma p| \leqslant C(1 + x^{2h} \eta^2)\, \eta^{2m-\beta-\gamma+\alpha/h} . \tag{3.28}$$

If $\beta \neq 0$, or $\alpha > 2h$, or $\gamma > 2m+2$, (3.28) is obvious. For $\alpha \leqslant 2h$, $\gamma \leqslant 2m+2$ we have

$$|D_x^\alpha D_\eta^\gamma p| \leqslant C|x|^{2h-\alpha} \eta^{2m+2-\gamma} ;$$

it will therefore be sufficient to prove

$$|x|^{2h-\alpha} \eta^2 \leqslant (1 + x^{2h} \eta^2)\, \eta^{\alpha/h} . \tag{3.29}$$

We check (3.29) separately in the two regions of Γ_R

$$A_1=\{|x|^h\eta > 1\}, \quad A_2=\{|x|^h\eta \leqslant 1\} .$$

We may re-write (3.29)

$$x^{2h}\eta^2 \leqslant (1+x^{2h}\eta^2)\,(|x|^h\eta)^{\alpha/h} ,$$

which is obviously satisfied in A_1; we may also re-write

$$1 \leqslant (1+x^{2h}\eta^2)\,(|x|^h\eta)^{-(2h-\alpha)/h} ,$$

which is satisfied in A_2 since $2h-\alpha \geqslant 0$.

Proof of Lemma 3.4. We have to prove that $(x_0, \xi_0) \notin WF_s Pu$ implies $(x_0, \xi_0) \notin WF_s u$, for $\xi_0 = \xi_{x_0,1}, \ldots, \xi_{x_0,m}$.

In fact, if $(x_0, \xi_0) \notin WF_s Pu$, there exists $\epsilon > 0$ and a conic neighborhood Λ of ξ_0 such that

$$(\{|x-x_0| < \epsilon\}\times\Lambda) \cap WF_s Pu=\phi .$$

In view of (3.23), if Λ is chosen sufficiently small, we have

$$\{x_0\}\times(\Lambda\backslash\{t\xi_0, t > 0\}) \cap WF_s u=\phi . \tag{3.30}$$

Take now $\chi(\xi) \in G^s(\mathbf{R}^n)$, homogeneous of order zero for large $|\xi|$, $\chi(\xi)=1$ in a conic neighborhood of ξ_0, $\operatorname{supp}\chi \subset \Lambda$ for large $|\xi|$. We have $\chi(D)\,Pu \in G^s$ in a neighborhood of x_0 (see for example Rodino-Zanghirati [23], p. 691). Let us prove that also $P\chi(D)\,u$ is G^s in a neighborhood of x_0; in fact

$$P\chi(D)\,u=\chi(D)\,Pu+[P, \chi(D)]\,u ,$$

where $[P, \chi(D)]\,u \in G^s$ in a neighborhood of x_0, since the s-microsupport of the symbol of $[P, \chi(D)]$ is included in

$$\Omega\times(\Lambda\backslash\{t\xi_0, t > 0\})$$

and therefore from (3.30) we obtain

$$(\{x_0\}\times\mathbf{R}^n) \cap WF_s\,[P, \chi(D)]\,u=\phi$$

(use Theorem 3.22 and Theorem 3.27 in [23]).

In view of the s-hypoellipticity of P, we have $\chi(D)\,u \in G^s$ in a neighborhood of x_0, and this implies $(x_0, \xi_0) \notin WF_s u$ (see again p. 691 in [23]).

This concludes the proofs of Lemma 3.4 and Theorem 3.1.

Theorem 3.5. *The operator in* $\mathbf{R}^2$ *of order* $2m+1$, $m \geqslant 1$,

$$P=(D_x+ix^{2k}D_y)^{2m+1}-ixD_y^{2m} \tag{3.31}$$

is analytic-micro-hypoelliptic and s-micro-hypoelliptic if $1 < s < 1+1/(2m)$ *for any integer* $k \geqslant 1$, *whereas it is neither hypoelliptic for any* $k \geqslant 1$, *nor s-micro-hypoelliptic if*

$$1+1/(2m)+\epsilon(k) < s < +\infty , \tag{3.32}$$

where $\epsilon(k) > 0$, $\epsilon(k) \to 0$ *as* $k \to +\infty$.

Since analytic-micro-hypoellipticity implies homogeneous hypoellipticity and homogeneous s-hypoellipticity for any $s \geqslant 1$, Theorem 3.5 provides an example of a homogeneous hypoelliptic operator P which is not hypoelliptic; moreover for s as in (3.32), P is homogeneous s-hypoelliptic but not s-micro-hypoelliptic.

Proof. The analytic-micro-hypoellipticity and the s-micro-hypoellipticity for $1 < s < 1+1/2m$ follow from Theorem 2.5. To obtain the other conclusions, we first find a formal solution of the equation $Pu=0$ by Fourier transforming with respect to the variable y:

$$(Pu)^\wedge(x, \eta)=\hat{P}\hat{u}(x, \eta)=0 , \tag{3.33}$$

where

$$\hat{P}(x, D_x, \eta)=(D_x+ix^{2k}\eta)^{2m+1}-ix\eta^{2m} . \tag{3.34}$$

For $\eta > 0$ a solution of $\hat{P}\hat{u}=0$ is given by

$$\hat{u}(x, \eta)=\psi(\eta^{m/(m+1)}x)\exp[x^{2k+1}\eta/(2k+1)] , \tag{3.35}$$

where $\psi(t)$ is a solution of

$$D_t^{2m+1}\psi-it\psi=0 . \tag{3.36}$$

We may assume $\psi(t)$ is a non-trivial function in $\mathscr{S}(\mathbf{R})$; in fact by Fourier transform we obtain for $\hat{\psi}(\tau)$

$$iD_\tau\hat{\psi}+\tau^{2m+1}\hat{\psi}=0$$

and therefore we may define $\psi(t)$ as the inverse Fourier transform of

$$\hat{\psi}(\tau)=\exp[-\tau^{2m+2}/(2m+2)] . \tag{3.37}$$

From (3.36) and the theory of the asymptotic integration, or else directly from (3.37), we have for $\psi(t)$ the estimates

$$|D^\alpha \psi(t)| \leqslant c_\alpha \exp[-\epsilon |t|^{(2m+2)/(2m+1)}] \tag{3.38}$$

with suitable positive constants c_α, ϵ.

From (3.35) one would obtain:

$$u(x, y) = (2\pi)^{-1} \int_0^{+\infty} \exp[iy\eta]\, \hat{u}(x, \eta)\, d\eta \,. \tag{3.39}$$

The solution $u(x, y)$ is purely formal, since for $x > 0$ the function $\hat{u}(x, \eta)$ is not in the Schwartz space $\mathscr{S}'(\mathbf{R}_\eta)$. However, we may take advantage of (3.38), which implies in particular for suitable $C > 0$, $\epsilon > 0$:

$$|\hat{u}(x, \eta)| \leqslant C \exp \varphi(x, \eta) \qquad \textit{for} \quad \eta > 0, \ \ x > 0 \tag{3.40}$$

with

$$\varphi(x, \eta) = x^{2k+1}\eta/(2k+1) - \epsilon\, x^{(2m+2)/(2m+1)}\, \eta^{2m/(2m+1)} \,. \tag{3.41}$$

Estimates of the type (3.40) are satisfied also by $D_x^\alpha \hat{u}(x, \eta)$ for all α, in view of (3.38).

We have $\varphi(x, \eta) \gtreqless 0$ inside the domains

$$A_\pm = \{x > 0,\ \eta > 0;\ x\eta^\varrho \gtreqless c\} \tag{3.42}$$

with

$$\varrho = 1/[2k(2m+1) - 1]\,, \qquad c = [(2k+1)\,\epsilon]^{(2m+1)\varrho} \,. \tag{3.43}$$

Take then $\chi(t) \in C^\infty(\mathbf{R})$, $\chi(t) = 1$ for $t < c'$, $\chi(t) = 0$ for $t > c''$, with $0 < c' < c'' < c$, and set

$$\tilde{u}(x, y) = (2\pi)^{-1} \int_0^{+\infty} \exp[iy\eta]\, \chi(x\eta^\varrho)\, \hat{u}(x, \eta)\, d\eta \,. \tag{3.44}$$

Now $\tilde{u}(x, y)$ is well defined, at least as distribution in $\mathscr{S}'(\mathbf{R}^2)$. On the other hand, sing supp $\tilde{u}$ is non-empty; in fact

$$\hat{u}(0, \eta) = \psi(0) = (2\pi)^{-1} \int \exp[-\tau^{2m+2}/(2m+2)]\, d\tau \neq 0$$

and therefore $\tilde{u}(x, y)$ is singular at the origin.

Compute now

$$P\tilde{u}(x, y) = (2\pi)^{-1} \int_0^{+\infty} \exp[iy\eta]\, \hat{P}(x, D_x, \eta)\, (\chi(x\eta^\varrho)\, \hat{u}(x, \eta))\, d\eta \,. \tag{3.45}$$

The function in the integral has support in the region

$$A_0 = \{x > 0,\ \eta > 0;\ c' \leqslant x\eta^{\varrho} \leqslant c''\}$$

which is included in A_-. In the same region A_0 we have in view of (3.41), (3.43)

$$\varphi(x, \eta) \leqslant -\epsilon' \eta^{\delta}$$

for some $\epsilon' > 0$, with

$$\delta = (4mk-1)/(4mk+2k-1)\ . \tag{3.46}$$

Therefore for $(x, \eta) \in A_0$

$$|\hat{u}(x, \eta)| \leqslant C \exp\,[-\epsilon\, \eta^{\delta}] \tag{3.47}$$

and estimates of the same type are satisfied by

$$\eta^{\alpha} D_x^{\beta} \hat{P}(x, D_x, \eta)\,(\chi(x\eta^{\varrho})\,\hat{u}(x, \eta)) \tag{3.48}$$

for all α and β, in view of (3.35), (3.38). Therefore from (3.45) we conclude $P\tilde{u} \in C^{\infty}(\mathbf{R}^2)$. This proves that P is not hypoelliptic.

As for the last part of the statement, consider the point $z = (x=0,\ y=0,\ \xi=0,\ \eta=1)$. We have easily from (3.44) that z is in $WF_s\,\tilde{u}$ for all s, whereas using in (3.45) the estimates (3.47) we deduce $z \notin WF_s\, P\tilde{u}$ for

$$s \geqslant 1/\delta = 1 + 1/(2m) + \epsilon(k)$$

with

$$\epsilon(k) = 1/[2m(4mk-1)]\ .$$

This ends the proof of Theorem 3.5.

Remark 3.6. A natural question is whether the operator P in (3.31) is s-hypoelliptic for $1 + 1/(2m) \leqslant s < +\infty$. It follows from the results of Gramchev [6] that P keeps s-micro-hypoelliptic if $1 + 1/(2m) \leqslant s \leqslant s'$, for a suitable $s' < 1 + 1/(2m) + \epsilon(k)$. On the contrary, it seems possible to prove that P is not s-hypoelliptic if $s'' < s < +\infty$ for a suitable $s'' > 1 + 1/(2m) + \epsilon(k)$, by using more precise estimates for the terms in (3.48).

Remark 3.7. Theorem 3.5 provides also an example of analytic-hypoelliptic operator which is not hypoelliptic (other examples are in Baouendi-Trèves [1], Rodino [20], [21], Cattabriga-Rodino-Zanghirati [4], Okaji [15], Gramchev [6]). In the opposite direction, it is well known that there exist operators which are hypoelliptic, but not analytic-hypoelliptic; see the heat operator in (2.18) and Theorem 2.4.

References

[1] M.S. Baouendi-F. Trèves, *A property of the functions and distributions annihilated by a locally integrable system of complex vector fields*, Ann. of Math., **113**, (1981), 387-421.

[2] P. Bolley-J. Camus-G. Métivier, *Regularité Gevrey et itérés pour une classe d'opérateurs hypoelliptiques*, Rend. Sem. Mat. Università Politecnico Torino, fasc. spec. Convegno «Linear partial and pseudo differential operators», 1982 (1983), 51-74.

[3] P. Bolley-J. Camus-L. Rodino, *Hypoellipticité analytique-Gevrey et itérés d'opérateurs*, Rend. Sem. Mat. Università Politecnico Torino, **45**, 1987 (1989), 1-61.

[4] L. Cattabriga-L. Rodino-L. Zanghirati, *Analytic-Gevrey hypoellipticity for a class of pseudo differential operators with multiple characteristics*, Comm. Partial Differential Equations, **15** (1990), 81-96.

[5] I.J. Chazarain, *Propagation des singularités pour une classe d'opérateurs à charactéristiques multiples et resolubilité locale*, Ann. Inst. Fourier, Grenoble, **24** (1974), 209-223.

[6] T. Gramchev, *Powers of Mizohata type operators in Gevrey classes*, Boll. Un. Mat. Ital., Sez. VII, **5-B** (1991), 135-156.

[7] S. Hashimoto-T. Matsuzawa-Y. Morimoto, *Opérateurs pseudo-différentiels et classes de Gevrey*, Comm. Partial Differential Equations, **8** (1983), 1277-1289.

[8] L. Hörmander, *Pseudo-differential operators and hypoelliptic equations*, Proc. of Symp. in Pure Math. (Amer. Math. Soc., Providence, R.I.) **10** (1967), 138-183.

[9] L. Hörmander, *Uniqueness theorems and wave front sets for solutions of linear differential equations with analytic coefficients*, Comm. Pure Appl. Math., **24** (1971), 671-704.

[10] L. Hörmander, *The analysis of linear partial differential operators*, I. II, III, IV, Springer-Verlag, Berlin 1983-1985.

[11] H. Komatsu, *Ultradistributions*, I: *Structure theorems and a characterization*; II: *The kernel theorem and ultradistributions with support in a submanifold*; III: *Vector valued ultradistributions and the theory of kernels*, J. Fac. Sci. Univ. Tokyo, Sect. IA, **20** (1973), 25-105; **24** (1977), 607-628; **29** (1982), 653-717.

[12] O. Liess-L. Rodino, *Inhomogeneous Gevrey classes and related pseudo-differential operators*, Boll. Un. Mat. Ital., Ser. VI, C-**3** (1984), 133-223.

[13] S. Mizohata, *Solutions nulles et solutions non analytiques*, J. Math. Kyoto Univ., **1** (1961-62), 271-302.

[14] Y. Morimoto-T. Morioka, *Some remarks on hypoelliptic operators which are not microhypoelliptic*, preprint 1991.

[15] T. Okaji, *Gevrey hypoelliptic operators which are not C^∞-hypoelliptic*, J. Math. Kyoto Univ., **28** (1988), 311-322.

[16] C. Parenti-L. Rodino, *Examples of hypoelliptic operators which are not microhypoelliptic*, Boll. Un. Mat. Ital., **17**-B (1980), 390-409.

[17] C. Parenti-L. Rodino, *Parametrices for a class of pseudo differential operators*, I, II, Annali Mat. Pura ed Appl., **125** (1980), 221-278.

[18] L. Rodino, *Gevrey hypoellipticity for a class of operators with multiple characteristics*, Astérisque, **89-90** (1981), 249-262.

[19] L. Rodino, *On the Gevrey wave front set of the solutions of a quasi-elliptic degenerate equation*, Rend. Sem. Mat. Università Politecnico Torino, fasc. spec. Convegno «Linear partial and pseudo differential operators» 1982 (1983), 221-234.

[20] L. Rodino, *On linear partial differential operators with multiple characteristics*, Symposium «Partial differential equations», Holzhau 1988, Teubner-Texte zur Mathematik, **112** (1989), 238-249.

[21] L. Rodino, *Analytic-hypoelliptic operators which are not C^∞-hypoelliptic*, Proceedings International Workshop «Methods of real analysis and partial differential equations», Capri 1990, to appear.

[22] L. Rodino, *Linear partial differential operators in Gevrey spaces*, World Scientific, Singapore; in progress.

[23] L. Rodino-L. Zanghirati, *Pseudo-differential operators with multiple characteristics and Gevrey singularities*, Comm. Partial Differential Equations, **11** (1986), 673-711.

[24] K. Taniguchi, *On multi-products of pseudo differential operators in Gevrey classes and its applications to Gevrey hypoellipticity*, Proc. Japan Acad., Ser. A, **61** (1985), 291-293.

[25] L. Schwartz, *Théorie des distributions*, Hermann 1957, Paris.

[26] W. Wasow, *Asymptotic exapansions for ordinary differential equations*, Interscience Publishers, 1965, John Wiley and Sons.

[27] L. Zanghirati, *Pseudodifferential operators of infinite order and Gevrey classes*, Ann. Univ. Ferrara, Sez. VII, Sc. Mat., **31** (1985), 197-219.

ULTRA WAVE FRONT SETS AND FOURIER INTEGRAL

OPERATORS OF INFINITE ORDER WITH AN APPLICATION

Kazuo Taniguchi

Department of Mathematics
University of Osaka Prefecture
Sakai, Osaka 591, Japan

Introduction. The fundamental solution of the Cauchy problem is given in the form of Fourier integral operator, and when the problem is not C^∞ well-posed, Cattabriga-Zanghirati[2] have proved that the symbols of Fourier integral operators in the fundamental solution are not able to be estimated by finite order but able to be estimated by infinite order of the following type

(1) $\quad C\exp[c\xi^{1/\kappa}], \quad c > 0.$

These symbols are estimeted from above by (1) for some constant κ satisfying $\kappa > 1$. Foureir integral operators and pseudo-differential operators with symbols estimated by (1) are investigated in Zanghirati[26], Rodino-Zanghirati[16], Cattabriga-Zanghirati[2] and Aoki[1], and it is proved that they have the following properties: i) They map the ultradistributions into themselves; ii) pseudo-differential operators with symbols estimated from above by (1) make algebras; and iii) they propagate Gevrey wave front sets canonically.

On the other hand, Shinkai[18] proves that, for the case of the not C^∞ well-posed hyperbolic operator $L = D_t^2 - t^{2j}D_x^2 + ait^kD_x$ $(k < j-1)$, the symbols of Fourier integral operators in its fundamental solution are estimated not only from above but also from below by (1), that is, they are estimated both from above and from below by

(2) $\quad C\exp[c\xi^{1/\kappa}], \quad c > 0.$

Fourier integral operators with such symbols in the fundamental solution give the existence of ultra wave front sets of the solutions of the Cauchy problem for not C^∞ well-posed hyperbolic operators. Especially, the constant κ in (2) for the symbols of the fundamental solution determines exactly the order of the Gevrey classes, in which the Cauchy problem in question is well-posed or not. We note that, for the well-posedness in Gevrey classes, Ivrii[7] determines the index of Gevrey classes where the Cauchy problem of the hyperbolic operator is well-posed, and this index corresponds the index of ultra wave front sets of the fundamental solutions.

Ultra wave front sets (UWF) are first defined by Wakabayashi[25] by the name "generalized wave front sets", and they contain both UWF and Gevrey wave front sets and are denoted, in Wakabayashi[25], $WF^{\{\kappa\}}$ and $WF_{\{\kappa\}}$, respectively. He also tried to get non-trivial inner estimates for UWF, but got only a lemma ("not really satisfactory" in his words) and he gave two examples concerning operators with constant coefficients.

In the present paper, we report the joint work with Professor Kenzo Shinkai and the author which is investigated in Shinkai - Taniguchi[20], Shinkai[18] and Taniguchi[24]. First, we give the definition of ultradistributions and the one of symbol classes and the classes of phase functions for Fourier integral operators with their properties. And then, we give the definition of UWF and give the propagation of UWF by Fourier integral operators with infinite order. Finally, we give examples of degenerate hyperbolic operators whose fundamental solutions are represented by the sum of Fourier integral operators with infinite order.

§1. Ultradistributions and Fourier integral operators of infinite order. Throughout this paper we always assume $\kappa > 1$. For positive constants h and ε we define a class $\mathcal{S}\{\kappa;h,\varepsilon\}$ of ultradifferentiable functions by a set of functions u(x) satisfying

$$(1.1) \qquad |\partial_x^\alpha u(x)| \leqq Ch^{-|\alpha|}\alpha!^\kappa \exp(-\varepsilon\langle x\rangle^{1/\kappa}).$$

Then, $\mathcal{S}\{\kappa;h,\varepsilon\}$ is a Banach space with the norm

$$\|u;\mathcal{S}\{\kappa;h,\varepsilon\}\| = \inf\{C \text{ of } (1.1)\}.$$

Definition 1.1. We define a class $\mathcal{S}\{\kappa\}$ by

$$\mathcal{S}\{\kappa\} = \underset{h\to 0,\varepsilon\to 0}{\text{ind}\quad\lim}\ \mathcal{S}\{\kappa;h,\varepsilon\}$$

and denote by $\mathcal{S}\{\kappa\}'$ the dual space of $\mathcal{S}\{\kappa\}$.

The class $\mathcal{S}\{\kappa\}'$ is a class of ultradistributions (cf. Gel'fand - Shilov[3] and Komatsu[11]) and is characterized as

Proposition 1.2. For $u \in \mathcal{S}\{\kappa\}'$ and $\chi \in \mathcal{S}\{\kappa\}\cap C_0^\infty$ the Fourier transform $\mathcal{F}[\chi u](\xi) \equiv \widehat{\chi u}(\xi)$ of χu is a measurable function and has an estimate

$$|\widehat{\chi u}(\xi)| \leqq C_\varepsilon \exp(\varepsilon\langle\xi\rangle^{1/\kappa})$$

for any ε.

Proof. We may assume that $u \in \mathcal{S}\{\kappa\}'$ has a compact support and we shall prove that, for any fixed ε, $\exp(-\varepsilon\langle\xi\rangle^{1/\kappa})\hat{u}$ is a functional on L^1 and has the following estimate

$$(1.2) \qquad |<\exp(-\varepsilon\langle\xi\rangle^{1/\kappa})\hat{u}\,,\,v>| \leqq C\|v\|_{L^1}$$

for $v \in L^1$. Then, (1.2) yields $\exp(-\varepsilon\langle\xi\rangle^{1/\kappa})\hat{u} \in L^\infty$ and

$$|\hat{u}| \leqq C_\varepsilon \exp(\varepsilon\langle\xi\rangle^{1/\kappa})$$

for any ε. This proves the propsition. So, it remains to prove (1.2). First, we assume $v \in \mathcal{S}\{\kappa;h,1\}$ with $h = \varepsilon^\kappa\kappa^{-\kappa}/2$. Denote by $\tilde{v}(x)$ the inverse

Fourier transform of $v(\xi)$ and take a function $\chi(x)$ in $\mathcal{S}\{\kappa;h,1\}\cap C_0^\infty(R^n)$ with the same $h = \varepsilon^\kappa\kappa^{-\kappa}/4$ such that $\chi(x) = 1$ on the support of u. Then, we have

$$\begin{aligned} <\exp(-\varepsilon\langle\xi\rangle^{1/\kappa})\hat{u}\,,\,v> &= <\hat{u}\,,\,\exp(-\varepsilon\langle\xi\rangle^{1/\kappa})v> \\ &= <u\,,\,\exp(-\varepsilon\langle D\rangle^{1/\kappa})\tilde{v}> \\ &= <u\,,\,\chi(x)\exp(-\varepsilon\langle D\rangle^{1/\kappa})\tilde{v}>. \end{aligned}$$

Here, we used Proposition 1.5 below for well-posedness of the third and fourth members of the above equation. Then, by the definition of the duality and the fact that $u \in \mathcal{S}\{\kappa\}'$ we have

$$(1.3)\qquad <\exp(-\varepsilon\langle\xi\rangle^{1/\kappa})\hat{u}\;,\;v> \leqq C\|\chi(x)\exp(-\varepsilon\langle D\rangle^{1/\kappa})\tilde{v}\;;\;\mathcal{S}\{\kappa;h,1\}\|.$$

Write

$$\chi(x)\exp(-\varepsilon\langle D\rangle^{1/\kappa})\tilde{v}(x) = \int e^{ix\cdot\xi}\chi(x)\exp(-\varepsilon\langle\xi\rangle^{1/\kappa})v(\xi)đ\xi,$$

where $đ\xi = (2\pi)^{-n}d\xi$. Then, from $h = \varepsilon^\kappa\kappa^{-\kappa}/4$, we have

$$|\partial_x^\alpha(\chi(x)\exp(-\varepsilon\langle D\rangle^{1/\kappa})\tilde{v})| \leqq Ch^{-|\alpha|}\alpha!^\kappa\exp(-\langle x\rangle^{1/\kappa})\|v\|_{L_1}$$

and hence we have

$$\|\chi(x)\exp(-\varepsilon\langle D\rangle^{1/\kappa})\tilde{v}\;;\;\mathcal{S}\{\kappa,h,1\}\| \leqq C\|v\|_{L_1}.$$

This and (1.3) yields (1.2) for $v \in \mathcal{S}\{\kappa;h,1\}$. For the general v in $L^1(R^n)$ we use the limiting process. Then, we have (1.2) for any v in $L^1(R^n)$. This concludes the proof of the proposition.

Now, we define classes of symbols and phase functions for Fourier integral operators. Let ρ and δ be real numbers satisfying $0 \leqq \delta \leqq 1/2 \leqq \rho \leqq 1$, $\kappa(1-\delta) \geqq 1$ and $\kappa\rho \geqq 1$.

Definition 1.3. Let $w(\theta)$ be a positive and non-decreasing function in $[1,\infty)$ or a function the type θ^m for a real m. We say that a symbol $p(x,\xi)$ belongs to a class $S_{\rho,\delta,G(\kappa)}[w]$ if $p(x,\xi)$ satisfies

$$(1.4)\qquad \begin{aligned} |p_{(\beta)}^{(\alpha)}(x,\xi)| \leqq\; & CM^{-|\alpha+\beta|}(\alpha!^\kappa + \alpha!^{\kappa\rho}\langle\xi\rangle^{(1-\rho)|\alpha|}) \\ & \times(\beta!^\kappa + \beta!^{\kappa(1-\delta)}\langle\xi\rangle^{\delta|\beta|})\langle\xi\rangle^{-|\alpha|}w(\langle\xi\rangle) \\ & \qquad\text{for all } x \text{ and } \xi, \end{aligned}$$

where $p_{(\beta)}^{(\alpha)} = \partial_\xi^\alpha(-i\partial_x)^\beta p$. We call the above function $w(\theta)$ an order function.

This definition originates from Taniguchi[21] for the case $\rho = 1$ and $\delta = 0$, and Iwasaki[8] and Métivier[14] for the general case of ρ and δ.

Remark 1. When $w(\theta) = \exp(C\theta^\sigma)$ for a $\sigma > 0$, the class $S_{\rho,\delta,G(\kappa)}[w]$ is a symbol class of exponential type, and this corresponds to the class investigated in Zanghirati[26], Rodino - Zanghirati[16], Cattabriga - Zanghirati[2] and Aoki[1]. We also remark that the classes of symbols in Gevrey classes are investigated in Liess - Rodino[13], Hashimoto - Matsuzawa - Morimoto[4] and Taniguchi[23].

Remark 2. When $w(\theta) = \theta^m$ for a real m we denote $S_{\rho,\delta,G(\kappa)}[w]$ by $S^m_{\rho,\delta,G(\kappa)}$. We also set $S^m_{\rho,G(\kappa)} = S^m_{\rho,1-\rho,G(\kappa)}$ and $S^m_{G(\kappa)} = S^m_{1,0,G(\kappa)}$.

Example. For $a(x,\xi) \in S^m_{G(\kappa)}$ the symbol $p(x,\xi) = a(x,\xi)\exp(\langle\xi\rangle^\sigma)$ belongs to $S_{1,0,G(\kappa)}[\exp(2\theta^\sigma)]$.

Definition 1.4. i) Let $0 \leqq \tau < 1$ and $1/2 \leqq \rho \leqq 1$. We say that a phase function $\phi(x,\xi)$ belongs to a class $\mathcal{P}_{\rho,G(\kappa)}(\tau)$ if $\phi(x,\xi)$ is a real-valued function satisfying

$$(1.5) \qquad \sum_{|\alpha|+|\beta| \leqq 2} |\partial_\xi^\alpha \partial_x^\beta (\phi(x,\xi) - x\cdot\xi)|/\langle\xi\rangle^{1-|\alpha|} \leqq \tau$$

and

$$(1.6) \qquad |\partial_\xi^\alpha \partial_x^\beta (\phi(x,\xi) - x\cdot\xi)| \leqq \tau M^{-(|\alpha+\beta|)} \times (|\alpha+\beta|!^\kappa + |\alpha+\beta|!^{\kappa\rho}\langle\xi\rangle^{(1-\rho)(|\alpha+\beta|-2)})\langle\xi\rangle^{1-|\alpha|} \quad \text{for } |\alpha+\beta| \geqq 2$$

for a constant M independent of α and β.

ii) Let $0 \leqq \tau < 1$, $0 \leqq \omega < 1/\kappa$ and $1/2 \leqq \rho \leqq 1$. We say that a (complex) phase function $\phi(x,\xi)$ belongs to $\mathcal{CP}_{\rho,G(\kappa)}(\tau;\omega)$ if $\mathrm{Re}\,\phi(x,\xi) \in \mathcal{P}_{\rho,G(\kappa)}(\tau)$ and $\mathrm{Im}\,\phi(x,\xi)$ is a symbol in $S^\omega_{\rho,G(\kappa)}$ satisfying

$$|(\mathrm{Im}\,\phi(x,\xi))^{(\alpha)}_{(\beta)}| \leqq \tau M^{-|\alpha+\beta|}(|\alpha+\beta|!^\kappa + |\alpha+\beta|!^{\kappa\rho}\langle\xi\rangle^{(1-\rho)|\alpha+\beta|})\langle\xi\rangle^{\omega-|\alpha|}.$$

We set

$$\mathcal{P}_{\rho,G(\kappa)} = \bigcup_{0 \leqq \tau < 1} \mathcal{P}_{\rho,G(\kappa)}(\tau), \quad \mathcal{CP}_{\rho,G(\kappa)} = \bigcup_{\substack{0 \leqq \tau < 1 \\ 0 \leqq \omega < 1/\kappa}} \mathcal{CP}_{\rho,G(\kappa)}(\tau;\omega)$$

and $\mathcal{P}_{G(\kappa)} = \mathcal{P}_{1,G(\kappa)}$.

Proposition 1.5. i) Let $w(\theta)$ be an order function satisfying

$$(1.7) \qquad w(\theta) \leqq \exp[C\theta^\sigma]$$

for a constant σ with $0 \leqq \sigma < 1/\kappa$. For a phase function $\phi(x,\xi) \in \mathcal{CP}_{\rho,G(\kappa)}$ and a symbol $p(x,\xi) \in S_{\rho,\delta,G(\kappa)}[w]$ we define a Fourier integral operator P_ϕ and a conjugate Fourier integral operator P_{ϕ^*} by

$$P_\phi u(x) = \int e^{i\phi(x,\xi)} p(x,\xi)\hat{u}(\xi)đ\xi,$$

$$P_{\phi^*}u(x) = \int e^{ix\cdot\xi}\{\int e^{-i\phi(y,\xi)}p(y,\xi)u(y)dy\}d\!\!\!{}^{-}\xi,$$

Then, the operators P_ϕ and P_{ϕ^*} map $\mathcal{S}\{\kappa\}$ to $\mathcal{S}\{\kappa\}$ continuously.

ii) Let $\mathcal{B}\{\kappa\}$ be a class of Gevrey functions u(x) satisfying

$$|\partial_x^\alpha u(x)| \leqq CM^{-|\alpha|}\alpha!^\kappa \qquad \text{for any } x$$

with constants C and M. Then, under the same condition of i) the operators P_ϕ and P_{ϕ^*} also map $\mathcal{B}\{\kappa\}$ to $\mathcal{B}\{\kappa\}$ continuously.

Proof. Write

$$f(x) = P_\phi u(x) \equiv \int e^{i\phi(x,\xi)}p(x,\xi)\hat{u}(\xi)d\!\!\!{}^{-}\xi$$

for $u(x) \in \mathcal{S}\{\kappa\}$ and define $L = \{1+|\nabla_\xi\,\phi(x,\xi)|^2\}^{-1}\{1 - i\overline{\nabla_\xi\phi(x,\xi)}\cdot\nabla_\xi\}$. Then, we have $Le^{i\phi(x,\xi)} = e^{i\phi(x,\xi)}$ and

$$f(x) = \int e^{i\phi(x,\xi)}(L^t)^N\{p(x,\xi)\hat{u}(\xi)\}d\!\!\!{}^{-}\xi,$$

where L^t is the transposed operator of L. By the induction on N we can prove

$$(1.8) \qquad \begin{aligned} |\partial_\xi^\alpha\partial_x^\beta(L^t)^N\{p(x,\xi)\hat{u}(\xi)\}| &\leqq CM_1^{-N}M_2^{-|\alpha+\beta|}(|\alpha|+N)!^\kappa\langle x\rangle^{-N} \\ &\times(\beta!^\kappa + \beta!^{\kappa(1-\delta)}\langle\xi\rangle^{\delta|\beta|} + \beta!^{\kappa\rho}\langle\xi\rangle^{(1-\rho)|\beta|}) \\ &\times\exp(C'\langle\xi\rangle^\sigma - \varepsilon\langle\xi\rangle^{1/\kappa}) \end{aligned}$$

for positive constants C, M_1, M_2, C' and ε, since $\hat{u}(\xi)$ belongs to $\mathcal{S}\{\kappa\}$. We note that the term $\beta!^{\kappa\rho}\langle\xi\rangle^{(1-\rho)|\beta|}$ in the above estimate comes from the derivatives of $\phi(x,\xi)$. Assume that x satisfies $C_0N^\kappa \leqq \langle x\rangle \leqq C_0(N+1)^\kappa$ (N = 0, 1,2,...) for a constant C_0 to be determined later. Then, setting $\phi_\beta = e^{-i\phi(x,\xi)}\partial_x^\beta e^{i\phi(x,\xi)}$ we have for any $\tilde{\varepsilon}$

$$(1.9) \qquad \begin{aligned} |\phi_\beta| &\leqq CM_3^{-|\beta|}\sum_{j=1}^{|\beta|}\{(|\beta|-j)!^\kappa + (|\beta|-j)!^{\kappa\rho}\langle\xi\rangle^{(1-\rho)(|\beta|-j)}\}\langle\xi\rangle^j \\ &\leqq CM_4^{-|\beta|}\beta!^\kappa\exp(\tilde{\varepsilon}\langle\xi\rangle^{1/\kappa}), \end{aligned}$$

since we have $\omega+1-\rho < 1$. Hence, using (1.8) with $\alpha = 0$ and (1.9) with sufficiently small $\tilde{\varepsilon}$ we have

$$(1.10) \qquad \begin{aligned} |\partial_x^\beta f(x)| &= |\sum_{\beta'+\beta''=\beta}\binom{\beta}{\beta'}\int e^{i\phi(x,\xi)}\phi_{\beta'}\partial_x^{\beta''}(L^t)^N\{p(x,\xi)\hat{u}(\xi)\}d\!\!\!{}^{-}\xi| \\ &\leqq C\sum_{\beta'+\beta''=\beta}\binom{\beta}{\beta'}\int M_4^{-|\beta'|}\beta'!^\kappa\exp(\tilde{\varepsilon}\langle\xi\rangle^{1/\kappa}) \end{aligned}$$

$$\times M_1^{-N} M_2^{-|\beta''|} N!^{\kappa} \langle x\rangle^{-N}$$
$$\times (\beta''!^{\kappa} + \beta''!^{\kappa(1-\delta)} \langle\xi\rangle^{\delta|\beta''|} + \beta''!^{\kappa\rho} \langle\xi\rangle^{(1-\rho)|\beta''|})$$
$$\times \exp(C_1\langle\xi\rangle^{\omega} + C_2\langle\xi\rangle^{\sigma} - \varepsilon\langle\xi\rangle^{1/\kappa}) d\xi$$
$$\leqq C' M_5^{-|\beta|} \beta!^{\kappa} M_1^{-N} N!^{\kappa} \langle x\rangle^{-N}$$
$$\leqq C'' M_5^{-|\beta|} \beta!^{\kappa} M_1^{-N} N!^{\kappa} (C_0 N^{\kappa})^{-N} \exp(\varepsilon_1 C_0^{1/\kappa}(N+1))$$
$$\times \exp(-\varepsilon_1 C_0^{1/\kappa} \langle x\rangle^{1/\kappa})$$

for any positive constant ε_1. Now, take C_0 and ε_1 satisfying

$$C_0 \geqq 2M_1^{-1}, \quad \exp(\varepsilon_1 C_0^{1/\kappa}) \leqq 2.$$

Then, f(x) satisfies (1.1) with $h = M_5$ and $\varepsilon = \varepsilon_1 C^{1/\kappa}$. Consequently, we have proved that P_ϕ maps $\mathcal{S}\{\kappa\}$ to $\mathcal{S}\{\kappa\}$ continuously. Similarly, we can prove that P_{ϕ^*} maps $\mathcal{S}\{\kappa\}$ to $\mathcal{S}\{\kappa\}$ continuously.

Next, we prove the continuity of P_ϕ in $\mathcal{B}\{\kappa\}$. Set $f = P_\phi u$ for $u \in \mathcal{B}\{\kappa\}$. Then, f(x) can be written in the form

$$f(x) = O_s\text{-}\iint e^{i(\phi(x,\xi) - x'\cdot\xi)} p(x,\xi) u(x') dx' d\xi. \tag{1.11}$$

Set $L = (1+|\nabla_\xi \phi(x,\xi) - x'|^2)^{-1}(1 - i(\overline{\nabla_\xi \phi(x,\xi) - x'}) \cdot \nabla_\xi\}$ and $\phi_\beta = e^{-i\phi(x,\xi)} \times \partial_x^\beta e^{i\phi(x,\xi)}$. Then, we have

$$\partial_x^\beta f(x) = \partial_x^\beta [O_s\text{-}\iint e^{i(\phi(x,\xi) - x'\cdot\xi)} (L^t)^{n+1} p(x,\xi) u(x') dx' d\xi]$$
$$= \sum_{\beta'+\beta''=\beta} \binom{\beta}{\beta'} [O_s\text{-}\iint e^{i(\phi(x,\xi) - x'\cdot\xi)} \phi_{\beta'} \partial_x^{\beta''} \{(L^t)^{n+1} p(x,\xi)\} u(x') dx' d\xi]$$
$$= \sum_{\beta'+\beta''=\beta} \binom{\beta}{\beta'} [\iint_{|\xi| \leqq \tilde{c}} e^{i(\phi(x,\xi) - x'\cdot\xi)} \phi_{\beta'} \partial_x^{\beta''} \{(L^t)^{n+1} p(x,\xi)\} u(x') dx' d\xi$$
$$+ \sum_{N=1}^{\infty} \iint_{\tilde{c}N^{\kappa} \leqq |\xi| \leqq \tilde{c}(N+1)^{\kappa}} e^{i(\phi(x,\xi) - x'\cdot\xi)}$$
$$\times \phi_{\beta'} \partial_x^{\beta''} \{(L^t)^{n+1} p(x,\xi)\}$$
$$\times (-i|\xi|^{-2} \xi \cdot \nabla_{x'})^N u(x') dx' d\xi].$$

Hence from (1.9) we have

$$|\partial_x^\beta f(x)| \leqq C \sum_{\beta'+\beta''=\beta} \binom{\beta}{\beta'} \sum_{N=0}^{\infty} \int_{\tilde{c}N^\kappa \leqq |\xi| \leqq \tilde{c}(N+1)^\kappa} M_4^{-|\beta'|} \beta'!^\kappa \exp(\tilde{\varepsilon}\langle\xi\rangle^{1/\kappa})$$
$$\times M_1^{-(n+1)} M_2^{-|\beta''|} (n+1)!^\kappa \langle x-x'\rangle^{-(n+1)}$$
$$\times (\beta''!^\kappa + \beta''!^{\kappa(1-\delta)}\langle\xi\rangle^{\delta|\beta''|} + \beta''!^{\kappa\rho}\langle\xi\rangle^{(1-\rho)|\beta''|})$$
$$\times M_6^{-N}|\xi|^{-N} N!^\kappa \exp(C_1\langle\xi\rangle^\omega + C_2\langle\xi\rangle^\sigma)\,dx'd\xi.$$

From $\omega < 1/\kappa$ and $\sigma < 1/\kappa$ we have for any $\tilde{\varepsilon}'$

$$(\beta!^\kappa + \beta!^{\kappa(1-\delta)}\langle\xi\rangle^{\delta|\beta|} + \beta!^{\kappa\rho}\langle\xi\rangle^{(1-\rho)|\beta|})$$
$$\times \exp(C_1\langle\xi\rangle^\omega + C_2\langle\xi\rangle^\sigma)$$
$$\leqq C_{\tilde{\varepsilon}'} M_7^{-|\beta|} \beta!^\kappa \exp(\tilde{\varepsilon}'\langle\xi\rangle^{1/\kappa}).$$

Hence, using $\langle\xi\rangle \leqq 2\tilde{c}(N+1)^\kappa$ on $\{\xi;\, \tilde{c}N^\kappa \leqq |\xi| \leqq \tilde{c}(N+1)^\kappa\}$ for $\tilde{c} \geqq 1$, we have

$$|\partial_x^\beta f(x)| \leqq C' \sum_{\beta'+\beta''=\beta} \binom{\beta}{\beta'} \sum_{N=0}^{\infty} M_4^{-|\beta'|} \beta'!^\kappa \exp(\tilde{\varepsilon}(2\tilde{c})^{1/\kappa}(N+1))$$
$$\times (M_2M_7)^{-|\beta''|} \exp(\tilde{\varepsilon}'(2\tilde{c})^{1/\kappa}(N+1))$$
$$\times M_6^{-N}\{N!^\kappa/(\tilde{c}N^\kappa)^N\}(N+1)^n$$
$$\leqq C'' M_8^{-|\beta|} \beta!^\kappa$$

by taking constants $\tilde{\varepsilon}$ and $\tilde{\varepsilon}'$ small enough and a constant $\tilde{c}$ large enough. This implies $f(x) \in \mathcal{B}\{\kappa\}$. Consequently, we have proved that P_ϕ maps $\mathcal{B}\{\kappa\}$ to itself. In the same way we can prove that P_{ϕ^*} maps $\mathcal{B}\{\kappa\}$ to itself. This concludes the proof of the proposition.

Remark. In (1.11) the integral is an oscillatory integral, and this can be defined as in Section 6 of Chap. 1 in Kumano-go[12]. We also remark that oscillatory integrals for amplitudes with infinite order are studied in Section 1.2 of Cattabriga - Zanghirati[2].

From the continuity of P_ϕ and P_{ϕ^*} in $\mathcal{S}\{\kappa\}$ the following is well-defined.

Definition 1.6. Let $w(\theta)$ be an order function satisfying (1.7), that is, it satisfies

$$w(\theta) \leqq \exp[C\theta^\sigma]$$

for a constant σ with $0 \leqq \sigma < 1/\kappa$. Then for $\phi(x,\xi) \in \mathcal{CP}_{\rho,G(\kappa)}$ and $p(x,\xi) \in S_{\rho,\delta,G(\kappa)}[w]$, the following operators

$$P_\phi \ :\ \mathcal{S}\{\kappa\}' \longrightarrow \mathcal{S}\{\kappa\}',$$
$$P_{\phi^*} \ :\ \mathcal{S}\{\kappa\}' \longrightarrow \mathcal{S}\{\kappa\}',$$

are defined by the principle of duality.

Example. For $a(x,\xi) \in S^m_{\rho,\delta,G(\kappa)}$ $(\kappa < 2)$ we consider a symbol $p(x,\xi) =$

$a(x,\xi)\exp(c\langle\xi\rangle^{1/2})$ with $c > 0$. Then, it belongs to $S_{\rho,\delta,G(\kappa)}[\exp(2c\theta^{1/2})]$ and for $1 < \kappa < 2$ the following maps are well-defined:

$$P_\phi \ : \ \mathcal{S}\{\kappa\}' \longrightarrow \mathcal{S}\{\kappa\}',$$
$$P_{\phi^*} \ : \ \mathcal{S}\{\kappa\}' \longrightarrow \mathcal{S}\{\kappa\}',$$

where ϕ is a phase function in $\mathcal{CP}_{\rho,G(\kappa)}$.

Concerning the algebra of pseudo-differential operators of infinte order we have the following (see Shinkai - Taniguchi[19] and Zanghirati[26]).

Proposition 1.7. Let $w_j(\theta)$, $j = 1,2$, be an order functions such that

(1.12) $$w_j(\theta) \leqq C_\varepsilon \exp(\varepsilon\theta^{1/\kappa}) \quad \text{for any } \varepsilon > 0 \ (j = 1,2)$$

and let $P_j = p_j(X,D_x)$ be pseudo-differential operators with symbols $p_j(x,\xi)$ in $S_{\rho,\delta,G(\kappa)}[w_j]$. Then, choosing an order function $w(\theta)$ satisfying $w(\theta) \geqq w_1(2\theta)w_2(\theta)$ there exists a symbol $q(x,\xi)$ in $S_{\rho,\delta,G(\kappa)}[w]$ such that the product P_1P_2 can be witten in the form

(1.13) $$P_1P_2 = q(X,D_x) + r(x,D_x)$$

with a symbol $r(x,\xi)$ satisfying

(1.14) $$|r^{(\alpha)}_{(\beta)}(x,\xi)| \leqq CM^{-|\alpha+\beta|}\alpha!^\kappa\beta!^\kappa\exp(-\varepsilon\langle\xi\rangle^{1/\kappa}).$$

with a positive constant ε independent of α and β.

We note that as in the Propositon 1.8 below the pseudo-differential operators with symbols satisfing (1.14) map $\mathcal{S}\{\kappa\}'$ to $\mathcal{B}\{\kappa\}$. So, we can call such pseudo-differential operators as regularizing operators.

Propsition 1.8. Let $R = r(X,D_x)$ be a pseudo-differential operator with a symbol $r(x,\xi)$ satisfying (1.14). Then, for $u \in \mathcal{S}\{\kappa\}'$ the operation Ru is well-defined and Ru belongs to $\mathcal{B}\{\kappa\}$.

Now, we prove Proposition 1.7 and Proposition 1.8.

Proof of Proposition 1.7. Write the symbol $\sigma(P_1P_2)(x,\xi)$ of P_1P_2 as

(1.15) $$\begin{aligned}\sigma(P_1P_2)(x,\xi) &= O_s\text{-}\iint e^{-iy\cdot\eta}p_1(x,\xi+\eta)p_2(x+y,\xi)dy\,đ\eta \\ &= O_s\text{-}\iint e^{-iy\cdot\eta}({}^tL_1)^{n+1}p_1(x,\xi+\eta)p_2(x+y,\xi)dy\,đ\eta,\end{aligned}$$

where $L_1 = (1 + \langle\xi+\eta\rangle^{2\delta}|y|^2)^{-1}(1 + i\langle\xi+\eta\rangle^{2\delta}y\cdot\nabla_\eta)$. Denote $\chi(\xi)$ a function in $\mathcal{S}\{\kappa\}$ satisfying

(1.16) $$0 \leqq \chi \leqq 1, \quad \chi = 1 \ (|\xi| \leqq 2/5), \quad \chi = 0 \ (|\xi| \geqq 1/2)$$

and divide (1.15) as

$$\sigma(P_1P_2)(x,\xi) = q(x,\xi) + r(x,\xi),$$

$$q(x,\xi) = O_s\text{-}\iint e^{-iy\cdot\eta}(L_1{}^t)^{n+1}p_1(x,\xi+\eta)\chi(\eta/\langle\xi\rangle)$$

$$\times p_2(x+y,\xi)dy đ\eta,$$

$$r(x,\xi) = O_s\text{-}\iint e^{-iy\cdot\eta}(L_1{}^t)^{n+1}p_1(x,\xi+\eta)(1-\chi(\eta/\langle\xi\rangle))$$

$$\times p_2(x+y,\xi)dy đ\eta.$$

Then, it is easy to prove $q(x,\xi) \in S_{\rho,\delta,G(\kappa)}[w]$ by writing

$$q(x,\xi) = \iint e^{-iy\cdot\eta}(L_2{}^t)^{\ell}(L_1{}^t)^{n+1}p_1(x,\xi+\eta)\chi(\eta/\langle\xi\rangle)$$

$$\times p_2(x+y,\xi)dy đ\eta,$$

with $L_2 = (1+\langle\xi\rangle^{2\delta}|\eta|^2)^{-1}(1-\langle\xi\rangle^{2\delta}\triangle_y)$ and $\ell = [n/2]+1$. In order to estimate $r(x,\xi)$ we write

$$r(x,\xi) = \iint_{|\eta|\leqq\tilde{c}} e^{-iy\cdot\eta}(L_2{}^t)^{\ell}(L_1{}^t)^{n+1}p_1(x,\xi+\eta)(1-\chi(\eta/\langle\xi\rangle))$$

$$\times p_2(x+y,\xi)dy đ\eta,$$

$$+\sum_{N=1}^{\infty}\iint_{\tilde{c}N^{\kappa}\leqq|\eta|\leqq\tilde{c}(N+1)^{\kappa}} e^{-iy\cdot\eta}(L_2{}^t)^{\ell}(L_1{}^t)^{n+1}\{p_1(x,\xi+\eta)$$

$$\times(1-\chi(\eta/\langle\xi\rangle))$$

$$\times(-i|\eta|^{-2}\eta\cdot\nabla_y)^N p_2(x+y,\xi)\}dy đ\eta.$$

Then, using (1.12) with sufficiently small ε we obtain (1.14) if we take $\tilde{c}$ sufficiently large.

Proof of Proposition 1.8. Let $R = r(X,D_x)$ for a symbol $r(x,\xi)$ satisfying (1.14). Then, as in the proof of Proposition 1.5 the operator R maps $\mathcal{S}\{\kappa\}$ to $\mathcal{S}\{\kappa\}$. Hence, by the principle of duality, the operator R also maps $\mathcal{S}\{\kappa\}'$ to $\mathcal{S}\{\kappa\}'$. So, it remains to prove, for $u\in\mathcal{S}\{\kappa\}'$, Ru belongs to $\mathcal{B}\{\kappa\}$. Set $f = Ru$. Then, for any α and $v\in\mathcal{S}\{\kappa\}$ we have

$$< D_x^{\alpha}f\ ,\ v > \ = (-1)^{|\alpha|} < Ru\ ,\ D_x^{\alpha}v > \ = (-1)^{|\alpha|} < u\ ,\ R^*D_x^{\alpha}v >$$

and

$$|< u\ ,\ R^*D_x^{\alpha}v >| \leqq C\|R^*D_x^{\alpha}v\ ;\ \ \mathcal{S}\{\kappa,h,\varepsilon\}\|$$

for positive constants h and ε such that $R^*D_x^{\alpha}v\in\mathcal{S}\{\kappa;h,\varepsilon\}$. Here, R^* is a formal conjugate operator of R. Note

$$(R^*D_x^{\alpha}v)(x) = O_s\text{-}\iint e^{i(x-x')\cdot\xi}r(x',\xi)(D_x^{\alpha}v)(x')dx'đ\xi.$$

Then, as in (1.10) we can prove that there exist positive constants h and ε such that $R^*D_x^{\alpha}v\in\mathcal{S}\{\kappa;h,\varepsilon\}$ and we have for some constant M

$$|\partial_x^\beta (R^* D_x^\alpha v)| \leq CM^{-|\alpha|}\alpha!^\kappa h^{-|\beta|}\beta!^\kappa \exp(-\varepsilon\langle x\rangle^{1/\kappa})||v||_{L_1}.$$

Hence, it follows that

$$||R^* D_x^\alpha v\ ;\ \mathcal{S}\{\kappa,h,\varepsilon\}|| \leq CM^{-|\alpha|}\alpha!^\kappa ||v||_{L_1}$$

and

$$|< u\ ,\ R^* D_x^\alpha v >| \leq CM^{-|\alpha|}\alpha!^\kappa ||v||_{L_1}.$$

This proves

$$(1.17)\qquad |< D_x^\alpha f\ ,\ v >| \leq CM^{-|\alpha|}\alpha!^\kappa ||v||_{L_1}$$

for v in $\mathcal{S}\{\kappa\}$. For the general v in L^1, by using the limiting process we have also (1.17). Consequently, for any α, $D_x^\alpha f$ belong to L^∞ and have

$$|\partial_x^\alpha f| \leq CM^{-|\alpha|}\alpha!^\kappa \qquad \text{for any } x.$$

This concludes the proof of the proposition.

For the asymptotic expansion of symbols of pseudo-differential operators we define

Definition 1.9. Let $w(\theta)$ be an order function as in Definition 1.3. Then, we say that a symbol $p(x,\xi)$ belongs to a class $Swf_{1,\delta,G(\kappa)}[w]$ if $p(x,\xi)$ belongs to a class $S_{1,\delta,G(\kappa)}[w]$ and there exists a formal sum $\Sigma p_j(x,\xi)$ of symbols $p_j(x,\xi)$ satisfying

$$(1.18)\qquad |p_{j(\beta)}^{(\alpha)}(x,\xi)| \leq CM^{-(|\alpha+\beta|+j)}\alpha! \times ((|\beta|+j)!^\kappa + (|\beta|+j)!^{\kappa(1-\delta)}\langle\xi\rangle^{\delta(|\beta|+j)})\langle\xi\rangle^{-j-|\alpha|}w(\langle\xi\rangle) \quad \text{for } |\xi| \geq c$$

with a constant c (≥ 1) and

$$(1.19)\qquad |\partial_\xi^\alpha \partial_x^\beta (p(x,\xi) - \sum_{j=0}^{N-1} p_j(x,\xi))| \leq CM^{-(|\alpha+\beta|+N)}\alpha! \times ((|\beta|+N)!^\kappa + (|\beta|+N)!^{\kappa(1-\delta)}\langle\xi\rangle^{\delta(|\beta|+N)}) \times \langle\xi\rangle^{-|\alpha|-N}w(\langle\xi\rangle) \quad \text{for } |\xi| \geq c(|\alpha|+N)^\kappa$$

for any N. In this case we say that a formal sum $\Sigma p_j(x,\xi)$ is the formal symbol associated with $p(x,\xi)$.

Then, we have

Proposition 1.10. Let $p_j(x,\xi)$ be a symbol in $Swf_{1,\delta,G(\kappa)}[w_j]$ (j = 1,2) with $w_j(\theta)$ satisfying (1.12). Then, taking an order function $w(\theta)$ satisfying $w(\theta) \geq w_1(\theta)w_2(\theta)$, there exists a symbol $q(x,\xi)$ in $Swf_{1,\delta,G(\kappa)}[$

w] such that (1.13) holds for a symbol $r(x,\xi)$ satisfying (1.14) and we have for any N

$$(1.20)\qquad |\partial_\xi^\alpha \partial_x^\beta (q(x,\xi) - \sum_{|\gamma|<N} \frac{1}{\gamma!} p_1^{(\gamma)}(x,\xi) p_{2(\gamma)}(x,\xi))|$$
$$\leqq CM^{-(|\alpha+\beta|+N)}$$
$$\times \alpha!((|\beta|+N)!^\kappa + (|\beta|+N)!^{\kappa(1-\delta)}\langle\xi\rangle^{\delta(|\beta|+N)})$$
$$\times \langle\xi\rangle^{-N-|\alpha|} w(\langle\xi\rangle)$$
$$\text{for } |\xi| \geqq c(|\alpha|+N)^\kappa.$$

For the proof of the above proposition we need following lemma, which are investigated in Hashimoto - Matsuzawa - Morimoto[4] for symbols with finite order and Zanghirati[26] for symbols with infinite order.

Lemma 1.11. Let $w(\theta)$ be an order function and let $\Sigma p_j(x,\xi)$ be a formal symbol satisfying (1.18) with a constant c ($\geqq 1$). Then, there exists a symbol $p(x,\xi)$ in $Swf_{1,\delta,G(\kappa)}[w]$ such that we have (1.19) for any N.

Outline of the proof. As in Hashimoto - Matsuzawa - Morimoto[4] we take a sequence $\{\psi_j(\xi)\}$ of functions $\psi_j(\xi)$ satisfying for a parameter R

$$(1.21)\qquad \begin{cases} \psi_j(\xi) = 1 \quad \text{if } \langle\xi\rangle \geqq Rj^\kappa, \quad \psi_j(\xi) = 0 \quad \text{if } \langle\xi\rangle \leqq Rj^\kappa/2\ , \\ |\partial_\xi^{\alpha+\beta}\psi_j(\xi)| \leqq CM_1^{-|\alpha+\beta|} j^{|\alpha|} \beta!^\kappa \langle\xi\rangle^{-|\alpha+\beta|} \quad \text{for } |\alpha| \leqq 2j \end{cases}$$

and define

$$p(x,\xi) = \sum_{j=0}^{\infty} p_j(x,\xi)\psi_j(\xi)(1 - \chi(\xi/(3c)))$$

with a function $\chi(\xi)$ in $\mathcal{S}\{\kappa\}$ satisfying (1.16). Here, in (1.21) the constants C and M_1 are independent of j and R. Then, for a fixed sufficiently large R we can prove

$$(1.22)\qquad |p_{(\beta)}^{(\alpha)}(x,\xi)| \leqq CM^{-|\alpha+\beta|}\alpha!(\beta!^\kappa + \beta!^{\kappa(1-\delta)}\langle\xi\rangle^{\delta|\beta|})\langle\xi\rangle^{-|\alpha|}w(\langle\xi\rangle)$$
$$\text{for } \langle\xi\rangle \geqq R|\alpha|^\kappa$$

and (1.19). So, by (1.22) an inequality (1.4) with $\rho = 1$ holds for $p_{(\beta)}^{(\alpha)}$ when $\langle\xi\rangle \geqq R|\alpha|^\kappa$ and it remains to prove (1.4) for $\langle\xi\rangle \leqq R|\alpha|^\kappa$ in order to prove $p(x,\xi) \in S_{1,\delta,G(\kappa)}[w]$. Note

$$j \leqq (2\langle\xi\rangle/R)^{1/\kappa} \leqq 2^{1/\kappa}|\alpha| \qquad \text{on supp}\,\psi_j$$

when $\langle\xi\rangle \leqq R|\alpha|^\kappa$. Then, we can write $p(x,\xi)$ in the form

$$p(x,\xi) = \sum_{j=0}^{2|\alpha|} p_j(x,\xi)\psi_j(\xi)(1 - \chi(\xi/(3c))) \quad \text{for } \langle\xi\rangle \leqq R|\alpha|^{\kappa}$$

and obtain the estimate (1.4) for $p^{(\alpha)}_{(\beta)}(x,\xi)$ in $\langle\xi\rangle \leqq R|\alpha|^{\kappa}$. This proves the lemma.

Now, we prove Proposition 1.10.

Proof of Proposition 1.10. Let $\Sigma p_{1,j}(x,\xi)$ and $\Sigma p_{2,j}(x,\xi)$ be formal symbols associated to $p_1(x,\xi)$ and $p_2(x,\xi)$, respectively. Define

$$q_j(x,\xi) = \sum_{j'+j''+|\gamma|=j} \frac{1}{\gamma!} p^{(\gamma)}_{1,j'}(x,\xi) p_{2,j''(\gamma)}(x,\xi).$$

Then, $q_j(x,\xi)$ satisfies (1.18) for an order function satisfying $w(\theta) \geqq w_1(\theta)w_2(\theta)$. Hence, from Lemma 1.11 there exists a symbol $q(x,\xi)$ in $S_{1,\delta,G(\kappa)}[w]$ with a formal symbol $\Sigma q_j(x,\xi)$ and $q(x,\xi)$ satisfies (1.20). Now, define

$$r(x,\xi) = O_s\text{-}\iint e^{-iy\cdot\eta} p_1(x,\xi+\eta)p_2(x+y,\xi)dy đ\eta - q(x,\xi).$$

Then, the equality (1.13) holds. In order to prove that $r(x,\xi)$ satisfies (1.14) we write $r(x,\xi)$ as

(1.23) $$r(x,\xi) = r_1(x,\xi) + r_2(x,\xi),$$

(1.24) $$r_1(x,\xi) = O_s\text{-}\iint e^{-iy\cdot\eta} p_1(x,\xi+\eta)\chi(\eta/\langle\xi\rangle)p_2(x+y,\xi)dy đ\eta - q(x,\xi),$$

(1.25) $$r_2(x,\xi) = O_s\text{-}\iint e^{-iy\cdot\eta} p_1(x,\xi+\eta)(1 - \chi(\eta/\langle\xi\rangle))p_2(x+y,\xi)dy đ\eta.$$

Then, as in the proof of Proposition 1.7 it easily follows that $r_2(x,\xi)$ satisfies (1.14). For the proof of (1.14) for $r_1(x,\xi)$, we fix a multi-index α and write $r_1^{(\alpha)}(x,\xi)$ as

(1.26) $$r_1^{(\alpha)}(x,\xi) = r_{1,\alpha}(x,\xi) + r'_{1,\alpha}(x,\xi) + r''_{1,\alpha}(x,\xi),$$

(1.27) $$r_{1,\alpha}(x,\xi) = \partial_\xi^\alpha\{\sum_{|\gamma|<N} \frac{1}{\gamma!} p_1^{(\gamma)}(x,\xi)p_{2(\gamma)}(x,\xi) - q(x,\xi)\},$$

(1.28) $$r'_{1,\alpha}(x,\xi) = \sum_{|\gamma|<N}\sum_{|\gamma'|=1} \frac{1}{\gamma!}\partial_\xi^\alpha\Big[\int_0^1 (1-\theta)^{|\gamma|}\{O_s\text{-}\iint e^{-iy\cdot\eta} p_1^{(\gamma)}(x,\xi+\eta) \times \chi^{(\gamma')}(\eta/\langle\xi\rangle)\langle\xi\rangle^{-1} p_{2(\gamma+\gamma')}(x+\theta y,\xi)dy đ\eta\}d\theta\Big],$$

(1.29) $$r''_{1,\alpha}(x,\xi)$$

$$= N \sum_{|\gamma|=N} \frac{1}{\gamma!}\partial_\xi^\alpha [\int_0^1 (1-\theta)^{N-1}\{O_s\text{-}\iint e^{-iy\cdot\eta} p_1^{(\gamma)}(x,\xi+\eta)\chi(\eta/\langle\xi\rangle)$$

$$\times p_{2(\gamma)}(x+\theta y,\xi)dy đ\eta\}d\theta].$$

We note that in the second term of (1.28) only the terms with $|\gamma'| = 1$ appear. Now, use $2\langle\xi\rangle/5 \leqq |\eta| \leqq \langle\xi\rangle/2$ on the support of $\chi^{(\gamma')}(\eta/\langle\xi\rangle)$ and rewrite (1.28) as

(1.30) $$r'_{1,\alpha}(x,\xi)$$

$$= \sum_{|\gamma|<N} \sum_{|\gamma'|=1} \frac{1}{\gamma!}\partial_\xi^\alpha [\int_0^1 (1-\theta)^{|\gamma|}\{O_s\text{-}\iint e^{-iy\cdot\eta} p_1^{(\gamma)}(x,\xi+\eta)$$

$$\times \chi^{(\gamma')}(\eta/\langle\xi\rangle)\langle\xi\rangle^{-1}$$

$$\times (-i|\eta|^{-2}\eta\cdot\nabla_y)^{N-|\gamma|-1} p_{2(\gamma+\gamma')}(x+\theta y,\xi)dy đ\eta\}d\theta]$$

Then, taking a small constant $\varepsilon > 0$, we can prove that an inequality

(1.31) $$|r_{1(\beta)}^{(\alpha)}(x,\xi)| \leqq CM^{-|\alpha+\beta|}\alpha!^\kappa(\beta!^\kappa + \beta!^{\kappa(1-\delta)}\langle\xi\rangle^{\delta|\beta|})\langle\xi\rangle^{-|\alpha|} \times \exp(-\varepsilon\langle\xi\rangle^{1/\kappa}).$$

holds for ξ satisfying $C_1(N+|\alpha|)^\kappa \leqq \langle\xi\rangle \leqq C_1(N+1+|\alpha|)^\kappa$ $(N = 0,1,\dots)$ if we take a constant C_1 large enough. Since $r_{1(\beta)}^{(\alpha)}(x,\xi)$ satisfies (1.31) for $\langle\xi\rangle \leqq C_1|\alpha|^\kappa$ from (1.24), we have proved that $r_1(x,\xi)$ satisfies (1.14). This concludes the proof of proposition.

§2. Wave front sets in Gevrey classes and ultra wave front sets.

Definition 2.1. Let $u \in \mathcal{S}\{\kappa\}'$ for $\kappa > 1$ and let μ satisfy $\mu \geqq \kappa$. Then, we say that a point $(x_0,\xi_0) \in T^*(R^n)\setminus\{0\}$ does not belong to a Gevrey wave front set $WF_{G(\mu)}(u)$ of u if there exists a symbol $a(x,\xi)$ in $S^0_{G(\kappa)}$ with $a(x_0,\theta\xi_0) \neq 0$ $(\theta \geqq 1)$ such that $a(X,D_x)u$ belongs to $\mathcal{B}\{\mu\}$.

Proposition 2.2 (cf. Theorem 3 of Taniguchi[22]). Let $u \in \mathcal{S}\{\kappa\}'$ and $\mu > \kappa$. Then, for a point (x_0,ξ_0) in $T^*(R^n)$ with $\xi_0 \neq 0$, $(x_0,\xi_0) \notin WF_{G(\mu)}(u)$ holds if and only if there exist a function $\chi(x)$ in $\mathcal{S}\{\kappa\}\cap C_0^\infty(R^n)$ with $\chi(x_0) \neq 0$ and a conic neighborhood Γ of ξ_0 such that for positive constants C and ε we have

(2.1) $$|\widehat{\chi u}(\xi)| \leqq C\exp(-\varepsilon\langle\xi\rangle^{1/\mu}) \quad \text{for } \xi \in \Gamma.$$

Proof. Let $u \in \mathcal{S}\{\kappa\}'$ and assume that $(x_0,\xi_0) \notin WF_{G(\mu)}(u)$. Then, from the definition there exists a symbol $a(x,\xi)$ in $S^0_{G(\kappa)}$ with

(2.2) $$a(x_0,\theta\xi_0) \neq 0 \quad \text{for } \theta \geqq 1$$

such that $a(X,D_X)u$ belongs to $\mathcal{B}\{\mu\}$. From (2.2) we can construct a microlocal parametrix A' of $A \equiv a(X,D_X)$ near the point (x_0,ξ_0) in the sence that there exist a neighborhood V of x_0 and a conic neighborhood of Γ such that we have

$$(2.3) \qquad |\partial_\xi^\alpha \partial_x^\beta (\sigma(A'A - I))| \leqq CM^{-|\alpha+\beta|}\alpha!^\kappa \beta!^\kappa \exp(-\varepsilon\langle\xi\rangle^{1/\kappa})$$

$$\text{for } x \in V \text{ and } \xi \in \Gamma.$$

for a positive constant ε. We note that such a microlocal parametrix is constructed by the method of multi-products of pseudo-differential operators. Now, take a function $\chi(x)$ in $\mathcal{S}\{\kappa\}$ and a symbol $\psi(\xi)$ in $S^0_{G(\kappa)}$ such that $\chi(x_0) \neq 0$ with $\operatorname{supp}\chi \subset V$ and $\psi(\theta\xi_0) \neq 0$ $(\theta \geqq 1)$ with $\operatorname{supp}\psi \subset \Gamma$. Then, since $\sigma(\psi(D_X)\chi(X)(A'A - I))(x,\xi)$ satisfies (1.14) from (2.3), we can prove from ii) of Proposition 1.5 and Proposition 1.8

$$\psi(D_X)\chi(X)u = (\psi(D_X)\chi(X)A')Au - (\psi(D_X)\chi(X)(A'A - I))u \in \mathcal{B}\{\kappa\}.$$

This yields (2.1). Conversely, Let $u \in \mathcal{S}\{\kappa\}'$ satisfy (2.1) for $\chi(x)$ in $\mathcal{S}\{\kappa\} \cap C_0^\infty(R^n)$ such that $\chi(x_0) \neq 0$ and a conic neighborhood Γ of ξ_0. Take a symbol $\psi(\xi)$ in $S^0_{G(\kappa)}$ satisfying $\operatorname{supp}\psi \subset \Gamma$ and $\psi(\theta\xi_0) \neq 0$ for $\theta \geqq 1$. Then, from (2.1) we have

$$\psi(D_X)\chi(X)u \in \mathcal{B}\{\kappa\}.$$

Let $\chi_0(\xi)$ be a function in $\mathcal{S}\{\kappa\}$ such that

$$0 \leqq \chi_0 \leqq 1, \qquad \chi_0 = 1 \ (|\xi| \leqq 2/5), \quad \chi_0 = 0 \ (|\xi| \geqq 1/2)$$

and define

$$a(x,\xi) = O_s\text{-}\iint e^{-iy\cdot\eta}\psi(\xi+\eta)\chi_0(\eta/(\delta\langle\xi\rangle))\chi_0(y/\delta)\chi(x+y)dy d\eta.$$

Then, for a small δ, we have $a(x,\xi) \in S^0_{G(\kappa)}$ with $a(x_0,\theta\xi_0) \neq 0$ $(\theta \geqq 1)$ and $r(x,\xi) \equiv \sigma(\psi(D_X)\chi(X))(x,\xi) - a(x,\xi)$ satisfies (1.14). Hence, from Proposition 1.8 we have $r(X,D_X)u \in \mathcal{B}\{\kappa\}$ and this yields $a(X,D_X)u = \psi(D_X)\chi(X))u - r(x,D_X)u \in \mathcal{B}\{\kappa\}$. This
concludes the proof of the theorem.

From this proposition the Gevrey wave front set $WF_{G(\kappa)}(u)$ defined in Definition 2.1 is the same as the one defined by Hörmander[5], and as shown below we can treat Gevrey wave front sets as the same way as the treatment of the wave front sets for the C^∞ case which are studied, for example, in Hörmander[6] and Kumano-go[12].

Proposition 2.3. Let $\kappa < \mu$ and let $\phi(x,\xi)$ be a phase function in $\mathcal{P}_{G(\kappa)}$ with positive homogeneity for large $|\xi|$ and $p(x,\xi) \in S_{\rho,\delta,G(\kappa)}[\exp(\exp(c\theta^\sigma)]$ for some σ with $\sigma < 1/\mu$. Then, for $u \in \mathcal{S}\{\kappa\}'$ and $(y_0,\eta_0) \in T^*(R^n) \setminus \{0\}$ with $|\eta_0| \gg 1$, we have

$$(2.4) \qquad (y_0,\eta_0) \notin WF_{G(\mu)}(u) \implies (x_0,\xi_0) \notin WF_{G(\mu)}(P_\phi u),$$

where

$$(2.5)\qquad \xi_0 = \nabla_x \phi(x_0,\eta_0), \quad y_0 = \nabla_\xi \phi(x_0,\eta_0).$$

Remark. This proposition is studied by Hörmander[5] for pseudo-differential operators, and Taniguchi[22] and Cattabriga-Zanghirati[2] for Fourier integral operators.

Proof. Let $u \in \mathcal{S}\{\kappa\}'$ and assume $(y_0,\eta_0) \notin WF_{G(\mu)}(u)$. Then, there exists a symbol $a(x,\xi) \in S^0_{G(\kappa)}$ such that (2.2) holds and Au belongs to $\mathcal{B}\{\mu\}$ with $A \equiv a(X,D_x)$. Now, Take a microlocal parametrix A' of A satisfing

$$|\partial_\xi^\alpha \partial_x^\beta (\sigma(A'A - I))| \leqq CM^{-|\alpha+\beta|}\alpha!^\kappa \beta!^\kappa \exp(-\varepsilon\langle\xi\rangle^{1/\kappa})$$

$$\text{for } x \in V_2 \text{ and } \xi \in \Gamma_2$$

for a neighborhood V_2 of y_0 and a conic neighborhood Γ_2 of η_0 and then, take neighborhoods V_1 and V_2' of x_0 and y_0, and conic neighborhoods Γ_1 and Γ_2' of ξ_0 and η_0 satisfying

$$V_2' \Subset V_2, \quad \Gamma_2' \cap S_\eta^{n-1} \Subset \Gamma_2 \cap S_\eta^{n-1}$$

and

$$(2.6)\qquad \begin{cases} \text{i)} \ \nabla_\xi \phi(x,\eta) \in V_2' & \text{for } x \in V_1, \ \eta \in \Gamma_2', \\ \text{ii)} \ \nabla_x \phi(x,\eta) \in \Gamma_1 & \text{for } x \in V_1, \ \eta \in \Gamma_2', \end{cases}$$

Here, we used (2.5). Now, let B_1 and B_2 are pseudo-differential operators with symbols $b_1(x,\xi)$ and $b_2(x,\xi)$ satisfying

$$(2.7)\qquad \operatorname{supp} b_1 \subset V_1 \times \Gamma_1, \quad b_1(x_0,\theta\xi_0) \neq 0 \quad \text{for } \theta \geqq 1,$$
$$(2.8)\qquad \operatorname{supp} b_2 \subset V_2 \times \Gamma_2, \quad b_2(x,\xi) = 1 \quad \text{on } V_2' \times \Gamma_2',$$

Then, from (2.6)-(2.8) we see that $\sigma(B_1P_\phi(I-B_2))$ satisfies (1.14) and hence we have

$$B_1P_\phi u = B_1P_\phi B_2A'(Au) + B_1P_\phi B_2(I-A'A)u + B_1P_\phi(1-B_2)u \in \mathcal{B}\{\mu\}.$$

Consequently, we have (2.4).

Corollary 2.4. Let $\kappa < \mu$ and let P be a hypoelliptic pseudo-differential operator with a symbol in $S_{\rho,\delta,G(\kappa)}[\exp(c\theta^\sigma)]$ for some σ satisfying $\sigma < 1/\mu$ and assume that it has a parametrix Q with a symbol in $S_{\rho,\delta,G(\kappa)}[\exp(c\theta^{\sigma'})]$ for some σ' satisfying $\sigma' < 1/\mu$. Then, we have

$$(2.9)\qquad WF_{G(\mu)}(u) = WF_{G(\mu)}(Pu).$$

Proof. The existence of the parametrix and Proposition 2.3 yields

$$WF_{G(\mu)}(u) \subset WF_{G(\mu)}(Pu).$$

Hence, we have (2.9) since we have from $WF_{G(\mu)}(u) \supset WF_{G(\mu)}(Pu)$ from (2.4) for $\phi(x,\xi) = x\cdot\xi$.

Next, we define a ultra wave front set $UWF^{(\mu)}(u)$ for $u \in \mathcal{S}\{\kappa\}'$.

Definition 2.5. Let κ and μ satisfy $\kappa \leqq \mu$. For $u \in \mathcal{S}\{\kappa\}'$ we define a ultra wave front set $UWF^{(\mu)}(u)$ of u as follows: We say that a point (x_0, ξ_0) in $T^*(R^n) \setminus \{0\}$ does not belong to $UWF^{(\mu)}(u)$ if there exist a function $\chi(x)$ in $\mathcal{S}\{\kappa\} \cap C_0^\infty(R^n)$ with $\chi(x_0) \neq 0$, a conic neighborhood Γ of ξ_0, and for any positive constant ε there exists a constant $C \equiv C_\varepsilon$ such that

$$|\widehat{\chi u}(\xi)| \leqq C\exp[\varepsilon\langle\xi\rangle^{1/\mu}] \quad \text{for } \xi \in \Gamma.$$

Remark 1. As stated in Introduction this definition is the same as that of Wakabayashi[25]. (See Definition 1.3.2 in Wakabayashi[25]).

Remark 2. Let $u \in \mathcal{S}\{\kappa\}'$ and let $\kappa < \mu$. Then, $(x_0, \xi) \notin UWF^{(\mu)}(u)$ for all ξ is equivalent to that $\chi u \in \mathcal{S}\{\mu\}'$ for some $\chi \in \mathcal{S}\{\kappa\}$ with $\chi(x_0) \neq 0$. (See Lemma 1.3.3 of Wakabayashi[25]). Especially, from Proposition 1.2 we have $UWF^{(\kappa)}(u) = \phi$ for $u \in \mathcal{S}\{\kappa\}'$.

Theorem 2.6. Let $\kappa < \mu$ and let $\phi(x,\xi) \in \mathcal{P}_{G(\kappa)}$ and $p(x,\xi) \in S_{\rho,\delta,G(\kappa)}[\exp(c\theta^\sigma)]$ for some σ with $\sigma < 1/\mu$. Assume that $\phi(x,\xi)$ is positively homogeneous for large $|\xi|$. Then, for $u \in \mathcal{S}\{\kappa\}'$ and $(y_0, \eta_0) \in T^*(R^n) \setminus \{0\}$ with $|\eta_0| \gg 1$, $(y_0, \eta_0) \notin UWF^{(\mu)}(u)$ yields

$$(x_0, \xi_0) \notin UWF^{(\mu)}(P_\phi u), \tag{2.10}$$

where (x_0, ξ_0) and (y_0, η_0) are canonically related by the equations (2.5), that is, they are related by

$$\xi_0 = \nabla_x \phi(x_0, \eta_0), \quad y_0 = \nabla_\xi \phi(x_0, \eta_0). \tag{2.11}$$

Proof. Assume $(y_0, \eta_0) \notin UWF^{(\mu)}(u)$. Then, from the definition we can take a neighborhood V_2 of y_0 and a conic neighborhood Γ_2 of η_0 such that for any ε and $\chi \in \mathcal{S}\{\kappa\}$ with $\text{supp}\,\chi \subset V_2$ an inequality

$$|\widehat{\chi u}(\eta)| \leqq C_\varepsilon \exp[\varepsilon\langle\eta\rangle^{1/\mu}] \quad \text{for } \eta \in \Gamma_2 \tag{2.12}$$

holds. Next, using (2.11) we take neighborhoods V_1 and V_2' of x_0 and y_0, and conic neighborhoods Γ_1 and Γ_2' of ξ_0 and η_0 satisfying

$$V_2' \Subset V_2, \quad \Gamma_2' \cap S_\eta^{n-1} \Subset \Gamma_2 \cap S_\eta^{n-1}$$

and

$$\begin{cases} \text{i)} \ \nabla_\xi \phi(x,\eta) \in V_2' & \text{for } x \in V_1, \quad \eta \in \Gamma_2', \\ \text{ii)} \ \nabla_x \phi^{-1}(x,\xi) \in \Gamma_2' & \text{for } x \in V_1, \quad \xi \in \Gamma_1, \end{cases} \tag{2.13}$$

where $\eta = \nabla_x \phi^{-1}(x,\xi)$ is the inverse function of $\xi = \nabla_x \phi(x,\eta)$. Let $\chi_1(x)$ and $\chi_2(x)$ be functions in $\mathcal{S}\{\kappa\}$ and $\psi_1(\xi)$ and $\psi_2(\xi)$ be symbols in $S^0_{G(\kappa)}$ satisfying

$$\text{supp}\,\chi_1 \subset V_1, \quad \chi_1(x_0) \neq 0, \tag{2.14}$$

(2.15) $\quad \operatorname{supp}\chi_2 \subset V_2, \quad \chi_2(y) = 1 \quad \text{for } y \in V_2',$

(2.16) $\quad \operatorname{supp}\psi_1 \subset \Gamma_1, \quad \psi_1(\xi) = 1 \quad \text{for } \xi \in \Gamma_1^0$

with some conic neighborhood Γ_1^0 of ξ_o, and

(2.17) $\quad \operatorname{supp}\psi_2 \subset \Gamma_2, \quad \psi_2(\eta) = 1 \quad \text{for } \eta \in \Gamma_2'.$

Now, write $\chi_1(x)P_\phi u$ as

$$(2.18) \qquad \chi_1 P_\phi u = \chi_1 P_\phi \psi_2(D)\chi_2 u + \chi_1 P_\phi \psi_2(D)(1-\chi_2)u + \chi_1 P_\phi(1-\psi_2(D))u$$
$$\equiv f_1(x) + f_2(x) + f_3(x).$$

From (2.13) and (2.16)-(2.17) we can show that $\sigma(\psi_1(D)\chi_1 P_\phi(1-\psi_2(D)))$ satisfies (1.14) and hence from Proposition 1.8 we have

$$\psi_1(D)f_3 = \psi_1(D)\chi_1 P_\phi(1-\psi_2(D))u \in \mathcal{B}\{\kappa\}.$$

Note that, by virtue of the compactness of the support of f_3 and Proposition 1.2, $\hat{f}_3(\xi)$ is a measurable function. Then, we have

$$(2.19) \qquad |\hat{f}_3(\xi)| \leqq C \qquad \text{for } \xi \in \Gamma_1^0 .$$

Similarly, from (2.13)-(2.15) we obtain that $\sigma(\chi_1 P_\phi \psi_2(D)(1-\chi_2))$ satisfies (1.14) and hence we get

$$f_2(x) = \chi_1 P_\phi \psi_2(D)(1-\chi_2)u \in \mathcal{B}\{\kappa\}.$$

This and the compactness of the support of f_2 yield $f_2(x) \in \mathcal{S}\{\kappa\}$ and

$$(2.20) \qquad |\hat{f}_2(\xi)| \leqq C \qquad \text{for all } \xi.$$

Next, we consider $f_1(x)$. Let τ be a constant satisfying (1.5)-(1.6) and write

$$(2.21) \qquad \hat{f}_1(\xi) = \iint e^{i(-x\cdot\xi+\phi(x,\eta))}\chi_1(x)p(x,\eta)\psi_2(\eta)\widehat{\chi_2 u}(\eta)\,d\eta dx$$
$$= \iint_{|\xi-\eta| \leqq \lambda\langle\eta\rangle} e^{i(-x\cdot\xi+\phi(x,\eta))}\chi_1(x)p(x,\eta)\psi_2(\eta)$$
$$\times \widehat{\chi_2 u}(\eta)\,d\eta dx$$
$$+ \iint_{|\xi-\eta| \geqq \lambda\langle\eta\rangle} e^{i(-x\cdot\xi+\phi(x,\eta))}\chi_1(x)p(x,\eta)\psi_2(\eta)$$
$$\times \widehat{\chi_2 u}(\eta)\,d\eta dx$$
$$\equiv I_1 + I_2$$

with $\lambda = (1 + \tau)/2$. Since the absolute value of the integrand of I_1 is estimated by

$$C\exp[c\langle\eta\rangle^\sigma + \varepsilon\langle\eta\rangle^{1/\mu}] \leqq C'\exp[2\varepsilon\langle\eta\rangle^{1/\mu}] \leqq C'\exp[2\varepsilon\{2/(1-\tau)\}^{1/\mu}\langle\xi\rangle^{1/\mu}]$$

from (2.12) and $\sigma < 1/\mu$, we have

(2.22) $\qquad |I_1| \leqq C'' \exp[2\varepsilon\{2/(1-\tau)\}^{1/\mu}\langle\xi\rangle^{1/\mu}]$.

Let $L = -i|-\xi + \nabla_x \phi(x,\eta)|^{-2}(-\xi + \nabla_x \phi(x,\eta)) \cdot \nabla_x$. Then, we have $L \exp[i(-x\cdot\xi + \phi(x,\eta))] = \exp[i(-x\cdot\xi+\phi(x,\eta))]$. Hence, using the integration by parts and $|-\xi + \nabla_x \phi(x,\eta)| \geqq C(\langle\xi\rangle + \langle\eta\rangle)$ on the support of the integrand of I_2 we can obtain

(2.23) $\qquad |I_2| \leqq C$.

Combining (2.18)-(2.23) we obtain

$$|\mathcal{F}[\chi_1 P_\phi u](\xi)| \leqq C\exp[2\varepsilon\{2/(1-\tau)\}^{1/\mu}\langle\xi\rangle^{1/\mu}] \qquad \text{for } \xi \in \Gamma_1^0.$$

Since we can take ε arbitrary, we obtain (2.10).

This theorem corresponds to Proposition 2.3 for the propagation of Gevrey wave front sets. For Fourier integral operators with complex phase functions we have

Corolarry 2.7 (cf. Proposition 1.7 of Taniguchi[24]). Let $\kappa < \mu$ and let $\phi(x,\xi) \in \mathcal{CP}_{\rho,G(\kappa)}(\omega;\tau)$ with $\omega < 1/\mu$ and $p(x,\xi) \in S_{\rho,\delta,G(\kappa)}[\exp(c\theta^\sigma)]$ with $\sigma < 1/\mu$. Assume that there exists a real phase function $\phi_0(x,\xi)$ in $\mathcal{P}_{G(\kappa)}$ with positive homogeneity for large $|\xi|$ such that

$$\phi(x,\xi) - \phi_0(x,\xi) \in S^\omega_{\rho,G(\kappa)} \qquad (\omega < 1/\mu).$$

Then, for $u \in \mathcal{S}\{\kappa\}'$ and $(y_0,\eta_0) \in T^*(R^n) \setminus \{0\}$ with $|\eta_0| \gg 1$, we have

(2.24) $\qquad (y_0,\eta_0) \notin UWF^{(\mu)}(u) \quad \Longrightarrow \quad (x_0,\xi_0) \notin UWF^{(\mu)}(P_\phi u)$,

where

$$\xi_0 = \nabla_x \phi_0(x_0,\eta_0), \qquad y_0 = \nabla_\xi \phi_0(x_0,\eta_0).$$

Proof. Let $q(x,\xi) = p(x,\xi)\exp[i(\phi(x,\xi) - \phi_0(x,\xi))]$. Then, the Fourier integral operator P_ϕ is a fourier integral operator Q_{ϕ_0} with the real phase function $\phi_0(x,\xi)$ and the symbol $q(x,\xi)$. Set $\sigma' = \max(\sigma, \omega)$. Then, $q(x,\xi)$ satisfies

$$\begin{aligned} |q^{(\alpha)}_{(\beta)}(x,\xi)| \leqq\ & CM^{-|\alpha+\beta|}(\alpha!^\kappa + \alpha!^{\kappa\rho}\langle\xi\rangle^{(1-\rho)|\alpha|}) \\ & \times (\beta!^\kappa + \beta!^{\kappa(1-\delta)}\langle\xi\rangle^{\delta|\beta|} + \beta!^{\kappa\rho}\langle\xi\rangle^{(1-\rho)|\beta|}) \\ & \times \langle\xi\rangle^{-|\alpha|}\exp[c\langle\xi\rangle^{\sigma'}]. \end{aligned}$$

Hence, we get (2.24) by almost the same way as the proof of Theorem 2.6.

§3. Ultra wave front sets for degenerate hyperbolic equations. The propagation of Gevrey wave front sets are investigated in Kajitani-Wakabayashi[10], Morimoto-Taniguchi[15], Shinkai[17] and Cattabriga-Zanghirati[2] for the solutions of not C^∞ well-posed Cauchy problem of hyperbolic operators and it is proved that Gevrey wave front sets of the solutions propagate along broken bicharacteristics. In these papers, the authors prove this result by using Fourier integral operators with symbols in Gevrey classes. In this section, by constructing the exact form of fundamental solutions, we give the propagation of ultra wave front sets for the fundamental solutions of the Cauchy problem

(3.1) $$Lu(t) = 0, \quad u(s) = 0, \quad \partial_t u(s) = u_0$$

for degenerate hyperbolic operators in $[s,T] \times R^1_t$:

$$L = D_t^2 - t^{2j} D_x^2 + ait^k D_x$$

with $k < j-1$ and

$$L = D_t^2 - g(x)^{2j} D_x^2 + aiD_x$$

with a function $g(x) \in \mathcal{B}\{\kappa\}$ satisfying

(3.2) $$\begin{cases} g(x) \leqq -1 & \text{for } x \leqq -1, \\ g(x) = x & \text{for } |x| \leqq 1, \\ g(x) \geqq 1 & \text{for } x \geqq 1. \end{cases}$$

Here, $D_t = -i\partial_t$ and $D_x = -i\partial_x$.

First, we consider the former degenerate hyperbolic operator

(3.3) $$L = D_t^2 - t^{2j} D_x^2 + ait^k D_x.$$

By investigating an ordinary differential equation

$$\frac{d^2}{dz^2} - \lambda(z^{2j} + bz^k)$$

with a parameter λ, Shinkai[18] proves

Theorem 3.1. Consider a Cauchy problem (3.1) for (3.3) with $s < 0 < < t$. Assume $k < j-1$. Then the fundamental solution $E(t,s)$ of (3.1) are constructed in the form

(3.4) $$E(t,s) = \sum_{m,n=1}^{2} E_{m,n,\phi_{m,n}}(t,s)$$

with Fourier integral operators $E_{m,n,\phi_{m,n}}(t,s)$ whose phase functions $\phi_{m,n}(t,s) \equiv \phi_{m,n}(t,s;\xi)$ are defined by

$$\phi_{m,n}(t,s;\xi) = x\xi + \{(-1)^m t^{j+1} + (-1)^n s^{j+1}\}\xi/(j+1)$$

and whose symbols are

(3.5) $$e_{m,n}(t,s;\xi) = a_{m,n} \exp[C_{m,n}\xi^{\sigma}]\xi^{-1}(1 + o(1)), \quad \xi \to +\infty$$

with

$$\sigma = (j - k - 1)/(2j - k).$$

Furthermore, he calculates the sign of $\operatorname{Re} C_{m,n}$ by using Stokes multipliers and show that some of them are positive. When $\operatorname{Re} C_{m,n} > 0$,

$E_{m,n,\phi_{m,n}}(t,s)$ is a Fourier integral operator of infinite order. So, UWF for the solution of (3.1) appears as in the following theorem.

Theorem 3.2. Assume $k < j-1$. Let $u(t) \equiv u(t,x)$ be the solution of (3.1) for (3.3) with $u_0(x) = \delta(x)$ (Dirac function). Let $\Gamma_{m,n}$ be the trajectory associated to $\phi_{m,n}$ for $t > 0$. Then, we get

$$(3.6) \qquad UWF^{(1/\sigma)}(u(t)) = \bigcup_{P} \Gamma_{m,n},$$

where $P = \{(m,n) \ ; \ \mathrm{Re}\, C_{m,n} > 0\}$.

Remark. The result (3.6) shows that if $k < j-1$, then (3.1) for (3.3) is not C^∞ well-posed and is $\gamma^{(\kappa)}$-well-posed for $1 < \kappa < (2j-k)/(j-k-1)$ (for the $\gamma^{(\kappa)}$-well-posedness see also Ivriĭ[7]).

Next, we consider a degenerate hyperbolic operator with respect to the space variable:

$$(3.7) \qquad L = D_t^2 - g(x)^{2j} D_x^2 + aiD_x$$

with a positive constant a, where j is an even number and $g(x)$ is a function in $\mathcal{B}\{\kappa\}$ satisfying (3.2). Let $\phi_\pm(t,s;x,\xi)$ be phase functions corresponding to the characteristic roots $\pm g(x)^j \xi$ of (3.7). Then, we have

Theorem 3.3 (Taniguchi[24]). Assume

$$2j/(2j-1) \leqq \kappa \leqq 2j/(j+1)$$

and denote $\delta = 1/(2j)$. Then, the fundamental solution of the Cauchy problem (3.1) for (3.7) is constructed in the form

$$(3.8) \qquad E(t,s) = E_{+,\phi_+}(t,s) + E_{-,\phi_-}(t,s) + (\text{regularizing operator})$$

and the symbols $e_\pm(t,s;x,\xi)$ of the Fourier integral operators $E_{\pm,\phi_\pm}(t,s)$ can be written in the form

$$(3.9) \qquad e_\pm(t,s;x,\xi) = \exp[f_\pm(t,s;x,\xi)]\, e'_\pm(t,s;x,\xi)$$

with symbols $f_\pm(t,s;x,\xi)$ in $S^{1/2}_{1-\delta,\delta,G(\kappa)}$ and elliptic symbols $e'_\pm(t,s;x,\xi)$ in $S^0_{1-\delta,\delta,G(\kappa)}$. Moreover, when $s < t$, the symbols $f_\pm(t,s;x,\xi)$ in (3.9) satisfy

$$(3.10) \qquad \mathrm{Re}\, f_+(t,s;x,\xi) \geqq C(t-s)\langle\xi\rangle^{1/2}/(|x|^j\langle\xi\rangle^{1/2}+1),$$

$$(3.11) \qquad \mathrm{Re}\, f_-(t,s;x,\xi) \leqq -C(t-s)\langle\xi\rangle^{1/2}/(|x|^j\langle\xi\rangle^{1/2}+1),$$

for a positive constant C.

We note, from (3.10), $E_{+,\phi_+}(t,s)$ is a Fourier integral operator with infinite order. Hence, using the fundamental solution (3.8) we have

Theorem 3.4. Let $u(t) \equiv u(t,x)$ be the solution of the Cauchy problem

(3.1) of the operator (3.7) for u_0 in $\mathcal{S}\{\kappa\}'$ with compact support. For a conic set V in $T^*(R^1)$ we denote, by $\Gamma(t,s;V)$, $\cup_\pm \{(x,\xi)$; (x,ξ) is a point at t of the bicharacteristic strip of $\pm g(x)^j\xi$ emanating from a point in V at s$\}$. Then, when μ satisfies $\kappa < \mu < 2$ we have

$$UWF^{(\mu)}(u(t)) = \Gamma(t,s;UWF^{(\mu)}(u_0))$$

and when $\mu \geqq 2$ we have

$$UWF^{(\mu)}(u(t)) \subset \Gamma(t,s;UWF^{(\mu)}(u_0)) \setminus T_0^*R ,$$

especially, we have

$$UWF^{(\mu)}(u(t)) \setminus T_0^*R = \Gamma(t,s;UWF^{(\mu)}(u_0) \setminus T_0^*R)$$

where $T_0^*R = \{(0,\xi)\,;\, \xi \in R \setminus \{0\}\}$. In particular, when $u_0 = \delta(x)$ (Dirac function) we have

$$(0,1) \in UWF^{(2)}(u(t)) .$$

For the construction of the fundamental solution (3.8) we use finite order Fourier integral operators with complex phase functions $\phi_\pm(t,s;x,\xi)$ - $if_\pm(t,s;x,\xi)$ as in Kajitani[9] instead of using Fourier integral operators of infinite order. Then, we can give the estimate (3.10) from below.

Remark. In the above we assumed a > 0. But, if we assume a < 0 we can also construct the fundamental solution E(t,s) for (3.6) in the same form (3.8) with (3.10)-(3.11) replaced by

$$\mathrm{Re}\, f_-(t,s;x,\xi) \geqq C(t-s)\langle\xi\rangle^{1/2}/(|x|^j\langle\xi\rangle^{1/2} + 1) ,$$
$$\mathrm{Re}\, f_+(t,s;x,\xi) \leqq -C(t-s)\langle\xi\rangle^{1/2}/(|x|^j\langle\xi\rangle^{1/2} + 1) .$$

REFERENCES

1. T. Aoki, Symbols and formal symbols of pseudodifferential operators, Advanced Studies Pure Math., 4:181 (1984).
2. L. Cattabriga and L. Zanghirati, Fourier integral operators of infinite order on Gevrey spaces, application to the Cauchy problem for certain hyperbolic operators, J. Math. Kyoto Univ., 30:149 (1990).
3. I. M. Gel'fand and G. E. Shilov, "Generalized functions," Academic, Press, New York and London (1964).
4. S. Hashimoto, T. Matsuzawa and Y. Morimoto, Opérateurs pseudodifférentiels et classes de Gevrey, Comm. Partial Differential Equations, 8:1277 (1983).
5. L. Hörmander, Uniqueness theorems and wave front sets for solutions of linear partial differential equations with analytic coefficients, Comm. Pure Appl. Math., 24:671 (1971).
6. L. Hörmander, Fourier integral operators I, Acta Math., 127:79 (1971).
7. Ja. V. Ivriĭ, Correctness of the Cauchy problem in Gevrey classes for nonstrictly hyperbolic operators, Math. USSR Sb. 25:365 (1975).
8. C. Iwasaki, Gevrey-hypoellipticity and pseudo-differential operators on Gevrey class, in: "Pseudo-differential operators" (Lecture note in Math. Vol. 1256), H. O. Cordes, B. Gramsch and H. Widom, ed., Springer-Verlag, Berlin, Heidelberg, New York, London, Paris and Tokyo (1987).

9. K. Kajitani, Fourier integral operators with complex valued phase function and the Cauchy problem for hyperbolic operators, to appear in Lecture Note in Pisa.
10. K. Kajitani and S. Wakabayashi, Microhyperbolic operators in Gevrey classes, Publ. RIMS Kyoto Univ., 25:169 (1989).
11. H. Komatsu, Ultradistributions, I, Structure theorems and a characterization, J. Fac. Sci., Univ. Tokyo, Sec. IA, 20:25 (1973).
12. H. Kumano-go, "Pseudo-differential operators," The MIT Press, Cambridge and London, (1982).
13. O. Liess and L. Rodino, Inhomogeneous Gevrey classes and related pseudo-differential operator, Bollettino U.M.I. Analisi Funzionale e Applicazioni Ser Vi, 3:233 (1984).
14. G. Métivier, Analytic hypoellipticity for operators with multiple characteristics, Comm. Partial Differential Equations, 6:1 (1982).
15. Y. Morimoto and K. Taniguchi, Propagation of wave front sets of solutions of the Cauchy problem for hyperbolic equations in Gevrey classes, Osaka J. Math., 23:765 (1986).
16. L. Rodino and L. Zanghirati, Pseudo differential operators with multiple characteristics and Gevrey singularities, Comm. Partial Differential Equations, 11:673 (1986).
17. K. Shinkai, Gevrey wave front sets of solutions for a weakly hyperbolic operator, Math. Japon., 30:701 (1985).
18. K. Shinkai, Stokes multipliers and a weakly hyperbolic operator, Comm. Partial Differential Equations, 16:667 (1991).
19. K. Shinkai and K. Taniguchi, Fundamental solution for a degenerate hyperbolic operator in Gevrey classes, to appear in Publ. RIMS Kyoto Univ. Vol. 28 No. 2.
20. K. Shinkai and K. Taniguchi, On ultra wave front sets and Fourier integral operators of infinite order, Osaka J. Math., 27:709 (1990).
21. K. Taniguchi, Fourier integral operators in Gevrey class on R^n and the fundamental solution for a hyperbolic operator, Publ. RIMS, Kyoto Univ. 20:491 (1984).
22. K. Taniguchi, Pseudo-differential operators acting on ultradistributions, Math. Japon., 30:719 (1985).
23. K. Taniguchi, On multi-products of pseudo-differential operators in Gevrey classes and its application to Gevrey hypoellipticity, Proc. Japan Acad., 61:291 (1985).
24. K. Taniguchi, A fundamental solution for a degenerate hyperbolic operator of second order and Fourier integral operators of complex phase, to appear in Comm. Partial Differential Equations.
25. S. Wakabayashi, The Cauchy problem for operators with constant coefficient hyperbolic principal part and propagation of singularities, Japan J. Math., 6:179 (1980).
26. L. Zanghirati, Pseudodifferential operators of infinite order and Gevrey classes, Ann. Univ. Ferrara Sez. VII Sc. Mat., 31:197 (1985).

ON MATHEMATICAL TOOLS FOR STUDYING PARTIAL DIFFERENTIAL EQUATIONS OF CONTINUUM PHYSICS: H-MEASURES AND YOUNG MEASURES

Luc Tartar

Department of Mathematics & Center for Nonlinear Physics
Carnegie Mellon University, Pittsburgh, PA 15213, USA

LEARNING FROM THE PAST

While preparing my lecture for a conference celebrating the 600th anniversary of the University of Ferrara, I thought about its title "New Developments in Partial Differential Equations and Applications to Mathematical Physics" and I wondered what could have meant Mathematical Physics six hundred years ago. In my understanding, Physics is mostly now concerned about Light and Matter in their different forms and, for a mathematician like me, it means a lot of questions in Partial Differential Equations, some much more difficult than others. I knew a question about Light which had been the subject of discussions five hundred years ago, as I had learned about a solution proposed by Leonardo da VINCI[1][(1)], and this was directly related to the subject of my talk. For what concerned Matter, I chose the question of motion of celestial bodies as adequate for that period, and thought the pioneer to be COPERNICUS.[(2)] Only after having prepared my lecture was I told that he had studied in Ferrara: he had mostly studied in Bologna and Padua, but he also obtained the degree of doctor of canon law in Ferrara in 1503.

Although not much of our Mathematics and Physics was known at that time, it was certainly not easy in those days to introduce with success a new idea. Not so long ago, PLANCK[(3)] did not perceive the situation to be much better when he wrote[(4)] *A new scientific truth does not triumph by convincing its opponents and making them see the light, but rather because its opponents finally die, and a new generation grows up that is familiar with it.*

To us the system of COPERNICUS seems but a small improvement on the old Ptolemaic system with its circles rolling on circles for explaining the apparent motion of the planets: the Sun indeed had been put at the center, instead of the Earth. This seems unnecessarily complicated, compared to the laws that KEPLER[(5)] derived using the precise observations of BRAHE.[(6)] One was still far from imagining mathematical equations for describing physical phenomena or designing experiments for discovering physical laws in the spirit of GALILEO,[(7)] but in order to give a rational derivation of the motion of the planets and explain the efficiency of KEPLER's laws, one had to wait for NEWTON's[(8)] law of gravitation and a crucial addition to Mathematics by LEIBNIZ[(9)] and NEWTON: Infinitesimal Calculus. Actually, the system of COPERNICUS was efficient enough, and we explain now its accuracy

(1) Leonardo da VINCI, 1452 – 1519.
(2) Nicolaus COPERNICUS (Mikołaj KOPERNIK), 1473 – 1543.
(3) Max Karl Ernst Ludwig PLANCK, 1858 – 1947
(4) Quoted by Clifford TRUESDELL, *Rational Thermodynamics*, second edition, Springer, 1984.
(5) Johannes KEPLER, 1571 – 1630.
(6) Tycho BRAHE, 1546 – 1601.
(7) Galileo GALILEI, 1564 – 1642.
(8) Sir Isaac NEWTON, 1643 – 1727.
(9) Gottfried Wilhelm LEIBNIZ, 1646 – 1716.

Developments in Partial Differential Equations and Applications to Mathematical Physics, Edited by G. Buttazzo *et al.*, Plenum Press, New York, 1992

in relation with FOURIER's[(10)] result that any periodic motion is the sum of circular motions, but the accuracy of Celestial Mechanics based on the law of gravitation finally led to a further advance: Uranus was found in 1781 by a systematic survey of the sky by HERSCHEL,[(11)] and its irregular motion led ADAMS[(12)] and Le VERRIER[(13)] to apply the theoretical work of LAGRANGE[(14)] and to discover the position of Neptune, observed in 1846. If Pluton was discovered in 1930 in the same way, it went otherwise for the anomalies in the motion of Mercury that Le VERRIER had tried in 1855 to explain in a similar way: the 1919 expedition to the island of Príncipe led by EDDINGTON[(15)] for observing a total eclipse of the sun and measuring of how much the light coming from Mercury would be bent near the Sun confirmed the computations of EINSTEIN[(16)] based on his general theory of relativity.

For a mathematician, if the measurement in a physical experiment compares with accuracy to the prediction of a mathematical model, it does not prove that Nature follows that precise model. A mathematician knows that every continuous function on a compact interval can be approximated uniformly by polynomials, but he does not deduce that polynomials are important, as they could be replaced by many other classes of functions: he embeds the question into a more general framework, the theory of approximation. Even if physicists transform into dogma a set of rules which has given good results on a list of interesting physical questions, mathematicians should remain skeptical, and this has been well expressed by PENROSE when he wrote[(17)] *Quantum theory, it may be said, has two things in its favour and only one against it. First, it agrees with all the experiments. Second, it is a theory of astonishing and profound mathematical beauty. The only thing to be said against the theory is that it makes absolutely no sense. Indeed, such a bizarre collection of ideas would hardly have been put forward had it not been the case that an equally bizarre and seemingly contradictory collection of experimental facts had forced themselves on the attention of the physics community.*

It is important to notice that some complicated rule can be transformed into some quite simple result once a new mathematical theory has been developed. COPERNICUS having studied canon law at Ferrara, was aware of the complicated laws of a necessarily human Church, in opposition with God's laws for the motion of celestial bodies which he thought probably simple, but it was probably difficult for him to imagine that they were even simpler when expressed using 18th century or even 20th century Mathematics. Having this example in mind, it is then surprising to find so many physicists and even mathematicians who think that the world is described by ordinary differential equations, preferably in hamiltonian form, as if God did not know better. Why is it that they would not learn about partial differential equations and look for developing the 21th century Mathematics which will certainly simplify most of what we think we have understood? Probably because their new religion forbids them to do so.

In those remote days one had to interpret natural phenomena without the help of fancy mathematical models; I learned of such an example in 1982 while I was spending a few days at Scuola Normale Superiore in Pisa. I was advised to go to Florence to see the exhibit of a manuscript of Leonardo da VINCI[1] and one of Leonardo's idea which struck me was related to a question, which I was not even aware of, concerning the light of the Sun reflected by the Moon: if we apply the law of reflection there is only one point of the Moon that can reflect the light from the Sun directly into our eye, and therefore we should only see a bright spot on the Moon and not the totality of the illuminated part. Leonardo's explanation was that there were seas on the Moon and that because of the waves, there was always the possibility to receive light from every illuminated point on the Moon. Of course there were skeptics arguing that there were no waves because there was no wind or, as we would say now, that there were no seas there, but the point is not that we know his hypothesis to be inaccurate, but that it contains the seed for an important improvement: if Leonardo had noticed that the size of the waves did not matter and that only the angles made by the waves were important, he might have discovered that the same result was expected for infinitesimal waves and deduced that rough surfaces reflect the light in every direction. Of course, we know all these facts because we have been told about

---

(10) Baron Jean Baptiste Joseph FOURIER, 1768 – 1830.

(11) Sir William Frederick HERSCHEL (Friedrich Wilhelm), 1738 – 1822.

(12) John Couch ADAMS, 1819 – 1892.

(13) Urbain Jean Joseph Le VERRIER, 1811 – 1877.

(14) Comte Joseph Louis LAGRANGE (Giuseppe Luigi LAGRANGIA), 1736 – 1813.

(15) Sir Arthur Stanley EDDINGTON, 1882 – 1944.

(16) Albert EINSTEIN, 1879 – 1955.

(17) From Roger PENROSE's review of "The Quantum World" by J. C. POLKINGHORNE, The Times Higher Education Supplement, March 23, 1984. I am grateful to John M. BALL for having sent me a copy of that review.

them, not because we are more intelligent than Leonardo was.

Geometrical optics gave the impression that Light was a question for geometers, and certainly reflection of light or SNELL's[18] law of refraction of light is a matter of sines. DESCARTES,[19] to whom this law is attributed in France because he published it first, went further than finding it experimentally because he tried an explanation of that law from more basic principles, but his analogy with the propagation of sound in solids was wrong and justly criticized by FERMAT,[20] whose own derivation needed a finite propagation speed for light, a fact only accepted after RØMER's[21] explanation in 1676 of the anomalies in the eclipses of the moons of Jupiter. HUYGENS[22] later showed the wave nature of light, but MALUS's[23] discovery of polarized light cannot be explained in the same framework of a scalar wave equation. MAXWELL's[24] system of equations, introduced for unifying electromagnetism, does explain polarization of light, but does not explain some more recent discoveries.

Should one believe the theory of quanta of light imagined by PLANCK, the ondulatory nature of electrons shown by de BROGLIE,[25] the spin of the electron and the existence of the positron explained through DIRAC's[26] system of equations, or even accept the rules of quantum mechanics and SCHRÖDINGER's[27] equation, as it is not even an hyperbolic system. Indeed, is it still reasonable to prefer NEWTON's idea of action at distance and a world described by ordinary differential equations to EINSTEIN's idea where a particle only feels a local field but tells of its presence through a system of partial differential equations, hyperbolic and probably semilinear for having only the speed of light as characteristic velocity?

Should one ask again the obvious question: what is a particle, anyway? Or in a simpler way, what is the meaning of the physicists' saying that an electron cannot be a point because a point would radiate energy? What was meant in mathematical terms was that there is no solution of Maxwell's equation corresponding to a point mass, stationary or moving around. Was it then assumed that Maxwell's equation could also describe electrons, although it had been only designed for describing electromagnetic effects, i.e. Light? One could argue now that DIRAC's equation is considered more appropriate for discussing about electrons.

There is not so much reason to be surprised when one remembers that a ray of light is not a solution of the wave equation or of MAXWELL's equation either, and that it is only an approximation valid for high frequencies. With the new mathematical tool of H-measures which I have developed for questions of homogenization, one can give a precise meaning to what a beam of light is for the wave equation, or a beam of polarized light for MAXWELL's equation. I have not checked what is the analogous result for DIRAC's equation, but the result will still be incomplete anyway because my approach cannot yet explain completely what happens for semilinear systems. In that spirit, as ray of lights are ideal objects which are useful for describing the solutions of the wave equation, electrons could be ideal objects useful for describing the solutions of DIRAC's equation, without being solutions themselves. If the obstacles due to the semilinear character were overcome, the equation of propagation for these objects would probably involve the mass and the spin of such an electron, the mass being probably entirely made of pure electromagnetic energy,[28] with EINSTEIN's relation $e = mc^2$, of course.

Quantum mechanics was invented for explaining the surprising effects of absorption and spontaneous emission at specific frequencies in experiments of spectroscopy. Because physicists thought that they had to find a list of numbers, which were thought to be proportional to $1/n^2 - 1/m^2$ in the case of hydrogen, they were quite happy when the spectrum of an operator related to their problem

---

[18] Willebrord van SNEL van ROYEN, 1580 – 1626.

[19] René DESCARTES, 1596 – 1650.

[20] Pierre de FERMAT, 1601 – 1665.

[21] Ole Christensen RØMER, 1644 – 1710.

[22] Christiaan HUYGENS, 1629 – 1695.

[23] Etienne Louis MALUS, 1775 – 1812.

[24] James Clerk MAXWELL, 1831 – 1879.

[25] Louis Victor Pierre Raymond, duc de BROGLIE, 1892 – 1987.

[26] Paul Adrien Maurice DIRAC, 1902 – 1984.

[27] Erwin SCHRÖDINGER, 1887 – 1961.

[28] I have thought for a long time that mass should only be a side effect of electromagnetism, but the first written argument which I read in that direction was an article of W. BOSTICK, based on MAXWELL's equation coupled with some rules about quantum mechanics; in my opinion one should work with DIRAC's equation instead.

appeared to give all the $1/n^2$, and they invented an argument about eigenvalues being levels of energy together with a recipe for creating the desired operator in other situations. We know now that with a more accurate experimental setting one finds a density of absorption for large bands of frequency: there are indeed peaks, but they are quite far from being localized at well defined frequencies, and therefore the numbers that the physicists were trying to recover do not even exist.

A more reasonable approach for a mathematician would be to say that an experiment of spectroscopy consists in sending a wave into a gas which contains objects having a size comparable to the wavelength used, so that one expects some resonance effects to occur, an extra difficulty being that these objects move and that one does not even know what shapes they have. The mathematical problem is then to study the solutions of an hyperbolic system in an heterogeneous material when the characteristic size of the inhomogeneities is comparable to the wavelength. Needless to say, mathematicians do not yet know how to solve such a general question of homogenization, but the partial results already obtained show some analogy with what physicists say; for instance, the mathematical meaning for the absorption and spontaneous emission rules is that effective equations often have extra nonlocal terms in space and time. In some instances, the corrections to be added to macroscopic equations can be computed by integrating H-measures, and are therefore quadratic corrections, very similar to some which are computed by using the rules of quantum mechanics; in other cases, H-measures only enter into the first correction of an expansion, with some similarities with the summation of diagrams in quantum field theory.

I have avoided an important trap that many like falling into, which is the postulate that we cannot understand what happens at a microscopic level and that the laws of Physics are probabilistic by nature. In studying oscillating solutions of partial differential equations, many nonnegative measures do appear in a quite natural way but normalizing them and talking about probabilities will not change the perfectly deterministic framework implied by dealing with hyperbolic systems. Of course, with only a partial information about the oscillations at time zero, there is some uncertainty about what can be said about these oscillations at a later time, but the mathematical understanding of the question should also tell us a way to obtain more information.

There is still a lot to be done but once such a mathematical theory will be more developed, it should become the natural framework for discussing many of the physical phenomena which have puzzled physicists in the last century. Once this goal attained, physicists will probably have found new puzzling experimental facts, and a new mathematical theory may have to be developed, which will render elementary this one which I am trying to create, and the quest for simplicity will continue at a higher level of understanding.

AN EXAMPLE OF HOMOGENIZATION OF A WAVE EQUATION

In order to describe in a simple way the mathematical tools of YOUNG[2] measures and H-measures[3] we will consider a simple homogenization problem corresponding to the propagation of waves in a material with periodic microstructure, but we will consider a situation where the wave length is long compared to the period. The space variable will belong to R^N, and mathematicians like to use an arbitrary value for N, at least as long as the amount of work is not too excessive compared to the interest of the question. The period cell will be denoted by $Y = \{y = \sum_{i=1}^{N} \theta_i y^i,\ 0 \leq \theta_i \leq 1,\ i = 1, \ldots, N\}$ where $y^1, \ldots, y^N$ are linearly independent vectors of R^N, and its volume will be denoted $|Y|$; we will say that a measurable function f is Y-periodic if for almost all $z \in R^N$ and all $i = 1, \ldots, N$, one has $f(z + y^i) = f(z)$. The density ρ is assumed to be a measurable Y-periodic function satisfying

$$0 < \rho_- \leq \rho(x) \leq \rho_+ < \infty \text{ almost everywhere} \tag{1}$$

and the acoustic tensor a is assumed to be a measurable Y-periodic symmetric tensor satisfying

$$\alpha|\xi|^2 \leq \sum_{i,j=1}^{N} a_{ij}(y)\xi_i\xi_j \leq \beta|\xi|^2 \text{ for all } \xi \in R^N \text{ and almost all } y \in Y, \tag{2}$$

with

$$0 < \alpha \leq \beta < \infty. \tag{3}$$

For a characteristic length $\varepsilon > 0$, we look for a solution u^ε of the wave equation

$$\frac{\partial}{\partial t}(\rho(\frac{x}{\varepsilon})\frac{\partial u^\varepsilon}{\partial t}) - \sum_{i,j=1}^{N} \frac{\partial}{\partial x_i}(a_{ij}(\frac{x}{\varepsilon})\frac{\partial u^\varepsilon}{\partial x_j}) = 0, \tag{4}$$

where the equation is taken in the sense of distributions as our coefficients may be discontinuous, and we ask u^ε to satisfy the initial conditions

$$u^\varepsilon(x,0) = v(x), \frac{\partial u^\varepsilon}{\partial t}(x,0) = w(x) \tag{5}$$

where the data v and w are independent of ε and correspond to a finite energy $E(0)$,

$$\frac{1}{2}\int_{R^N}(|w|^2 + \sum_{i,j=1}^{N} a_{ij}\frac{\partial v}{\partial x_i}\frac{\partial v}{\partial x_j})dx = E(0) < \infty, \tag{6}$$

so that the solution will have the same finite energy

$$\frac{1}{2}\int_{R^N}(|\frac{\partial u^\varepsilon}{\partial t}|^2 + \sum_{i,j=1}^{N} a_{ij}\frac{\partial u^\varepsilon}{\partial x_i}\frac{\partial u^\varepsilon}{\partial x_j})dx = Constant = E(0). \tag{7}$$

The case of data depending upon ε and such that part of the energy is sent into wavelengths of order ε, for example $v(x)$ being replaced by $\varepsilon^{1-N/2}v(x/\varepsilon)$ and $w(x)$ being replaced by $\varepsilon^{-N/2}w(x/\varepsilon)$, is not entirely understood.

The question is to understand what the solution u^ε looks like and to identify its limit u^0 as ε tends to 0, usually in a weak topology. This particular problem offers no surprise and u^0 satisfies an effective equation of the same type

$$\frac{\partial}{\partial t}(\rho^{eff}\frac{\partial u^0}{\partial t}) - \sum_{i,j=1}^{N}\frac{\partial}{\partial x_i}(a_{ij}^{eff}\frac{\partial u^0}{\partial x_j}) = 0, \tag{8}$$

with the same initial data

$$u^0(x,0) = v(x), \frac{\partial u^0}{\partial t}(x,0) = w(x), \tag{9}$$

where ρ^{eff} and a^{eff} are independent of x, due to the periodicity hypothesis, and satisfy

$$0 < \rho_- \leq \rho^{eff} \leq \rho_+ < \infty \tag{10}$$

and

$$\alpha|\xi|^2 \leq \sum_{i,j=1}^{N} a_{ij}^{eff}\xi_i\xi_j \leq \beta|\xi|^2 \text{ for all } \xi \in R^N. \tag{11}$$

Of course, the effective density ρ^{eff} is the average of ρ,

$$\rho^{eff} = \frac{1}{|Y|}\int_Y \rho(y)dy, \tag{12}$$

but the effective acoustic tensor a^{eff} is not in general the average of a.

Because of the periodicity hypothesis, one can give a simple algorithm for computing the effective acoustic tensor a^{eff}: it requires solving N elliptic problems on the unit cell Y.

For each choice of a vector $\lambda \in R^N$, there is a unique Y-periodic function $z_\lambda \in H^1_{loc}(R^N)$ solution of

$$-\sum_{i,j=1}^{N}\frac{\partial}{\partial y_i}(a_{ij}(y)(\frac{\partial z_\lambda}{\partial y_j} + \lambda_j)) = 0 \tag{13}$$

and satisfying the normalization condition

$$\int_Y z_\lambda(y)dy = 0. \tag{14}$$

Then for $i = 1, \ldots, N$ one has

$$(a^{eff}\lambda)_i = \frac{1}{|Y|}\int_Y (\sum_{j=1}^{N} a_{ij}(y)(\frac{\partial z_\lambda}{\partial y_j} + \lambda_j))dy. \tag{15}$$

Repeating this computation for N linearly independent vectors λ determines a^{eff}, which is positive definite (and symmetric as $a(y)$ is symmetric almost everywhere in Y).

WEAK CONVERGENCE, HOMOGENIZATION, YOUNG MEASURES

In the preceding example with a periodic structure, averaging a function on a period εY should be considered analogous to making a macroscopic measurement in an experiment where something happens at a microscopic level; in our example one sees the microscopic level by looking at a length scale of the order of ε. In a nonperiodic situation, averaging is replaced by weak convergence: a sequence f^ε converges weakly to f^0 as ε tends to 0 if $\int f^\varepsilon(x)\varphi(x)dx \to \int f^0(x)\varphi(x)dx$ for a suitable class of functions φ : ε is usually a length (or time) scale and f^0 will be called the macroscopic quantity corresponding to the microscopic quantity f^ε. Of course, there are macroscopic quantities without microscopic analog: there is no function H such that if f^ε converges weakly to f^0 and $g^\varepsilon = (f^\varepsilon)^2$ converges weakly to $g^0 \geq (f^0)^2$ one can deduce that $h^\varepsilon = H(f^\varepsilon)$ converges weakly to $g^0 - (f^0)^2$. This is analog to the situation of a gas when the microscopic velocity is not equal to the macroscopic velocity: the averaged kinetic energy is more than the kinetic energy computed from the macroscopic velocity and the difference is then called the internal energy, usually related to temperature which only has a macroscopic meaning.

Some effective quantities are obtained by taking weak limits: potentials, electric or magnetic or velocity fields, induction or vorticity fields, densities of charge or mass or energy (they are coefficients of differential forms), while others are not obtained by taking weak limits: electrical or thermal conductivity, electric or magnetic permittivity, elastic properties, sound speed, and this happens for mixtures, composite materials, polycrystals.

Physicists often tend to disagree when mathematicians let some physical parameter ε converge to 0, but this is only a first step which consists in identifying what is the right topology for the various quantities involved in order to find a limiting equation whose solution will be near the physical one. One should also remember that there are infinitely many ways to imbed a given problem into a sequence of such problems, and that various scalings may correspond to different physical questions, each having its own limiting behaviour.

Homogenization is a mathematical theory whose first goal is to derive for each situation of interest what are the effective equations valid at a macroscopic level, assuming that one has a complete information about the microscopic level. Its second goal is to deduce what can be said under partial information about the microscopic level. Uncovering new mathematical objects which carry in a concise way some important information about the microstructure is a reward of that approach. The term microstructure is used when dealing with sequences converging only weakly in order to express that something is happening at a microscopic level; one also talks about an oscillating sequence of functions.

The term microgeometry is used when dealing with many oscillating sequences of functions constructed on the same geometrical pattern: an open set Ω of R^N is decomposed as a countable union of disjoint measurable subsets of Ω

$$\Omega = \cup_i \omega_i^\varepsilon, \tag{16}$$

and one only considers sequences U^ε with values in R^p of the form

$$U^\varepsilon(x) = \sum_i \chi_i^\varepsilon(x)V^i \tag{17}$$

where the functions χ_i^ε are the characteristic functions of ω_i^ε and V^i are elements of R^p, often belonging to a closed bounded set K.

The main question is to study how weak limits of U^ε, or effective quantities generated from it, depend upon the values V^i and on the information on the decomposition of Ω in ω_i^ε.

YOUNG measures describe the weak limits of U^ε for all possible choices of V^i. If

$$\chi_i^\varepsilon \text{ converges weakly * to } \theta_i \text{ in } L^\infty(\Omega) \tag{18}$$

so that one has

$$0 \leq \theta_i(x) \leq 1 \text{ and } \sum_i \theta_i(x) = 1 \text{ almost everywhere in } \Omega, \tag{19}$$

then for every continuous function F on R^p, one has

$$F(U^\varepsilon) \text{ converges weakly * to } \sum_i \theta_i F(V^i) \text{ in } L^\infty(\Omega). \tag{20}$$

If one defines the probability measure ν_x on R^p by

$$\langle \nu_x, G \rangle = \sum_i \theta_i(x) G(V^i) \tag{21}$$

for every continuous function G, then the YOUNG measure associated with the sequence U^ε is the measurable family of all these ν_x.

For a general uniformly bounded sequence of functions U^ε taking their values in a closed subset K of R^p, there is a subsequence and a measurable family of ν_x which are probability measures on K such that for every continuous function F on R^p, one has

$$F(U^\varepsilon) \text{ converges weakly * to } f \text{ in } L^\infty(\Omega) \tag{22}$$

with

$$f(x) = \langle \nu_x, F \rangle \text{ almost everywhere in } \Omega. \tag{23}$$

What the YOUNG measures do for a mixture is to know what are the local proportions of all the materials used in that mixture.

SMALL AMPLITUDE HOMOGENIZATION I

YOUNG measures only see statistics and they cannot, except in dimension 1, help computing effective coefficients like the algorithm (13)-(14)-(15) for a periodic case. In a layered material, they lack the knowledge of an important geometric parameter, the direction of the layers. In order to compute some second order corrections in homogenization, I have introduced a new tool, which I naturally called H-measures[3], which is a measure in (x,ξ), where ξ is a unit direction of an hyperplane. In the periodic case, they can be described by using the Fourier expansion of the coefficients: if we assume that

$$a_{ij}(y) = A_{ij} + \gamma b_{ij}(y) \tag{24}$$

where A is symmetric positive definite and γ is small, then a^{eff} is analytic in γ

$$a^{eff} = A + \gamma \frac{1}{|Y|} \int_Y b(z)dz + \gamma^2 B^{(2)} + \gamma^3 B^{(3)} + \ldots\ldots, \tag{25}$$

as well as the functions z_λ solutions of (13)-(14), and therefore by an easy induction one can compute all the correctors $B^{(r)}$ by using the Fourier coefficients of b.

In the case where Y is the unit cube, one has

$$b_m = \int_Y b(z) e^{-2i\pi(m.z)} dz, \quad m \in Z^N \tag{26}$$

and

$$b(y) = \sum_{m \in Z^N} b_m e^{2i\pi(m.y)} \text{ in } L^2(Y) \tag{27}$$

and an easy computation gives

$$B^{(2)} = \sum_{m \in Z^N \setminus 0} b_{-m} \frac{m \otimes m}{(Am.m)} b_m. \tag{28}$$

Using (26), the formula (28) can be expressed in terms of the 2-point correlation function

$$C(h) = \frac{1}{|Y|} \int_Y b(z+h)b(z)dz, \tag{29}$$

but it uses a singular integral with kernel

$$K(h) = \sum_{m \in Z^N \setminus 0} e^{2i\pi(m.h)} \frac{m \otimes m}{(Am.m)}. \tag{30}$$

For a sequence which is not periodic, one cannot define n-point correlation functions without the knowledge of a characteristic length. The H-measures which I have introduced do not use any characteristic length in their definition, and therefore one cannot deduce from them the 2-point correlation function; however, there are situations where the complete knowledge of the 2-point correlation function is not necessary and where H-measures contain all the desired information.

H-MEASURES

A similar formula for the corrector $B^{(2)}$ exists in the case of nonperiodic microstructures and uses H-measures, but at the moment there is no general formula for expressing the correctors $B^{(r)}$ with $r \geq 3$.

If Ω is an open subset of R^N and U^ε is a sequence converging to 0 in $(L^2(\Omega))^p$ weak, then after extraction of a subsequence one can define a hermitian nonnegative $p \times p$ matrix of Radon measures μ in (x, ξ), with $\xi \in S^{N-1}$ the unit sphere in R^N; μ is called the H-measure associated to the subsequence and it enables to compute the weak * limits of products of the type $L_1(U_i^\varepsilon)\overline{L_2(U_j^\varepsilon)}$ where L_1 and L_2 are some "pseudo-differential" operators of order 0, U^ε being extended by 0 outside Ω.

The class of symbols of these "pseudo-differential" operators of order zero have the form

$$s(x,\xi) = \sum_{n=1}^{\infty} a_n(\xi) b_n(x) \tag{31}$$

with $a_n \in C(S^{N-1})$, the space of continuous functions on the unit sphere and $b_n \in C_0(R^N)$, the space of continuous functions converging to 0 at infinity, with

$$\sum_{n=1}^{\infty} ||a_n||.||b_n|| < \infty \tag{32}$$

where the norms are sup norms.

The standard operator S with symbol s is defined by

$$\mathbf{F}(Su)(\xi) = \sum_{n=1}^{\infty} a_n(\frac{\xi}{|\xi|}) F(b_n u)(\xi), \text{ almost everywhere in } \xi \in R^N, \text{ for } u \in L^2(R^N) \tag{33}$$

where $\mathbf{F}$ denotes the Fourier transform. A linear continuous operator L from $L^2(R^N)$ into itself is said to have symbol s if $L - S$ is a compact operator from $L^2(R^N)$ into itself.

With these notations, if L_1 and L_2 are operators with symbols s_1 and s_2, and

$$L_1(U_i^\varepsilon)\overline{L_2(U_j^\varepsilon)} \text{converges weakly to a measure } \nu, \tag{34}$$

one has

$$\langle \nu, \varphi \rangle = \langle \mu^{ij}, \varphi s_1 \overline{s_2} \rangle \text{ for every } \varphi \in C_c(\Omega), \tag{35}$$

the space of continuous functions with compact support in Ω.

An important consequence is the localization principle, which expresses how the H-measure is constrained by any differential information on the sequence U^ε: if A_{ij} are continuous functions in Ω and

$$\sum_{i=1}^{N} \sum_{j=1}^{p} \frac{\partial}{\partial x_i}(A_{ij} U_j^\varepsilon) \to 0 \text{ in } H^{-1}_{loc}(\Omega) \tag{36}$$

then

$$\sum_{i=1}^{N}\sum_{j=1}^{p}\xi_i A_{ij}(x)\mu^{jk} = 0 \text{ for } k = 1,\dots, p. \tag{37}$$

An early version of the theory, the compensated compactness theory, could only handle the case of constant coefficients and discuss the possible weak limits of quadratic quantities; H-measures give new results in other directions.

SMALL AMPLITUDE HOMOGENIZATION II

We do not consider the case of periodic coefficients anymore. Let A be positive definite, and let $B^\varepsilon \rightharpoonup B^0$ in $L^\infty(\Omega; L(R^N, R^N))$ weak * and assume that $B^\varepsilon - B^0$ corresponds to a H-measure μ. For γ small and $f \in H^{-1}(\Omega)$, we solve

$$-div((A + \gamma B^\varepsilon)gradu^\varepsilon) = f \text{ in } \Omega \text{ with } u^\varepsilon \in H_0^1(\Omega). \tag{38}$$

After extraction of a subsequence (independent of f), u^η converges weakly to u^0 solution of

$$-div((A^{eff}(x;\gamma)gradu^0) = f \text{ in } \Omega \text{ with } u^0 \in H_0^1(\Omega), \tag{39}$$

where A^{eff} is analytic in γ

$$A^{eff}(x;\gamma) = A + \gamma B^0 - \gamma^2 M + O(\gamma^3) \tag{40}$$

and the correction M can be computed from the H-measure μ: for $i,\ j = 1,\dots,\ N$ and $\varphi \in C_c(\Omega)$, one has

$$\int_\Omega M_{ij}(x)\varphi(x)dx = \sum_{k,l=1}^{N}\langle\mu^{ik,lj}, \varphi(x)\frac{\xi_k\xi_l}{(A\xi.\xi)}\rangle. \tag{41}$$

In the particular case of a mixture of isotropic materials, with $A = a_0(x)I$ and $B^\varepsilon(x) = b^\varepsilon(x)I$, with

$$b^\varepsilon \rightharpoonup b_0 \text{ and } (b^\varepsilon - b_0)^2 \rightharpoonup \beta^2 \text{ in } L^\infty(\Omega) \text{ weak } *, \tag{42}$$

one deduces

$$Trace(M) = \frac{\beta^2}{a_0}, \tag{43}$$

and therefore if the effective material is isotropic, i.e. $A^{eff}(x;\gamma) = a^{eff}(x;\gamma)I$, or only isotropic at order 2 in γ, i.e. if $M(x) = m(x)I$, then one has

$$a^{eff} = a_0 + \gamma b_0 - \gamma^2\frac{\beta^2}{Na_0} + O(\gamma^3). \tag{44}$$

This latter result was previously known under additional hypotheses of symmetry.

We will discuss later a similar result for isotropic linearized elasticity.

PROPAGATION OF OSCILLATIONS AND CONCENTRATION EFFECTS I

H-measures can be used to describe both oscillations and concentration effects.

If a scalar function u^ε is defined by

$$u^\varepsilon(x) = v(x, \frac{x}{\varepsilon}) \tag{45}$$

with $v(x, y)$ periodic in y with average 0 (and smooth enough), the period Y being the unit cube for simplification, and if the Fourier expansion of v in y is

$$v(x, y) = \sum_{m\in Z^N\setminus 0} v_m(x)e^{2i\pi(m.y)}, \tag{46}$$

then u^ε corresponds to the H-measure μ defined by

$$\langle \mu, \varphi \rangle = \sum_{m \in Z^N \setminus 0} \int_\Omega |v_m(x)|^2 \varphi(x, \frac{m}{|m|}) dx \tag{47}$$

for all continuous functions φ with compact support in $\Omega \times S^{N-1}$. Such a sequence will be called a periodically modulated oscillating sequence, a more general oscillating sequence being a sequence u^ε converging weakly but not necessarily strongly in $L^2(\Omega)$, and such that $(u^\varepsilon)^2$ converges weakly to a function in $L^1(\Omega)$; for such a general sequence the H-measure need not be atomic in ξ.

If a scalar function u^ε is defined by

$$u^\varepsilon = \varepsilon^{-\frac{N}{2}} f(\frac{x-z}{\varepsilon}) \text{ with } z \in R^N \text{ and } f \in L^2(R^N) \tag{48}$$

then u^ε corresponds to the H-measure μ defined by

$$\langle \mu, \varphi \rangle = \int_{R^N} |\mathbf{F} f(\xi)|^2 \varphi(z, \frac{\xi}{|\xi|}) d\xi \tag{49}$$

for all continuous functions φ with compact support in $R^N \times S^{N-1}$. Such a sequence will be called a concentration effect at the point z, a more general concentration effect being a sequence u^ε converging weakly but not strongly to u^0 in $L^2(\Omega)$ and such that $(u^\varepsilon - u^0)^2$ converges weakly * to a measure which is singular with respect to the Lebesgue measure.

Of course, a general weakly converging sequence may show both oscillations and concentration effects.

For some partial differential equations of hyperbolic nature, one can measure in a quantitative way the propagation of oscillations and concentration effects, and this is done by deriving a partial differential equation in (x, ξ) for the H-measure μ. In the case of a first order scalar equation, let u^ε converge to 0 weakly in $L^2(\Omega)$, correspond to a H-measure μ and satisfy

$$\sum_{i=1}^N b_i(x) \frac{\partial u^\varepsilon}{\partial x_i} = f^\varepsilon \tag{50}$$

with f^ε converging strongly to 0 in $H^{-1}_{loc}(\Omega)$, the coefficients b_i being of class C^1, $i = 1, \ldots, N$. The localization principle implies then that μ satisfies

$$P(x, \xi)\mu = 0 \tag{51}$$

where

$$P(x, \xi) = \sum_{i=1}^N b_i(x) \xi_i. \tag{52}$$

In order to study the propagation properties for oscillations and concentration effects present in the sequence u^ε, we assume moreover that the coefficients b_i are real, and that f^ε converges weakly to 0 in $L^2(\Omega)$. Under these hypotheses, the H-measure μ satisfies the equation

$$\langle \mu, \{\varphi, P\} - \varphi div b \rangle = \langle 2 Re \mu^{12}, \varphi \rangle \tag{53}$$

for all C^1 test functions φ with compact support in x, where $\{g, h\}$ denotes the Poisson bracket

$$\{g, h\} = \sum_{i=1}^N (\frac{\partial g}{\partial \xi_i} \frac{\partial h}{\partial x_i} - \frac{\partial g}{\partial x_i} \frac{\partial h}{\partial \xi_i}). \tag{54}$$

Equation (53) expresses a propagation effect along the bicharacteristics associated to $P(x, \xi)$

$$\frac{dx_i}{dt} = \frac{\partial P}{\partial \xi_i}; \frac{d\xi_i}{dt} = -\frac{\partial P}{\partial x_i} \text{ for } i = 1, \ldots, N, \tag{55}$$

the equation in ξ being homogeneous in ξ and inducing therefore an equation on the unit sphere. In the propagation equation (53), the source term μ^{12} corresponds to the H-measure associated to the sequence $(u^\varepsilon, f^\varepsilon)$, so that μ^{11} is μ; if $f^\varepsilon = Lu^\varepsilon$ where L has symbol s, then $\mu^{12} = \bar{s}\mu$, but if f^ε is nonlinear in u^ε it is not known yet how to describe what μ^{12} can be for a given μ.

Equation (53) can be supplemented with an initial condition for the H-measure μ.

PROPAGATION OF OSCILLATIONS AND CONCENTRATION EFFECTS II

Let us consider now the question of propagation of oscillations and concentration effects for a wave equation. For this we consider a sequence u^ε converging weakly to 0 in $H^1(R^N \times (0,T))$ and satisfying

$$\frac{\partial}{\partial t}(\rho(x)\frac{\partial u^\varepsilon}{\partial t}) - \sum_{i,j=1}^{N} \frac{\partial}{\partial x_i}(a_{ij}(x)\frac{\partial u^\varepsilon}{\partial x_j}) = f^\varepsilon \tag{56}$$

with $grad u^\varepsilon$ corresponding to a H-measure μ; the localization principle implies that μ has the form

$$\mu^{ij} = \xi_i \xi_j \nu \text{ for } i,j = 1,\dots, N \tag{57}$$

with a nonnegative measure ν. Assuming that the functions ρ and a_{ij}, i, $j = 1,\dots, N$ are continuous and that f^ε converges strongly to 0 in H^{-1}_{loc}, the localization principle implies that ν satisfies

$$Q(x,\xi)\nu = 0 \tag{58}$$

with

$$Q(x,\xi) = \rho(x)\xi_0^2 - \sum_{i,j=1}^{N} a_{ij}(x)\xi_i\xi_j \tag{59}$$

where, as usual, t is replaced by x_0 with dual variable ξ_0. Notice that (58) is a way to describe the principle of equipartition of energy.

If we assume now that ρ is real positive and of class C^1, that the acoustic tensor a is real symmetric positive definite and of class C^1, then ν satisfies the propagation equation

$$\langle \nu, \{\varphi, Q\}\rangle = \langle 2Re\nu^{12}, \varphi\rangle \tag{60}$$

for all C^1 test functions φ with compact support in x. In (60), ν^{12} corresponds to some components of the H-measure associated to $(grad u^\varepsilon, f^\varepsilon)$. Equation (60) expresses a propagation effect along the classical light rays, which are the bicharacteristics associated to $Q(x,\xi)$,

$$\frac{dx_i}{d\tau} = \frac{\partial Q}{\partial \xi_i}; \frac{d\xi_i}{d\tau} = -\frac{\partial Q}{\partial x_i} \text{ for } i = 1,\dots, N, \tag{61}$$

the equation in ξ being homogeneous in ξ and inducing therefore an equation on the unit sphere. This result gives a mathematical framework for what is meant by a light beam at a point x_0 pointing in a direction ξ_0. The H-measure is neither a solution of the wave equation, nor a formal asymptotic solution for high frequency: it describes, in the limit of infinite frequency, the way to decide where the energy goes for any oscillating sequence of initial data with finite energy.

Equation (60) can be supplemented with an initial condition for the H-measure ν.

SMALL AMPLITUDE HOMOGENIZATION III

Once the framework of an application of H-measures has been developed, it can be generalized to various equations or systems at the only expense of having to perform some often tedious computations of linear algebra. Let us describe for example the question of small amplitude homogenization for the system of linearized elasticity in the case of a mixture of isotropic materials[4]. The stress tensor σ^ε satisfies the equilibrium equation

$$\sum_{j=1}^{N} \frac{\partial \sigma_{ij}^\varepsilon}{\partial x_j} = f_i \text{ for } i = 1,\dots, N, \tag{62}$$

and is here related to the linearized strain tensor e^ε defined by

$$e_{ij}^\varepsilon = \frac{1}{2}\left(\frac{\partial u_i^\varepsilon}{\partial x_j} + \frac{\partial u_j^\varepsilon}{dx_i}\right) \text{ for } i,j = 1,\ldots, N, \tag{63}$$

where $u^\varepsilon(x)$ denotes the displacement of the point x. The particular constitutive relation corresponding to an isotropic material has the form

$$\sigma_{ij}^\varepsilon = 2\mu^\varepsilon e_{ij}^\varepsilon + \lambda^\varepsilon \delta_{ij} \sum_{k=1}^{N} e_{kk}^\varepsilon \tag{64}$$

and the hypothesis of small amplitude means that

$$\mu^\varepsilon = \mu_0 + \gamma\mu_1^\varepsilon; \lambda^\varepsilon = \lambda_0 + \gamma\lambda_1^\varepsilon, \tag{65}$$

where for simplification we assume that μ_1^ε and λ_1^ε converge to 0 in $L^\infty(\Omega)$ weak *. Of course, γ is assumed small enough so that μ^ε and λ^ε uniformly satisfy the usual ellipticity condition required for applying the theory of homogenization, i.e. $\mu^\varepsilon > 0$ and $2\mu^\varepsilon + N\lambda^\varepsilon > 0$. The homogenized equation will have the same form, but may correspond to a general anisotropic material with constitutive relation

$$\sigma_{ij} = \sum_{k,l=1}^{N} C_{ijkl}^{eff} e_{kl} \tag{66}$$

with the usual symmetries in $ijkl$. Of course the effective elasticity tensor C^{eff} is analytic in γ

$$C_{ijkl}^{eff} = \mu_0(\delta_{ik}\delta_{jl} + \delta_{il}\delta_{jk}) + \lambda_0\delta_{ij}\delta_{kl} - \gamma^2 D_{ijkl} + O(\gamma^3) \tag{67}$$

and the coefficients D_{ijkl} can be expressed in terms of the H-measure ν associated to the sequence $(\mu_1^\varepsilon, \lambda_1^\varepsilon)$. The formulae involve the moments of order 4 of ν^{11}, the moments of order 2 of ν^{12} and the moment of order 0 of ν^{22}. The contribution of ν^{11} to D_{ijkl} is obtained by integrating

$$\frac{(\delta_{ik}\xi_j\xi_l + \delta_{il}\xi_j\xi_k + \delta_{jk}\xi_i\xi_l + \delta_{jl}\xi_i\xi_k)}{\mu_0} - 4\frac{\xi_i\xi_j\xi_k\xi_l(\mu_0 + \lambda_0)}{\mu_0(2\mu_0 + \lambda_0)}, \tag{68}$$

the contribution of ν^{12} to D_{ijkl} is obtained by integrating

$$2\frac{(\delta_{kl}\xi_i\xi_j + \delta_{ij}\xi_k\xi_l)}{(2\mu_0 + \lambda_0)}, \tag{69}$$

and the contribution of ν^{22} to D_{ijkl} is obtained by integrating

$$\frac{\delta_{ij}\delta_{kl}}{(2\mu_0 + \lambda_0)}. \tag{70}$$

In the very special case where the effective material is isotropic, or simply isotropic at order 2 in γ, i.e. if D_{ijkl} has the form

$$D_{ijkl} = M(\delta_{ik}\delta_{jl} + \delta_{il}\delta_{jk}) + \Lambda\delta_{ij}\delta_{kl}, \tag{71}$$

then M and Λ can be computed using only the moments of order 0 of ν, i.e. from the weak * limits of the quantities $(\mu_1^\varepsilon)^2$, $\mu_1^\varepsilon\lambda_1^\varepsilon$ and $(\lambda_1^\varepsilon)^2$: one finds

$$M = \frac{4(N+1)\mu_0 + 2N\lambda_0}{N(N+2)\mu_0(2\mu_0 + \lambda_0)} weak * lim(\mu_1^\varepsilon)^2 \tag{72}$$

and

$$2M + N\Lambda = \frac{1}{N(2\mu_0 + \lambda_0)} weak * lim(2\mu_1^\varepsilon + N\lambda_1^\varepsilon)^2. \tag{73}$$

A PROBLEM MIXING YOUNG MEASURES AND H-MEASURES

The solutions of many important problems seem to require the use of a mathematical object, yet to be developed, which will encompass both the YOUNG measures and the H-measures. Partial results about the relations between YOUNG measures and H-measures have been obtained with François MURAT[5] and we can see how they can be used on an example.

We consider the model of micromagnetics of William BROWN[6], for a crystal occupying a bounded open domain Ω of R^3, as studied recently by Richard JAMES & David KINDERLEHRER[7]. After normalization, we consider the equation

$$-div(gradu + m\chi_\Omega) = 0 \text{ in } R^3 \tag{74}$$

where χ_Ω is the characteristic function of Ω and m satisfies the constraint

$$|m(x)| = 1 \text{ almost everywhere in } \Omega, \tag{75}$$

and we seek m minimizing the quantity $J(m)$ defined by

$$J(m) = \int_{R^3} |gradu|^2 dx + \int_\Omega (\varphi(m) - H_0.m)dx. \tag{76}$$

In that model the magnetic field H is $gradu$ and the magnetic induction field B is $H + m$, m corresponding to a spin effect, φ is an anisotropic energy due to the crystalline nature of the body and H_0 is an applied magnetic field.

An exchange energy, usually taken to be quadratic in $gradm$, has been neglected.

The mathematical difficulty comes from the fact that the functional J is not lower semicontinuous for the natural topology for m, the $(L^\infty(\Omega))^3$ weak * topology. Minimizing sequences might then develop oscillations, and this is in qualitative agreement with the experimentally observed formation of small magnetic domains, although there are still some quantitative discrepancies and it is not clear yet how good this model is. A better understanding of the relaxation of the functional J might shed some light upon this question.

If a sequence m^ε converges to m^0 in $(L^\infty(\Omega))^3$ weak *, the computation of the limit of $\varphi(m^\varepsilon)$ requires more than the weak * limit of m^ε (as φ is not affine) and can be obtained from the YOUNG measure ν associated to a subsequence; on the other hand the computation of the limit of $|gradu^\varepsilon|^2$ cannot be computed from the YOUNG measure ν alone, but can be computed from the H-measure μ associated to a subsequence of $m^\varepsilon - m^0$. One has

$$\int_\Omega \varphi(m^\varepsilon)\psi(x)dx \to \int_\Omega \langle\nu_x, \varphi\rangle\psi(x)dx \tag{77}$$

and

$$\int_{R^3} |gradu^\varepsilon|^2\psi(x)dx \to \int_{R^3} |gradu^0|^2\psi(x)dx + \sum_{i,j=1}^{3} \langle\mu^{ij}, \psi(x)\xi_i\xi_j\rangle \tag{78}$$

for every bounded continuous function ψ, where u^0 denotes the solution corresponding to m^0, which only satisfies $|m^0(x)| \leq 1$ for almost every $x \in \Omega$, as $m^0(x)$ is the center of mass of the probability ν_x which lives on the sphere S^2, i.e.

$$m_i^0(y) = \langle\nu_y, x_i\rangle \text{ almost everywhere in } \Omega \text{ for } i = 1,2,3. \tag{79}$$

The crucial question is then to understand what relations link the YOUNG measure ν and the H-measure μ. Without answering this question, we can only describe an abstract relaxation problem where we seek to minimize the functional J_1 defined in the following way. We let X be the space of all pairs (ν, μ) for which there exists a sequence m^ε satisfying the constraint (75), such that m^ε defines the YOUNG measure ν, and such that $m^\varepsilon - m^0$ defines the H-measure μ, where m^0 is defined by (79). We define the functional J_1 on X by the formula

$$J_1(\nu,\mu) = \int_{R^3} |gradu^0|^2 dx + \sum_{i,j=1}^{3} \langle\mu^{ij}, \xi_i\xi_j\rangle + \int_\Omega \langle\nu_x, \varphi\rangle dx - \int_\Omega H_0.m^0 dx, \tag{80}$$

where u^0 is the solution of

$$-div(gradu^0 + m^0\chi_\Omega) = 0 \text{ in } R^3. \tag{81}$$

By following the construction of ν and μ, one can put a topology on X that makes it compact and renders J_1 continuous, so that J_1 does attain its minimum on X. The initial problem is imbedded into this new one and corresponds to each ν_x being a Dirac mass and μ being 0. Of course, the preceding result is only a change of language for recasting the problem.

A precise description of X is not yet available, but a partial result shows that for a given YOUNG measure ν, there is a pair (ν, μ) belonging to X such that $\sum_{i,j=1}^{3}\langle \mu^{ij}, \xi_i\xi_j\rangle = 0$. This gives rise to a new relaxed problem defined on the set Y of all YOUNG measures, where one defines a functional J_2 by the formula

$$J_2(\nu) = \int_{R^3} |gradu^0|^2 dx + \int_\Omega \langle \nu_x, \varphi\rangle dx - \int_\Omega H_0.m^0 dx, \tag{82}$$

with m^0 and u^0 defined by (79) and (81). If Y is equipped with the weak * topology, then Y is compact and J_2 is lower semi-continuous and does attain its minimum. The initial problem is imbedded into this new one and corresponds to each ν_x being a Dirac mass.

We finally define Z to be the convex set of functions m^0 satisfying

$$|m^0(x)| \leq 1 \text{ almost everywhere in } \Omega, \tag{83}$$

and define the functional J_3 by

$$J_3(m^0) = \int_{R^3} |gradu^0|^2 dx + \int_\Omega (\psi(m^0) - H_0.m^0) dx, \tag{84}$$

where u^0 is defined by (81) and where ψ is the convex function defined on the unit ball by

$$\psi(m) = Inf_\nu \langle \nu, \varphi\rangle \text{ for all probability measures } \nu \text{ on } S^2 \text{ with center of mass m.} \tag{85}$$

If we equip Z with the weak * topology, then Z is compact and J_3 is lower semi-continuous and does attain its minimum. The initial problem is imbedded into this new one and corresponds to m^0 taking almost everywhere its values on the unit sphere.

If no solution of this last problem satisfies $|m^0(x)| = 1$ almost everywhere, then there are no classical solution mimimizing J, and minimizing sequences tend to create somewhere in Ω some tiny magnetic domains, the statistics of orientations for m being described by the YOUNG measure ν satisfying (85); the H-measure μ, sees another kind of information, like the orientations of the walls of these magnetic domains. Of course, having neglected the exchange energy, there is nothing to limit the size of the magnetic domains in this simplified model.

NONLOCAL EFFECTS INDUCED BY HOMOGENIZATION

For explaining some strange rules invented by physicists, like absorption and spontaneous emission of particles, it is important to realize that effective equations may contain nonlocal terms in space or/and time even when the microscopic level is described by classical partial differential equations; this phenomenon seems actually quite usual when dealing with hyperbolic equations. As a typical example, we consider the following problem, which has been studied by Youcef AMIRAT, Kamal HAMDACHE and Abdelhamid ZIANI[8], and by myself[9]:

$$\frac{\partial u^\varepsilon}{\partial t} + a^\varepsilon(y)\frac{\partial u^\varepsilon}{\partial x} = f(x,y,t); u^\varepsilon(x,y,0) = v(x,y), \tag{86}$$

where the sequence a^ε converges to a^0 in L^∞ weak *. If we assume that a^ε satisfies

$$0 < \alpha \leq a^\varepsilon(y) \leq \beta < \infty \text{ almost everywhere} \tag{87}$$

and if the sequence a^ε defines a YOUNG measure ν, then one can characterize the effective equation which the weak * limit u^0 of the sequence u^ε must satisfy (there is indeed only one such equation if one restricts attention to linear convolution equations in x and t). Another nonnegative measure π

will appear in the equation and it is obtained from ν through a nonlinear transformation. With data which are measurable and bounded, the effective equation is

$$\frac{\partial u^0}{\partial t} + a^0(y)\frac{\partial u^0}{\partial x} + M = f(x,y,t); u^0(x,y,0) = v(x,y) \tag{88}$$

where the nonlocal term M has the form

$$M(x,y,t) = -\int_0^t \int_{[\alpha,\beta]} \frac{\partial^2 u^0(x-\lambda(t-s),y,s)}{\partial x^2} d\pi_y(\lambda)ds. \tag{89}$$

One sees that a transport equation with velocity constant along the flow but fluctuating in another direction induces a nonlocal effect in space and time; of course the effective equation does possess the finite propagation speed property.

More general questions and in particular nonlinear effects should be understood in order to explain turbulence effects for example.

Quite intricate hierarchies of corrections can occur in nonlinear situations, as can be seen with the following example[10], related to the equation

$$\frac{\partial u^\varepsilon}{\partial t} + (a^0(x,t) + \gamma b^\varepsilon(x,t))(u^\varepsilon)^2 = f(x,t); u^\varepsilon(x,0) = v(x). \tag{90}$$

Let us assume that a^0 and b^ε are uniformly bounded measurable functions, that

$$0 < \alpha \leq a^0(x,t) \text{ almost everywhere,} \tag{91}$$

$$b^\varepsilon \text{ is uniformly equicontinuous in t and converges weakly * to 0,} \tag{92}$$

and that the functions f and v are nonnegative, measurable and bounded. Then when the parameter γ is small, the solutions u^ε stay nonnegative and are defined for all t. After extracting a subsequence one has

$$u^\varepsilon \text{ converges to } U_0 + \gamma^2 U_2 + \gamma^3 U_3 + O(\gamma^4) \text{ in } L^\infty \textit{ weak } * \tag{93}$$

where U_0 is the solution of

$$\frac{\partial U_0}{\partial t} + a^0(x,t)(U_0)^2 = f(x,t); U_0(x,0) = v(x), \tag{94}$$

and U_2 is the solution of

$$\frac{\partial U_2}{\partial t} + 2a^0(x,t)U_0U_2 + Z_1 + Z_2 = 0; U_2(x,0) = 0, \tag{95}$$

where the nonlinear memory effects Z_1 and Z_2 are defined in the following way. One first extracts a subsequence such that

$$b^\varepsilon(x,s)b^\varepsilon(x,t) \text{ converges to } M_2(x,s,t) \text{ in } L^\infty \textit{ weak}*, \tag{96}$$

and then one defines M* and R* by the formulae

$$M^*(x,s,t) = M_2(x,s,t)U_0(x,s)U_0(x,t) \tag{97}$$

$$R^*(x,s,t) = U_0(x,s)exp(-2\int_s^t a^0(x,\tau)U_0(x,\tau)d\tau). \tag{98}$$

With these notations one has

$$Z_1(x,t) = -2\int_0^t M^*(x,s,t)R^*(x,s,t)ds \tag{99}$$

$$Z_2(x,t) = a^0(x,t)\int_0^t\int_0^t M^*(x,s,\sigma)R^*(x,s,t)R^*(x,\sigma,t)dsd\sigma. \tag{100}$$

In the equation for U_3 appears the function M_3 such that $b_\varepsilon(x,t_1)b_\varepsilon(x,t_2)b_\varepsilon(x,t_3)$ converges in L^∞ weak * to $M_3(x,t_1,t_2,t_3)$, and so on. A complete analysis would probably lead to similar procedures than that followed by theoretical physicists when they deal with their beloved diagrams.

QUASI-CRYSTALS

By quickly cooling some Al-Mn alloys, experimental physicists[11] discovered in 1984 a resulting material whose X-ray diffraction pattern showed an unexpected five-fold symmetry or icosahedral symmetry. Tiling games following the work of Roger PENROSE have often been played by theoretical physicists to generate average five fold symmetries, but they cannot cast any light upon what could have happened inside the material to create this strange observation.

If one submits a material to a combination of heat and stress, the material will change its microstructure in order to adapt itself to these new constraints: this is in essence what the blacksmith's art is about. If the different components of a mixture are allowed to rearrange themselves locally in order to optimize some criterium, one should find the optimal configuration by studying all the effective coefficients corresponding to given proportions and then optimize the criterium on this set of effective coefficients. Unfortunately, even for two-component mixtures, there is not yet a complete description of such a set of effective coefficients.

A reasonable guess is that optimal effective coefficients are usually on the boundary of the set of effective coefficients. In the case of small variations of elastic properties (and using the oversimplification of linearized elasticity), we have seen that H-measures can be used for computing a better approximation of the effective coefficients and that the formula used the set of moments of order 4 of a nonnegative measure on the sphere S^{N-1}. Even if thin ribbons have often been considered, they should be first thought as three-dimensional bodies in order to understand what happens inside them and so we should consider the case $N = 3$, although many tiling games have been played in the plane.

Of course, the preceding analysis would be useful if it was true that the results of X-ray diffraction experiments were connected to H-measures; strictly speaking it cannot be so as H-measures are defined without using any characteristic length, while for X-ray diffraction experiments it is important to select a wavelength related to the characteristic atomic distances. A variant of H-measures using a characteristic length has been introduced by Patrick GERARD[12] under the name of semi-classical measures, and it may be more appropriate for that question. Nevertheless, it might be useful to understand the structure of the set of moments of order 4 of nonnegative measures on the sphere S^2, as it appeared in the formula for computing second order effects, and check if five fold symmetry or icosahedral symmetry can indeed be related to the structure of this set of moments, which I have studied then with Gilles FRANCFORT and François MURAT[13].

In dimension 3, there are 15 moments of order 4, so the set of moments of order 4 of nonnegative measures on S^2 is a closed convex cone of R^{15}. One can show that points on the boundary of this convex cone can be obtained as moments of at least one measure which is combination of at most 5 Dirac masses, while a minimum of 6 is required for points in the interior.

The 15 moments of an isotropic distribution do not correspond to a boundary point. If a nonnegative measure has this list of moments and is a combination of only 6 Dirac masses, then the 6 points (and their antipodes) must be the vertices of a regular icosaedron, a geometry which had already been used by Gilles FRANCFORT and François MURAT[14] for constructing isotropic mixtures. On the other hand, the 15 moments of a transversally isotropic distribution may correspond to a boundary point. In such a case, if a nonnegative measure has this list of moments and is a combination of only 5 Dirac masses, then the 5 points (and their antipodes) must be the vertices of two regular pentagons.

At the moment, there is no obvious reason why only H-measures with the least number of Dirac masses for a given list of moments would play a role in such a problem. Understanding the correction in γ^3, or introducing a characteristic length in the definition of H-measures, could shed some light on this question.

CONCLUSION

YOUNG measures is a simple mathematical tool for describing questions of local statistics, too simple for solving the important questions of homogenization which are so crucial for understanding Physics. The introduction of H-measures is just a step forward in the construction of new mathematical tools for understanding more of these questions of Physics; H-measures appear quite useful for the computation of many quadratic corrections, but a better mathematical tool still has to be developed.

ACKNOWLEDGEMENTS

This research work was supported in part by NSF grants DMS-8803317 and DMS-91-00834.

REFERENCES

1. The codex Hammer of Leonardo da Vinci, Catalogue, Florence, Palazzo Vecchio, 1982.
2. YOUNG Laurence C., *Lectures on the Calculus of Variations*, W.B. Saunders, Philadelphia, 1969.
3. TARTAR Luc, H-Measures, a new approach for studying homogenisation, oscillations and concentration effects in partial differential equations, Proc. Royal Soc. Edinburgh, 115A, 193-230, 1990.
4. TARTAR Luc, H-measures and small amplitude homogenization, in *Random Media and Composites*, Robert V. KOHN & Graeme W. MILTON eds., 89-99, S.I.A.M., Philadelphia, 1989.
5. MURAT François & TARTAR Luc, On the relation between YOUNG measures and H-measures, in preparation.
6. BROWN William F., *Micromagnetics*, Interscience, New York, 1963.
7. JAMES Richard D. & KINDERLEHRER David, Frustration in ferromagnetic materials, Cont. Mech. Therm. 2, (1990), 215-239.
8. AMIRAT Youcef & HAMDACHE Kamal & ZIANI Abdelhamid, Homogénéisation d'équations hyperboliques du premier ordre et applications aux écoulements miscibles en milieu poreux, Ann. Inst. Henri Poincaré, Vol. 6, no. 5, (1989), 397-417.
9. TARTAR Luc, Nonlocal effects induced by homogenization, in *Partial Differential Equations and the Calculus of Variations, Essays in Honor of Ennio De Giorgi*, II, 925-938, Birkhaüser, Boston, 1989.
10. TARTAR Luc, Memory effects and homogenization, Arch. Rational Mech. Anal., 111 (1990), 121-133.
11. SCHECHTMAN Dany & BLECH Ilan & GRATIAS Denis & CAHN John W., Metallic phase with long-range orientation order and no translational symmetry, Phys. Rev. Lett. 53, 1951-1953, (1984).
12. GÉRARD Patrick, Mesures semi-classiques et ondes de BLOCH, in Equations aux Dérivées Partielles, Exposé XVI, Séminaire 1990-1991, Ecole Polytechnique, Palaiseau.
13. FRANCFORT Gilles & MURAT François & TARTAR Luc, in preparation.
14. FRANCFORT Gilles & MURAT François, Homogenization and optimal bounds in linear elasticity, Arch. Rational Mech. Anal., 94 (1986), 307-334.

A VARIATIONAL APPROACH TO PROBLEMS OF PLASTIC DEFORMATION

Primo Brandi, Anna Salvadori

Dipartimento di Matematica
Università degli Studi di Perugia
06100 Perugia, Italy

1.INTRODUCTION

We present some aspects of a research we have developed in a joint project with L.Cesari and W.H.Yang[1] on a variational approach to problems of plastic deformation (3,4,7e,9,16).

About ten years ago, Yang started working on the formulation of a mathematical model in plasticity. More precisely he deals with the deformation of a structure (e.g.,beams and plates) under a load in various situations:for example,the load can be distributed or concentrated, fixed or moving and the structure can be subjected to different constraints (e.g., it can be clamped, cantilever or simply supported). Yang's model is new and different from well-known analogous formulations,it proves to be highly efficient as regards its applications and fits the physical meaning very well. It consists of two optimization problems,in duality, which present non-trivial peculiarities. For this reason an ad hoc research (7e) was necessary to settle the model in a rigorous and operative mathematical framework (see also (9,3c,4)). We herein illustrate the abstract optimization problem introduced in (7e) and emphasize the applications of Yang's model for plastic deformation.

We wish to mention that the formulation of this variational problem, which involves constraints of a distributional type, was inspired both by the abstract optimal control problem, with distributed and boundary controls, introduced by Cesari in (6b) (see also (6c,8)) and by the research regarding discontinuous solutions for simple and multiple integrals of the calculus of variations that we have developed over recent years (7a,b,c,d).

2.PRESENTATION OF THE MODEL

In order to illustrate Yang's model and point out the main mathematical difficulties it produces, a particular and very simple case is sufficient (see (16e)).

Let us consider a clamped beam of length 2L subjected to a uniform load q. The first (or primal) problem takes into account the static and the matherial properties, the dual problem deals with the kinematics, i.e., the geometry of deformation.

---

[1] W.H.Yang is Professor of Applied Mechanics at the University of Michigan

Let $M : [-L,L] \to \mathbb{R}$ denote the bending moment of the beam. The equilibrium equation has the form

1) $$\frac{d^2M}{dx^2}(x) = q \quad , \quad x \in [-L,L]$$

and no boundary condition is prescribed since the beam is clamped.

The asymptotic matherial behaviour is modeled by the inequality

2) $$|M(x)| \le M_o \qquad x \in [-L,L]$$

where M_o is a constant which depends on the matherial.

The variational formulation consits of maximizing load q under constraints 1) and 2), i.e.,

(P) $$\begin{aligned} &\sup q \\ &\frac{d^2M}{dx^2}(x) = q \quad , \quad x \in [-L,L] \\ &|M(x)| \le M_o \quad , \quad x \in [-L,L] \end{aligned}$$

Now it is easy to see that the solution of problem (P) is

$$\bar{M}(x) = 2M_o x^2 - M_o \qquad \bar{q} = 4M_o/L^2.$$

Let us now take into consideration the deformation of the beam under the maximum load. Denoted by $w : [-L,L] \to \mathbb{R}$ the transverse velocity of the beam, observe that the clamped beam has four kinematic boundary conditions

3) $$w(-L) = w(L) = 0 \quad , \quad \frac{dw}{dx}(-L) = \frac{dw}{dx}(L) = 0.$$

According to Yang's formulation, equation 1) is written in the weak form

$$\int_{-L}^{L} \frac{d^2M}{dx^2}(x)\, w(x)\, dx = q\int_{-L}^{L} w(x)\, dx.$$

Then, integrating by parts and taking into account 3) and 2), we obtain

4) $$q \le \int_{-L}^{L} \left|\frac{d^2w}{dx^2}(x)\right| dx$$

where we have assumed, without loss of generality, that $\int_{-L}^{L} w(x)\, dx = 1$.

Thus, searching for the best approximation in relation to 4), we obtain the dual problem

(P') $$\begin{aligned} &\inf M_o \int_{-L}^{L} \left|\frac{d^2w}{dx^2}(x)\right| dx \\ &\int_{-L}^{L} w(x)\, dx = 1 \\ &w(-L)=w(L)=0 \quad , \quad \frac{dw}{dx}(-L)=\frac{dw}{dx}(L)=0 \end{aligned}$$

Problem (P') seeks the minimizer $\bar{w}$ among all kinematically admissible functions. The physical meaning suggests that the following is the optimal solution:

$$\bar{w}(x) = -\frac{x}{L^2} + \frac{1}{L} \quad , \text{ for } x \in [0,L] , \quad \bar{w}(x) = \frac{x}{L^2} + \frac{1}{L} \quad , \text{ for } x \in [-L,0] .$$

Moreover, the two dual problems must have no gaps, i.e., max P = min P'. This implies that

$$\int_{-L}^{L} \left|\frac{d^2\bar{w}}{dx^2}(x)\right| dx = \frac{4}{L^2}.$$

Of course this result needs a suitable interpretation. First of all it may seem that the slope boundary conditions are violated. We can overcome this difficulty by admitting discontinuities in the first derivative and putting $\frac{d\bar{w}}{dx}(-L) = \frac{d\bar{w}}{dx}(L) = 0$. In other words, we interpret that $\bar{w}$ approaches the clamped boundary with a finite slope and then jumps to zero slope at the end-points, as required. In this context it could be natural to interpret the integral functional to minimize as the generalized variation of the first derivative. Unfortunately $V(\frac{dw}{dx}) = 2/L^2$. In order to obtain the requested value, i.e., $4/L^2$, we must replace V with a modified variation which takes into account the jumps at the end-points (see Section 4c)).

We wish to observe that in plasticity, unlike in elasticity or fluid mechanics, kinematically admissible non-smooth solutions are allowed. Furthermore, we shall merely state that in this case primal problem ($\mathcal{P}$) admits a smooth solution but, in the case of a concentrated load, equation 1) must be interpreted in a weak form and M, or its derivatives, may admit discontinuities.

3. THE ABSTRACT FORMULATION

The previous example points out that Yang's variational approach presents the following main mathematical difficulties: both the dual optimization problems involve differential elements of higher order and present constraints which can be of a distributional type; moreover, non-smooth solutions are expected with a consequent difficult interpretation of the boundary conditions.

We have chosen to frame all the different situations proposed by Yang (16) in a unitary and general formulation. As a theorical support for this topic, an abstract optimization problem, which involves constraints of a distributional type, has been taken into consideration in (7e).

More precisely, given a functional $I : W \to \mathbb{R} \cup (+\infty)$, where W is a subset of a reflexive, separable Banach space S, we denote by $J : \bar{W} \to \mathbb{R} \cup (+\infty)$ the Serrin-type weak extension of I to $\bar{W}$ (the weak closure of W). Moreover, let $\mathcal{F} : \bar{W} \to \mathcal{D}'$ be a given functional, where $\mathcal{D}'$ as usual denotes the space of the distributions over a bounded open set $G \subset \mathbb{R}$.

We consider the minimization problem

$$(\mathcal{P}) \qquad \min_{u \in \Omega} J(u)$$

where $\Omega \subset \bar{W}$ is a class of elements which satisfy the distributional relation

$$(c) \qquad \mathcal{F}u = 0.$$

Problem ($\mathcal{P}$) can be written $\min(J(x) + g(x))$, where $g: \bar{W} \to \{0, +\infty\}$ is the function defined by $g(u)=0$ if (c) is satisfied, $g(u)=+\infty$ elsewhere.

The existence result for problem ($\mathcal{P}$) can be proved as a suitable application of Tonelli's direct method of the calculus of variations.

THEOREM 1. Assume that the following conditions are satisfied[1]*:*

i) for any sequence $(u_k)_k$ *in* Ω_0 *which* w*-converges to a function* u*, a subsequence* $(j_k)_k$ *exists such that* $(\mathcal{F}u_{j_k})_k$ w* *converges to* $\mathcal{F}u$*;*

[1] As usual, we denote by w and w* the weak topology of S and the weak-star topology of $\mathcal{D}'$, respectively. Moreover, we put $\Omega_0 = \{u \in \Omega : \mathcal{F}u = 0\}$.

ii) functional I *is lower bounded and* w*-lower semicontinuous on* $\Omega\cap W$;

iii) class Ω *is bounded and if* $(u_k)_k$ *is a sequence in* $\Omega\cap W$ *which* w*-converges to a function* u, *then* $u\in\Omega$.

Then an element $u_\circ\in\Omega$ *exists such that*

$$J(u_\circ) + g(u_\circ) = \min_{\Omega} (J(u) + g(u)) = i$$

Moreover, if $i < +\infty$, *then we also have* $g(u_\circ) = 0$ *and*

$$J(u_\circ) = \min_{\Omega} (J(u) + g(u)) = \min_{\Omega_\circ} J(u) \geq \inf_{\Omega\cap W} I(u).$$

REMARK. Note that the w-continuity of *F* ensures assumption i). Moreover, assumption iii) is trivially satisfied provided Ω is bounded and w-closed.

4. ON SOME NOTEWORTHY CASES

Let us consider some particularly interesting cases of our abstract formulation.

4a) On functional I

Let $U\colon \overline{W} \to L_p(G)$ and $L\colon \overline{W} \to M$ be given operators with $L(W) \subset L_q(G)$, $p,q \geq 1$, where M denotes the space of the finite Borel measures over G. Let $F_\circ\colon A\times \mathbb{R}^N \to \mathbb{R}\cup(+\infty)$ be a function with the property that, for every $u\in L_p(G)$ and $v\in L_q(G)$, $F_\circ(\cdot,u(\cdot),v(\cdot))$ is measurable, where $A\subset\mathbb{R}^{\nu+n}$ is given.

Let us take into consideration the Lebesgue functional

$$I(u) = \int_G F_\circ(x, (Uu)(x), (Lu)(x)) \, dx$$

Note that it can be considered as a rather general version of the classical Lebesgue integral of the calculus of variations. We recall that, in the classical setting, functionals of this type have been studied by E.Rothe (13), G. Fichera (11) and by L.Cesari and D.E.Cowles (6b,c;8) in optimal control theory. For recent results concerning the w-lower semicontinuity of the present functional we refer to (3b).

Examples of possible operators *U* and *L* are the following:

1) $Uu=u$ $\quad Lu=Du$ with $W = W^{1,1}(G)$, where Du denotes the gradient;
2) $Uu=(u,Du,D^2u,\dots,D^nu)$, $Lu=D^{n+1}u$ with $W = W^{n+1,1}(G)$, where D^ju denotes the gradient of $D^{j-1}u$, $j=2,\dots,n$;
3) Uu= a linear combination of the derivatives up to order n, Lu= a linear combination of the derivatives of order n+1 with $W = W^{n+1}(G)$.

4b) On the constraint $Fu=0$

When integral functional I is of the type in 4a), then it is natural to take into consideration a distributed constraint which involves operator *L*. For example we can have:

1) $Lu = \mu$, where $\mu\in M$;
2) $(Lu)(x) = f(x,(Uu)(x))$ a.e. in G, where $f\colon G\times\mathbb{R}^n \to \mathbb{R}^N$ is a given function.

In order to satisfy distributional constraint 1), the solutions (or their differential elements) must be necessarily discontinuous, not Sobolev's. As we have shown a natural framework is BV, i.e., the space of the functions of bounded generalized variation (see (7e)).

Note that, when applied to problems of plasticity, equations 1) and 2) fit to model the equilibrium equation under all the possible loads. In particular, if μ is concentrated then it represents a concentrated load which will be fixed or moving according to whether the support of μ is fixed or not. Furthermore, equation 2) allows us to deal with a distributed load, which becomes uniform provided f is a constant function.

But our abstract formulation also considers different constraints. For example an integral constraint of the type

$$3) \quad \int_G (Lu)(x)\, dx = \text{const.}$$

which may be present in dual problem (P').

4c) On class Ω

Class Ω is usually characterized by distributed constraints of the type $(x,(Uu)(x))\varepsilon A$, a.e. in G, and by boundary conditions.

Since we are interested in applying our theory to problems where non-smooth solutions are expected, then the main difficulty regarding class Ω is a suitable interpretation of the boundary conditions. More precisely, these boundary conditions (as observed in Section 2) must take into account the values attained on the boundary and, moreover, they must be preserved under w-convergence. Of course these properties force us to modify the field of solutions. In fact, even if BV is a good setting to handle distributed constraints (as we have already observed in Section 4b)), it is not efficient for modeling boundary conditions. This is because a BV function neglects the subsets of null measures, as the boundary is in general.

For this reason a modified definition of generalized variation, which is computed taking note of the values attained on a given subset Γ of the boundary, has been introduced (see Section 5). The corresponding space BV_Γ proves to be a suitable setting to deal with both distributed and boundary constraints (see (7e)). In this framework the boundary conditions are interpreted exactly as boundary data, rather than a kind of trace. Note that this idea, even if not developed in detail , was adopted in our previous papers with Cesari (7a,b,c,d) to model boundary conditions for discontinuous BV solutions of optimization problems.

5. THE Γ-VARIATION

We briefly illustrate here the concept of Γ-variation which was introduced in (3a) and adopted in (7e) to frame the boundary conditions.

For the sake of semplicity, we herein consider only the one dimensional case. We start by recalling the definition of a spectrum of variations that was introduced in (2) for a summable function $u:[a,b] \to R$. Let N denote a family of subsets of null measure in $[a,b]$, which is closed under countable unions. We shall write N-a.e. when we refer to a neglected set of the family N. Moreover we put

$$V_N(u) = \min_{N \varepsilon N} V(u,N)$$

where V(u,N) is the variation computated by disregarding set N.

If $V_N(u) < +\infty$, we shall say that u is of bounded N-variation and we denote the space of such functions by N-BV. Of course two functions which coincide N-a.e. have the same N-variation, and $u \varepsilon N$-BV iff it is equal N-a.e. to a function of bounded variation .

Analogously, we shall denote by N-$W^{1,1}$ the space of the functions which are equal N-a.e. to a function in $W^{1,1}$.

Note that if $N_1 \subset N_2$, then $V_{N_2}(u) \le V_{N_1}(u)$. Moreover $V_\emptyset$ is the classical Jordan variation, while if N is the family of all the null sets, then V_N is the generalized variation (6,12). In other words $\{V_N(u), N\}$ is a spectrum of variations for u whose lower and upper lines are the generalized variation and the Jordan variation respectively.

Each variation has properties which are analogous to those of the generalized variation (see (2) for details). We only mention the lower semicontinuity result.

THEOREM 2. Let $(u_k)_{k\geq o}$ *be a sequence of summable functions such that* $u_k \to u_o$ N-a.e., *then* $\liminf_{k\to\infty} V_N(u_k) \geq V_N(u_o)$.

In particular we are interested here in the line of the spectrum which takes into account the values attained by u at the end-points. Let Γ be a given subset of boundary $\{a,b\}$, we denote by N_Γ the family of all the null sets in $[a,b]-\Gamma$. In brief we write $V_{N_\Gamma} = V_\Gamma$ and call it the Γ-variation. Moreover, we put $N_\Gamma\text{-}BV = BV_\Gamma$ and $N_\Gamma\text{-}W^{1,1} = W^{1,1}_\Gamma$.

Observe that, by virtue of Theorem 2, we have that V_Γ is lower semicontinuous with respect to the metric

$$d(u,v) = \|u - v\|_{L_1} + \max_{t\varepsilon\Gamma} |u(t) - v(t)|$$

Furthermore it can be proved that space BV_Γ is the d-closure of $W^{1,1}_\Gamma$.

We mention that the following comparison between V_Γ and the generalized variation V holds (see (3a)):

$$V_\Gamma(u) = V(u) + j(\Gamma)$$

where $j(\Gamma)$ is the "jump" at the boundary, more precisely $j(\Gamma) = \sum_{x\varepsilon\Gamma} |u(x) - \lim\text{ess}_{y\to x}\, u(y)|$. Of course $V_\Gamma(u) = V(u)$ provided $u\varepsilon W^{1,1}_\Gamma$.

Finally, we state a compacteness result which is analogous to the Cafiero-Fleming theorem (5).

THEOREM 3. Let $(u_k)_k$ *be a sequence of summable functions such that*

$$\sup_k \{d(u_k,0) + V(u_k)\} < +\infty,$$

then a subsequence (still called $(u_k)_k$ *) and a function* $u_o\varepsilon BV_\Gamma$ *exist such that*

$$\lim_{k\to\infty} d(u_k,u_o) = 0.$$

6. APPLICATIONS TO PROBLEMS OF PLASTIC DEFORMATION

Despite a theoretical interest, of course the main aim of the present research is to apply it to practical purpose. It finds its source in a recent study by Cesari-Yang (9a,b) on a variational approach to problems of plastic deformation, where some aspects of the theory are reconsidered in a new light. More precisely, our abstract formulation extends and generalizes the primal and dual problems considered in (9a) as a model for the plastic deformation of beams and plates. In particular, an application of our general existence Theorem 1 allows us to solve in detail the following models (see (7e)): clamped, cantilever and simply supported beams or plates under a distributed or concentrated load. We wish to mention that our solutions completely agree with the physical meaning of possible non-smooth solutions in plasticity.

In order to illustrate one of the above mentioned applications, let us go back to the example proposed in Section 2. Of course problem (P') is more interesting for our purpose since it involves boundary conditions. We have proposed a solution $\overline{w}$ of such a problem which agrees with the physical meaning and we have roughly suggested a possible interpretation which also fits the mathematical meaning. Now we are ready to prove that $\bar{w}$ is a rigorous solutions of our mathematical model.

We take this opportunity to point out the fundamental role of space BV_Γ in the applications.

Let $G = [-L,L]$, $\Gamma = \{-L,L\}$. We consider the space $S = W^{1,2}$ and the subset $W = \{w \varepsilon S : w \varepsilon W_\Gamma^{1,1}, \frac{dw}{dx} \varepsilon W_\Gamma^{1,1}, \|w\|_{W^{1,2}} \leq K, V(\frac{dw}{dx}) \leq K, |w(\pm L)| \leq K, |\frac{dw}{dx}(\pm L)| \leq K\}$, where K is a given constant. Note that for every $w \varepsilon W$ we also have $V_\Gamma(w) = V(w) \leq K$ and $V_\Gamma(\frac{dw}{dx}) = V(\frac{dw}{dx})$.

By virtue of the Rellich-Kondrachov theorem (1) and compacteness Theorem 3, the following result holds.

THEOREM 4. For every sequence $(w_k)_k$ *in* W *a subsequence (still called* $(w_k)_k$*) and a function* $w_o \varepsilon BV_\Gamma$ *exist such that*

$$\lim_{k\to\infty} \{d(w_k, w_o) + d(\frac{dw_k}{dx}, \frac{dw_o}{dx})\} = 0 .$$

As a consequence, we get that every $w \varepsilon \bar{W}$ satisfies the conditions:

$$w \varepsilon BV_\Gamma, \ \frac{dw}{dx} \varepsilon BV_\Gamma, \ \|w\|_{W^{1,2}} \leq K, \ V_\Gamma(\frac{dw}{dx}) \leq K, \ V_\Gamma(w) \leq K, \ |w(\pm L)| \leq K, \ |\frac{dw}{dx}(\pm L)| \leq K.$$

Now let $\Omega \subset \bar{W}$ be the class of the functions which satisfy the constraints

1) $\int_{-L}^{L} w(x)\, dx = 1$,

2) $w(-L) = w(L) = 0, \ \frac{dw}{dx}(-L) = \frac{dw}{dx}(L) = 0.$

Let us take $I : W \to \mathbb{R}$ defined by $I(w) = \int_{-L}^{L} |\frac{d^2w}{dx^2}(x)|\, dx = V(\frac{dw}{dx})$.

Of course I is w-lower semicontinuous and it can be proved that

$$J(w) = V_\Gamma(w) \quad , w \varepsilon \bar{W}.$$

An application of our existence Theorem 1, allows us to prove the existence of a function $w_o \varepsilon \Omega$ such that

$$J(w_o) = \min_{w \varepsilon \Omega} J(w).$$

Of course function $\bar{w} \varepsilon \Omega$ and it is easy to prove that

$$J(\bar{w}) = V_\Gamma(\frac{d\bar{w}}{dx}) = 4/L^2.$$

7. REFERENCES

1. R.A.Adams, Sobolev Spaces, Academic Press (1975).

2. P.Brandi, Uno spettro di variazioni con peso nei sottoinsiemi di $\tilde{\mathbb{R}}$. Teoremi di semicontinuità, *Rend.Circ.Mat.Palermo*, 26:165 (1977).

3. P.Brandi and A.Salvadori, (a) The Γ-variation, *Dip.Mat.Univ.Perugia preprint*, 23 (1991);
(b) On the lower semicontinuity for weak topologies, *Dip.Mat.Univ.Perugia preprint*, 22(1991);
(c) Existence theorems for the dual problem in plastic deformation, to appear;
(e) Discontinuous solutions in plastic deformation, *Atti Sem.Mat.Fis. Univ.Modena*, to appear.

4. P.Brandi, A.Salvadori and W.H.Yang, Second order problems in plasticity with general distributional input, to appear.

5. H.Brezis, Analyse Fonctionnelle. Theorie et Applications, Masson (1984).

6. L.Cesari, (a) Sulle funzioni a variazione limitata, *Ann Scuola Norm. Sup.Pisa*, 5:299 (1936);
(b) Existence theorems for abstract multidimentional control problems, *J.Opt.Theor.Appl.*, 6:21o (1970);
(c) Closure theorems for orientor fields and weak convergence,*Arch.Rat. Mech.Anal.*, 55:332 (1974);
(e) OptimizationTheory and Applications, Springer-Verlag (1983).

7. L.Cesari, P.Brandi and A.Salvadori, (a) Discontinuous solutions in problems of optimization, *Ann.Scuola Norm.Sup.Pisa*, 15:219 (1988);
(b) Existence theorems concerning simple integrals of the calculus of variations for discontinuous solutions,*Arch.Rat.Mech.Anal.*, 98:307 (1987);
(c) Existence theorems for multiple integrals of the calculus of variations for discontinuous solutions, *Ann.Mat.Pura Appl.*, 152:95 (1988);
(d) Seminormality conditions in calculus of variations for BV solutions, *Ann.Mat.Pura Appl.*, to appear;
(e) Existence theorems for multiple integrals with distributional constraints, *Dip.Mat.Univ.Perugia preprint*, 21 (1991).

8. L.Cesari and D.E.Cowles, Existence theorems in multidimensional problems of optimization with distributed and buondary controls,*Arch.Rat.Mech.Anal Anal.*, 46:321 (1972).

9. L.Cesari and W.H.Yang, (a) Serrin's integrals and second order problems of plasticity, *Proc.Royal Soc.Edimburg*, 117:193 (1991);
(b) BV functions and first order problems of plasticity, *J.Opt.Theor. Appl.*, to appear.

10.I.Ekeland and R.Temam, Analyse Convexe et Problemes Variationnels, Dunod, Gauthier-Villars (1974).

11.G.Fichera, (a) Semicontinuity of multiple integrals in ordinary form,*Arch. Rat.Mech.Anal.*, 17:339 (1964);
(b) Semicontinuità ed esistenza del minimo per una classe di integrali multipli, *Rev.Roum.Math.Pures Appl.*, 12:1217 (1967).

12.K.Krickeberg, Distributionen, functionen beschrankter variation und Lebesguescher inhalt nichterparameterischer flachen, *Ann.Mat.Pura Appl.*, 44: 105 (1957).

13.E.H.Rothe, An existence theorem in the calculus of variations based on Sobolev's imbedding theorems,*Arch.Rat.Mech.Anal.*, 21:151 (1965).

14.J.Serrin, On the definition and properties of certain variational integrals, *Trans.Amer.Math.Soc.*, 101:139 (1961).

15.R.Temam, Mathematical Problems in Plasticity, Gauthier-Villars (1983);

16.W.H.Yang,(a) Minimization approach to limit solutions of plates, *Comp. Meth.Appl.Mech.Eng.*, 28:265 (1981);
(b) A variational principle and an algorithm for limit analysis of beams and plates, *Comp.Meth.Appl.Mech.Eng.*, 33:575 (1982);
(c) A duality theorem for plastic plates, *Acta Math.*, 69:177 (1987);
(d) Asymptotic analysis for some limit solutions of plates with concentrated load, to appear;
(e) Calculus of Variations in plasticity, *Atti Sem.Mat.Fis.Univ.Modena*, to appear.

SOME REGULARITY RESULTS FOR A NONLINEAR ELLIPTIC EQUATION

Anna Mercaldo

Dipartimento di Matematica ed Applicazioni
Via Cintia - Complesso di Monte S. Angelo
80126 Napoli - Italia.

Let us consider the following nonlinear elliptic problem:

$$\begin{cases} Au = -\Delta_p u - (b_i(x)|u|^{p-2}u)_{x_i} + h(x,u) = (f_i)_{x_i} & \text{in } \Omega \\ u \in W_0^{1,p}(\Omega) \end{cases} \tag{1}$$

where $p > 1$, $\Omega \subset R^n$ is a bounded open set, $\Delta_p u = \operatorname{div}(|\nabla u|^{p-2}\nabla u)$, $b_i(x)$ and $f_i(x)$ are in suitable Lorentz spaces* and $h(x,u)$ is a nonlinear term satisfying the sign condition $h(x,u)u \geq 0$.

Regularity, existence and uniqueness results for a solution of (1) when $p = 2$, the coefficients $b_i(x)$ and the data $f_i(x)$ are in Lebesgue spaces or in Lorentz spaces are well known (see e.g. [8], [6], [1]).

When $p \neq 2$, regularity results for (1) with $b_i(x) = 0$ and $f_i(x)$ in Lebesgue spaces are in [4], while in [2] the case $b_i(x)$ and $f_i(x)$ in suitable Lorentz spaces is considered.

An existence result for the problem (1) is also proved in [2], where it is shown that a solution u exists if the norm of coefficients $b_i(x)$ are sufficiently "small".

In this work, a priori bounds for the solution u of (1) in Lorentz spaces, in Orlicz spaces and $L^\infty(\Omega)$ are presented.

The following theorem gives estimates for the Lorentz norm of u, that generalize known results ([2], [4]).

---

* If $1<p<\infty$ and $1\leq q\leq\infty$ the Lorentz space $L(p,q)$ is the class of functions u such that:

$$\|u\|_{p,q}=\left(\int_0^\infty [u^*(s)s^{1/p}]^q \frac{ds}{s}\right)^{1/q}<\infty,$$

$$\|u\|_{p,\infty}=\sup u^*(s)s^{1/p}<\infty.$$

where $u^*(s)$ is the decreasing rearrangement of u (for more detailes see [7]).

Developments in Partial Differential Equations and Applications to Mathematical Physics, Edited by G. Buttazzo *et al.*, Plenum Press, New York, 1992

Theorem 1. *Let $u \in W_0^{1,p}(\Omega)$ be a solution of the problem (1) and $s = [q(p-1)]^*$. Then:*

a) *if $b(x) \in L\left(\frac{n}{p-1}, \infty\right)$ that is $b^{\#}(x) \leq \frac{B^{p-1}}{|x|^{p-1}}$,*

where $b^{\#}(x)$ is the spherically rearrangement of $b(x) = \left(\sum_i |b_i(x)|^2\right)^{1/2}$,

and if $f(x) = \left(\sum_i |f_i(x)|^2\right)^{1/2} \in L\left(q, \frac{k}{p-1}\right)$ with

$$p' \leq q < \frac{n}{(p-1)(B+1)} \qquad\qquad p \leq k \leq q(p-1),$$

then $u \in L(s,k)$ and

$$\|u\|_{s,k} \leq C_1 \|f\|_{q,k/(p-1)}^{p'/p}$$

with C_1 constant depending on n, B, k, q and p;

b) *if $b(x) \in L\left(\frac{n}{p-1}, \frac{r}{p-1}\right)$ with $n \leq r < \infty$ and $f(x) \in L\left(q, \frac{k}{p-1}\right)$ with*

$$p' \leq q < \frac{n}{p-1} \qquad\qquad p \leq k \leq q(p-1),$$

then $u \in L(s,k)$ and

$$\|u\|_{s,k} \leq C_2 \|f\|_{q,k/(p-1)}^{p'/p}$$

with C_2 constant depending on $n, |\Omega|, k, r, q$ and p.

To obtain a priori bounds for the norm of u in Orlicz space $L_\phi(\Omega)$ with $\phi(t) = \exp\left(|t|^{n/(n-1)}\right) - 1$ and $L^\infty(\Omega)$, we assume that the coefficients b_i belong to $L^{r/(p-1)}(\Omega)$ with $r > n$ and that f is in $L^q(\Omega)$ with $q \geq n/(p-1)$.

Theorem 2. *Let $u \in W_0^{1,p}(\Omega)$ be a solution of the problem (1). If $b_i(x) \in L^{r/(p-1)}(\Omega)$ with $r > n$, then:*

1) *if $f(x) \in L^{n/(p-1)}(\Omega)$ then $u \in L_\phi(\Omega)$ with $\phi(t) = \exp(|t|^{n/(n-1)}) - 1$ and*

$$\|u\|_\phi \leq C_3 \|f\|_{n/(p-1)}^{p'/p}$$

with C_3 constant depending on $n, |\Omega|, r$ and $\|b\|_{r/(p-1)}$;

2) *if $f(x) \in L^q(\Omega)$ with $q > n/(p-1)$, then $u \in L^\infty(\Omega)$ and:*

$$\|u\|_\infty \leq C_4 \|f\|_q^{p'/p}$$

with C_4 constant depending on $n, |\Omega|, p$ and $\|b\|_{r/(p-1)}$.

Previous theorem is proved in [5] when $b_i(x) = 0$.

The proofs of theorems 1 and 2 are achieved by using, in different way, the following comparison result ([3]) in which the spherically rearrangement of the solution u of (1) is compared with the solution v of a suitable spherically symmetric elliptic problem. Using the expression of v, we can evaluate the norm of the solution of the problem (1) and hence to derive a priori bound for u.

Theorem 3. *Let* $u \in W_0^{1,p}(\Omega)$ *be a solution of* (1) *with the assumption that* $b(x) \in L\left(\frac{n}{p-1}, \infty\right)$ *and* $f(x) \in L^{p'}(\Omega)$. *Then:*

$$u^{\#}(x) \leq v(x) \qquad \text{a.e.} \quad x \in \Omega^{\#}, \tag{2}$$

where

$$v(x) = \int_{C_n|x|^n}^{|\Omega|} \frac{F(t)^{p'/p}}{nC_n^{1/n} t^{1-1/n}} \exp\left(\int_{C_n|x|^n}^{t} \frac{B(r)^{p'/p}}{nC_n^{1/n} r^{1-1/n}} dr\right) dt \tag{3}$$

is the weak solution in $W_0^{1,p}(\Omega^{\#})$ *of the linear equation:*

$$-\Delta v - \left(B(C_n|x|^n)^{p'/p} v \frac{x_i}{|x|}\right)_{x_i} = \left(F(C_n|x|^n)^{p'/p} \frac{x_i}{|x|}\right)_{x_i} \quad \text{in} \quad \Omega^{\#} \tag{4}$$

and $B(s)^{p'}$, $F(s)^{p'}$ *are the functions such that*

$$\int_{|u|>t} f(x)^{p'} dx = \int_0^{\mu(t)} F(s)^{p'} ds, \qquad \int_{|u|>t} b(x)^{p'} dx = \int_0^{\mu(t)} B(s)^{p'} ds.$$

These results hold also when the operator $-\Delta_p$ is replaced by a Leray-Lions operator. The details of their proofs are in [3].

REFERENCES

[1] A. Alvino, P. Buonocore, G. Trombetti, *On Dirichlet problem for second order elliptic equations,* Nonlinear Analisys T.M.A., (7) **14** (1990), pp. 559-570.

[2] M. F. Betta, A. Mercaldo, *Existence and regularity results for a nonlinear elliptic equation,* to appear on Rend. Mat. Appl.

[3] M. F. Betta, A. Mercaldo, *Comparison and regularity results for a nonlinear elliptic equation,* to appear on Nonlin. Anal. T.M.A.

[4] L. Boccardo, D. Giacchetti, *Alcune osservazioni sulla regolarità delle soluzioni di problemi fortemente non lineari e applicazioni,* Ricerche di Mat., **34** (1985), pp. 309-323.

[5] E. Giarrusso, D. Nunziante, *Regularity theorems in limit cases for solutions of linear and nonlinear elliptic equations,* Rend. Inst. Mat. Univ. Trieste, **20** (1988), pp. 39-58.

[6] E. Maz'ja, *On weak solution of the Dirichlet and Neumann problems,* Trans. Moscow Math. Soc., **20** (1965), pp. 135-171.

[7] R. O'Neil, *Integral tranforms and tensor products on Orlicz spaces and* $L(p,q)$ *spaces,* J. Analyse Math., **21** (1968), pp. 1-276.

[8] G. Stampacchia, *Le probléme de Dirichlet pour le équations elliptiques du second ordre à coefficients discontinus,* Ann. Inst. Fourier (Grenoble), **15** (1965), pp. 189-258.

ON THE STATIONARY OSEEN EQUATIONS WITH NONVANISHING DIVERGENCE IN EXTERIOR DOMAINS OF $\mathbb{R}^3$

Reinhard Farwig

Institut für Mathematik
RWTH Aachen
Templergraben 55
D–5100 Aachen, Germany

INTRODUCTION

Consider the Navier-Stokes equations describing the flow of an incompressible and viscous fluid past one or several obstacles in $\mathbb{R}^3$ such that the velocity converges to a vector u_∞ at infinity. In the most interesting case where $u_\infty \neq 0$ we know by experiments that the flow gets turbulent in a region behind the obstacle, in the so–called wake. Linearizing the Navier-Stokes equations in the velocity field with respect to the constant vector $u_\infty = k(1,0,0)$, $k > 0$, we get the Oseen equations [10]

$$(1) \qquad \begin{cases} -\nu\Delta u + k\partial_1 u + \nabla p = f & \text{in } \Omega \\ \operatorname{div} u = g & \text{in } \Omega \\ u = 0 & \text{on } \partial\Omega \end{cases}$$

where $u(x) \to 0$ at infinity. Here $\Omega \subset \mathbb{R}^3$ denotes an exterior domain with boundary $\partial\Omega$ of class C^2, and $\partial_1 = \partial/\partial x_1$. We include a prescribed nonvanishing divergence g, since a priori estimates of the generalized Oseen equations (1) are a crucial step in the study of the stationary Navier-Stokes equations of a compressible, viscous fluid in an exterior domain, see e.g. Matsumura, Nishida [9], when $u_\infty = 0$, and Padula [11], when $u_\infty \neq 0$. Contrary to the exterior Stokes problem there are only few papers on the exterior Oseen equations (with $g = 0$). Faxén [6] and Bemelmans [2] used the theory of hydrodynamical potentials to solve the homogeneous Oseen equations, i.e. when $f = 0$ and $g = 0$, while Finn [7] exploited the skew–symmetry of the operator ∂_1 and investigated Green's function. If $g = 0$ and $\Omega = \mathbb{R}^n$, let $u = Sf$ denote the solution of (1). Babenko [1] was the first to observe that $\partial_1 S$ is a bounded linear operator on $L^q(\mathbb{R}^3)^3$ for each $q \in (1, \infty)$. This consequence of the Lizorkin–Marcinkiewicz multiplier theorem is also used by Galdi [8] proving that for each $f \in L^q(\Omega)^n$, $\Omega \subset \mathbb{R}^n$ an exterior domain, there is a unique solution $(u, \nabla p)$ of (1) such that $D^2 u$ and $\partial_1 u$ are in $L^q(\Omega)$.

In our approach to the three–dimensional Oseen equations (1) we essentially exploit the anisotropic behavior of its fundamental solution [10]

$$(2) \qquad E(x) = \big(E_{ij}(x)\big) = \big((\delta_{ij}\Delta - \partial_i\partial_j)\Phi(x)\big), \qquad e(x) = -\nabla(1/4\pi r),$$

where

$$\Phi(x) = \frac{1}{4\pi k}\int_0^{ks(x)/2\nu} \frac{1-e^{-\alpha}}{\alpha}\,d\alpha \quad \text{with} \quad s(x) = r - x_1 \geq 0, \quad r = |x|.$$

Developments in Partial Differential Equations and Applications to Mathematical Physics, Edited by G. Buttazzo *et al.*, Plenum Press, New York, 1992

Thus $|E(x)| = O\big((1+r)^{-1}(1+s)^{-1}\big)$. In particular $|E(x)| \sim 1/r$ in the wake region $\{x \in I\!R^3;\ s(x) \le 1\}$, which asymptotically is a paraboloid, but $|E(x)| \sim 1/r^2$ for each ray $\{x = t\omega;\ t \ge 0\}, \omega \in S_2 \setminus \{(1,0,0)\}$. Therefore we introduce the anisotropic weights

$$\eta_b^a(x) = (1+r)^a(1+s)^b, \qquad a,\, b \in I\!R,$$

and the weighted L^2-spaces

$$L^2_{a,b}(\Omega) = \Big\{ u \in L^2_{loc}(\Omega);\ \big\|u\big\|_{a,b} := \Big(\int_\Omega dx\, \big|u\big|^2 \eta_b^a\Big)^{1/2} < \infty \Big\},$$

in which we want to solve (1). Note that $\eta_{-b}^{-a} \in L^1(I\!R^3)$ iff $a > 2$ and $a + b > 3$.

THE OSEEN EQUATIONS IN WEIGHTED SOBOLEV SPACES

Theorem 1. *Let $\Omega \subset I\!R^3$ be an exterior domain with boundary of class C^2. Further let $0 \le |\alpha| < \beta < 1$. Then for every $f \in L^2_{\alpha+1,\beta}(\Omega)^3$ and $g \in L^2_{\alpha+1,\beta}(\Omega)$ with $\nabla g \in L^2_{\alpha+1,\beta}(\Omega)^3$ there is a unique solution (u,p) of (1) with $u \in L^2_{\alpha-1,\beta}(\Omega)^3$. Further there is a constant $c > 0$ such that*

$$\big\|u\big\|_{\alpha-1,\beta} + \big\|Du\big\|_{\alpha,\beta} + \big\|(\partial_1 u,\, D^2 u,\, \nabla p)\big\|_{\alpha+1,\beta} + \big\|p\big\|_{\alpha,\beta-1} \le c\big\|(f,\, g,\, \nabla g)\big\|_{\alpha+1,\beta}. \tag{3}$$

Proof. Extending f and g onto $I\!R^3 \setminus \Omega$ we consider the Oseen equations

$$\begin{cases} -\nu\Delta u + k\partial_1 u + \nabla p = f & \text{in } I\!R^3 \\ \operatorname{div} u = g & \text{in } I\!R^3. \end{cases} \tag{4}$$

First we construct a solution v of the equation $\operatorname{div} v = g$ in the form $v = e * g$. Then $(U = u - v, p)$ is a solution of (4) with $g \equiv 0$ and f being replaced by $F = f + (-\nu\Delta + k\partial_1)v$. Since $U = E * F$, $p = e * F$, it suffices to study the convolution operators with kernels e and E in the anisotropically weighted spaces $L^2_{\alpha+1,\beta}(I\!R^3)$.

Lemma 1. *Let $0 < \beta < 1$, $|\alpha + \beta| < 2$ and $g \in L^2_{\alpha+1,\beta}(I\!R^3)$. Then $v = e * g \in L^2_{\alpha,\beta-1}(I\!R^3)^3$, $\nabla v \in L^2_{\alpha+1,\beta}(I\!R^3)^9$ and*

$$\big\|v\big\|_{\alpha,\beta-1} + \big\|\nabla v\big\|_{\alpha+1,\beta} \le c\big\|g\big\|_{\alpha+1,\beta}.$$

Proof. The convolution integral $v = e * g$ with kernel e which is weakly singular and homogeneous of order -2 is explicitly estimated in technical and lenghty calculations [5]. Further note that ∇e is a strongly singular kernel and homogeneous of order -3 such that the integral over the unit sphere vanishes. Thus the theory of Calderón-Zygmund operators in weighted L^p-spaces [12] yields the estimate of ∇v, since it is easily shown that $\eta_\beta^{\alpha+1}$ is a weight of the Muckenhoupt class A_2. Thus Lemma 1 is proved. ∎

Lemma 2. *Let $0 \le |\alpha| < \beta < 1$ and $F \in L^2_{\alpha+1,\beta}(I\!R^3)^3$. Then $U \in L^2_{\alpha-1,\beta}(I\!R^3)^3$, $DU \in L^2_{\alpha,\beta}(I\!R^3)^9$, $\partial_1 U \in L^2_{\alpha+1,\beta}(I\!R^3)^3$ and $|D^2U| \in L^2_{\alpha+1,\beta}(I\!R^3)$, and*

$$\big\|U\big\|_{\alpha-1,\beta} + \big\|DU\big\|_{\alpha,\beta} + \big\|(\partial_1 U,\, D^2 U)\big\|_{\alpha+1,\beta} \le c\big\|F\big\|_{\alpha+1,\beta}.$$

Proof. The convolution kernels E, DE and $\partial_1 E$ are weakly singular, but not homogeneous. Nevertheless the proof of the estimates of U, DU and $\partial_1 U$ follows the same lines as in the proof of Lemma 1, but for technical reasons we only get that $U \in L^2_{\alpha-1-\epsilon,\beta}(I\!R^3)^3$ with $\epsilon > 0$ arbitrarily small. However note that U_j is a solution of the strongly elliptic equation

$$-\nu\Delta U_j + k\partial_1 U_j = F_j - \partial_j p,$$

where $\nabla p = e * F \in L^2_{\alpha+1,\beta}(I\!R^3)^3$ by Lemma 1. Using a variational approach [4] we may remove the positive ϵ and prove the remaining parts of Lemma 2. ∎

To finish the proof of Theorem 1 we get by Lemma 1 that $|\nabla v|, |D^2 v| \in L^2_{\alpha+1,\beta}(I\!R^3)$. Thus $F \in L^2_{\alpha+1,\beta}(I\!R^3)^3$, and Lemma 2 may be applied. Finally we solve the Oseen problem (1) when $f = 0$ and $g = 0$, but $u = a \in H^{3/2}(\partial\Omega)$ on $\partial\Omega$, via the theory of hydrodynamical potentials [5]. Hence Theorem 1 is proved. ∎

Final Conclusions. i) By Lemma 1 and Bogovskii's results [3] we get that in an exterior domain $\Omega \subset I\!R^3$ the equation of continuity div $v = g$ for $g \in L^2_{\alpha+1,\beta}(\Omega)$, $0 < \beta < 1$, $|\alpha+\beta| < 2$, has a solution $v \in H^1_{loc}(\bar{\Omega})^3$ such that $v \in L^2_{\alpha,\beta-1}(\Omega)^3$, $\nabla v \in L^2_{\alpha+1,\beta}(\Omega)^9$, $v|_{\partial\Omega} = 0$ and such that $\|v\|_{\alpha,\beta-1} + \|\nabla v\|_{\alpha+1,\beta} \leq c\|g\|_{\alpha+1,\beta}$.
ii) By the definition of the anisotropically weighted spaces $L^2_{\alpha,\beta}(\Omega)$, $\beta > 0$, the solution u of (1) and all its derivatives exhibit a wake behavior. Further $D^2 u$ and $\partial_1 u$ are subject to the same estimates, cf. [1,8].
iii) The results of Theorem 1 together with embedding estimates of anisotropically weighted Sobolev spaces may be used to prove existence of solutions of the exterior Navier–Stokes problem.

REFERENCES

1. Babenko, K.I.: On stationary solutions of the problem of flow past a body of a viscous incompressible fluid. Math. USSR Sbornik **20**, 1–25 (1973).
2. Bemelmans, J.: Eine Außenraumaufgabe für die instationären Navier–Stokes–Gleichungen. Math. Z. **162**, 145–173 (1978).
3. Bogovskii, M.E.: Solution of the first boundary value problem for the equation of continuity of an incompressible medium. Soviet. Math. Doklady **20**, 1094–1098 (1979).
4. Farwig, R.: A variational approach in weighted Sobolev spaces to the operator $-\Delta + \partial/\partial x_1$ in exterior domains of $I\!R^3$. To appear in Math. Z.
5. Farwig, R.: The stationary exterior 3D–problem of Oseen and Navier–Stokes equations in anisotropically weighted Sobolev spaces. To appear in Math. Z.
6. Faxén, H.: Fredholm'sche Integralgleichungen zu der Hydrodynamik zäher Flüssigkeiten I. Ark. Mat. Astr. Fys. **21A**, No. 14 (1928/29).
7. Finn, R.: On the exterior stationary problem for the Navier–Stokes equations, and associated perturbation problems. Arch. Rat. Mech. Anal. **19**, 363–406 (1965).
8. Galdi, G.P.: On the Oseen boundary–value problem in exterior domains. Preprint.
9. Matsumura, A., Nishida, T.: Exterior stationary problems for the equation of motion of compressible viscous and heat–conductive fluids. Proc. EQUADIFF Xanthi 1987, Dafermos, Ladas, Papanicolaou (eds.), M. Dekker Inc. (1989), 473–479.
10. Oseen, C.W.: Neuere Methoden und Ergebnisse in der Hydrodynamik. Leipzig: Akademische Verlagsgesellschaft 1927.
11. Padula, M.: Asymptotic properties of steady solutions of compressible fluids in external regions. Preprint.
12. Torchinsky, A.: Real–variable methods in harmonic analysis. Orlando: Academic Press 1986.

SPECIAL BOUNDED HESSIAN AND PARTIAL REGULARITY OF EQUILIBRIUM FOR A PLASTIC PLATE

Franco Tomarelli

Dipartimento di Matematica - Politecnico

20133 Milano

INTRODUCTION

The aim of this lecture is to announce some results obtained in a joint research with M.Carriero e A.Leaci about variational problems with free gradient-discontinuity set, say problems where the unknown is a pair (u, C) such that C is a closed set and u is a function with smooth derivatives outside C; we enphasize that C is not necessarily the union of essential boundaries unlike the case of a free boundary problem.

A suitable relaxed form of a free gradient-discontinuity problem leads to the study of an integral functional depending on second derivatives, which has to be minimized among scalar functions whose hessian is a Radon measure without Cantor part. We discuss the existence of solutions for the weak problem together with properties of the free gradient-discontinuity set, and relationship with the strong formulation.

Problems with a free discontinuity set arise in various contexts like liquid crystals configuration, phase transition when taking into account surface energy, and image segmentation in computer vision[2,10]. To fix ideas, we deal here with a model for an elastic-perfectly plastic thin plate submitted to boundary conditions and possibly to unilateral constraints, when a transverse dead load is prescribed.

We consider a bounded connected open subset Ω of $\mathbf{R}^2$, with piecewise C^2 boundary $\partial\Omega$ satisfying the cone property, as the undeformed natural state of the plate.

In order to have equilibrium with gradient discontinuity supported by a 1-dimensional set, we introduce the following deformation energy functional

$$F(v,K) = \int_{\Omega\setminus K} Q(D^2 v)dx + \int_{K\cap\Omega} (1 + |[Dv]|)d\mathcal{H}^1 \tag{1}$$

where K is a closed set of $\mathbf{R}^2$ (the a priori unknown set of plastic yielding), v is the unknown vertical displacement assumed to be continuous in $\bar{\Omega}$ and smooth in $\Omega \setminus K$, Q is a given positive definite quadratic form, D and D^2 denote the distributional first and second derivatives, $|[Dv]|$ the jump of Dv along K and $\mathcal{H}^1$ is the 1-dimensional Hausdorff measure. This model corresponds to a linear elastic energy density for deformations in $\Omega \setminus K$, and to a weak elastic hinge energy density $|[Dv]|$ for bending the plate along K, while the plate yielding contribution to energy is the length $\mathcal{H}^1(K)$ of the plasticity set.

We give necessary conditions and sufficient conditions for the existence of equilibrium of this mechanical system.

Our approach is a higher order analogous to the one introduced by E. De Giorgi for problems with a free discontinuity set when minimizing functionals depending on the first derivatives [1,5,6,7].

We mention that equilibrium of an elastic-plastic plate has been studied also by F. Demengel, by assuming that the deformation energy is given by $\int_\Omega \psi(D^2 v)dx$ where $D^2 v$ is the distributional hessian of v, and ψ is real valued, convex with linear growth at infinity; the existence of a weak solution to the corresponding minimum problem is proved in[8,9].

SPECIAL BOUNDED HESSIAN FUNCTIONS

We introduce the set of admissible displacements as the space of Special Bounded Hessian functions, denoted by $SBH(\Omega)$.

For any $x \in \mathbf{R}^2$ we set $B_\rho(x) = \{y \in \mathbf{R}^2; |y-x| < \rho\}$, and $B_\rho = B_\rho(0)$; here and in the following $|\cdot|$ denotes the euclidean norm; $\mathcal{L}^2$ is the two dimensional Lebesgue measure.

For any Borel function $w : \Omega \to \mathbf{R}^k$, $x \in \Omega$ and $z \in \widetilde{\mathbf{R}}^k = \mathbf{R}^k \cup \{\infty\}$ we say that z is the approximate limit of w at x, and we write

$$z = \operatorname{ap}\lim_{y \to x} w(y),$$

if

$$g(z) = \lim_{\rho \to 0} \frac{\int_{B_\rho(x)} g(w(y))dy}{|B_\rho|} \qquad \forall g \in C^0(\widetilde{\mathbf{R}}^k).$$

The singular set

$$S_w = \{x \in \Omega;\ \operatorname{ap}\lim_{y \to x} w(y) \quad \text{does not exist}\}$$

is a Borel set of negligible Lebesgue measure; as usual we denote by $\tilde{w} : \Omega \setminus S_w \to \widetilde{\mathbf{R}}^k$ the function

$$\tilde{w}(x) = \operatorname{ap}\lim_{y \to x} w(y).$$

Let $x \in \Omega \setminus S_w$ be such that $\tilde{w}(x) \in \mathbf{R}^k$; we say that w is approximately differentiable at x if there exists a $k \times 2$ matrix $\nabla w(x)$ such that

$$\operatorname{ap}\lim_{y \to x} \frac{|w(y) - \tilde{w}(x) - \nabla w(x) \cdot (y-x)|}{|y-x|} = 0.$$

For any real euclidean finite dimensional space Y we denote by $L^p(\Omega, Y)$, $W^{1,p}(\Omega, Y)$ and $\mathcal{M}(\Omega, Y)$ respectively the Lebesgue and Sobolev function spaces and the bounded measures in Ω with values in Y. We denote by $|\cdot|_T$ the total variation of a measure in $\mathcal{M}(\Omega, Y)$. Moreover

$$BV(\Omega, \mathbf{R}^k) = \{w \in L^1(\Omega, \mathbf{R}^k) : Dw \in \mathcal{M}(\Omega, \mathbf{R}^{2k})\}$$

denotes the space of functions with bounded variation in Ω with values in $\mathbf{R}^k$; from now on Dv denotes the distributional gradient of v in Ω.

For every $w \in BV(\Omega, \mathbf{R}^k)$ the following properties hold:

1) S_w is countably $(\mathcal{H}^1; 1)$ rectifiable;

2) ∇w exists $\mathcal{L}^2$-a.e. on Ω and coincides with the Radon-Nikodym derivative of Dw with respect to $\mathcal{L}^2$;

3) for $\mathcal{H}^1$ almost all $x \in S_w$ there exist $n = n_w(x) \in \partial B_1$ and $w^+(x), w^-(x) \in \mathbf{R}^k$ (outer and inner trace, respectively, of w at x in the direction n) such that

$$\lim_{\rho \to 0} \rho^{-2} \int_{\{y \in B_\rho(x); y \cdot n > 0\}} |w(y) - w^+(x)| \, dy = 0,$$

$$\lim_{\rho \to 0} \rho^{-2} \int_{\{y \in B_\rho(x); y \cdot n < 0\}} |w(y) - w^-(x)| \, dy = 0.$$

Recently a class of special functions of bounded variation[1,6] has been considered with the aim of solving variational problems with free discontinuity set.

We introduce here the space of functions with special bounded hessian.

Definition 1 - We define $SBV(\Omega, \mathbf{R}^k)$ as the class of all functions $w \in BV(\Omega, \mathbf{R}^k)$ such that

$$|Dw|_T = \int_\Omega |\nabla w| dx + \int_{S_w \cap \Omega} |[w]| d\mathcal{H}^1,$$

where $[w] = w^+ - w^-$.

Definition 2 - We define $SBH(\Omega)$ as the class of all functions $v \in L^1(\Omega)$ such that $Dv \in SBV(\Omega, \mathbf{R}^2)$.

From now on we denote by $\nabla^2 v = \nabla Dv$ the approximate gradient of Dv, by $S_{Dv} = S_{D_{x_1}v} \cup S_{D_{x_2}v}$ the singular set of Dv and by $|[Dv]| = \left([D_{x_1}v]^2 + [D_{x_2}v]^2\right)^{1/2}$ the jump of Dv on S_{Dv}.

Remark 3 - By the definition the following properties of a function $v \in SBH(\Omega)$ can be deduced immediately:

1) The distributional derivative of v is absolutely continuous with respect to the Lebesgue measure, hence $Dv = \nabla v$;

2) $|D^2 v|_T = \int_\Omega |\nabla^2 v| dx + \int_{S_{Dv} \cap \Omega} |[Dv]| d\mathcal{H}^1$.

Remark 4 - By rank-1 properties of singular hessian of v in $SBH(\Omega)$ one gets that $\mathcal{H}^1$-a.e. in S_{Dv} there is n such that

$$|[Dv]| \, d\mathcal{H}^1 = \left| \left[\frac{\partial v}{\partial n} \right] \right| d\mathcal{H}^1 = \left| (D^2 v)^s \right| = |(\Delta v)^s|$$

where $(\Delta v)^s$ denotes the singular part of the distributional laplacian of v with respect to $\mathcal{L}^2$.

Proposition 5 - Let $\Omega \subset \mathbf{R}^2$ be a bounded open set C^2 uniformly regular. Then $SBH(\Omega) \subset C^0(\overline{\Omega}) \cap W^{1,q}(\Omega)$ for $1 \leq q \leq 2$, and the embedding in $W^{1,q}(\Omega)$ is compact if $q < 2$.
Traces can be defined as bounded linear maps

$$\gamma_0 : SBH(\Omega) \to W^{1,1}(\partial\Omega)$$
$$\gamma_1 : SBH(\Omega) \to L^1(\partial\Omega)$$

such that $\gamma_0(v) = v\Big|_{\partial\Omega}$ and $\gamma_1(v) = \frac{\partial v}{\partial N}\Big|_{\partial\Omega}$ for every $v \in C^2(\overline{\Omega})$, where N is the outward normal to $\partial\Omega$.

THE RELAXED PROBLEM

We choose the quadratic form Q (see (1)) by considering for any $v \in SBH(\Omega)$ the following functional

$$E(v) = \int_\Omega \left((1-\nu)|\nabla^2 v|^2 + \nu|(\Delta v)^a|^2\right) dx + \mathcal{H}^1(S_{Dv} \cap \Omega) + \int_{S_{Dv}\cap\Omega} |[Dv]| d\mathcal{H}^1 \quad (2)$$

where $\nu \in [0,1)$ is the Poisson coefficient and $(\Delta v)^a$ denotes the absolutely continuous part (with respect to $\mathcal{L}^2$) of the distributional Laplacian $\Delta v = D^2_{x_1} v + D^2_{x_2} v$.
Let us consider a transverse load L given by

$$\begin{cases} \langle L, v\rangle = \int_\Omega g v dx + \int_{\partial\Omega} l v d\mathcal{H}^1 \\ g \in L^q(\Omega), l \in L^s(\partial\Omega), \text{ with } q, s > 1. \end{cases} \quad (3)$$

The first result describes a free plate under a load L satisfying mechanical compatibility conditions.

Theorem 6 - There is a constant $\delta(\Omega) > 0$ such that, if we assume (3) and

$$\langle L, 1\rangle = \langle L, x_1\rangle = \langle L, x_2\rangle = 0 \ , \quad (4)$$

$$|L|_T < \delta(\Omega) \ , \quad (5)$$

then the functional

$$E(v) - \langle L, v\rangle \quad (6)$$

achieves its minimum over $SBH(\Omega)$.

The next case describes a clamped plate; the boundary datum of the gradient is relaxed as it is usual in nonreflexive problems.

Theorem 7 - Assume (3),(5). Then, for any given w in $SBH(\mathbf{R}^2)$ such that

$$\int_{\mathbf{R}^2} \left((1-\nu)|\nabla^2 v|^2 + \nu|(\Delta v)^a|^2\right) dx + \int_{S_{Dv}} (1 + |[Dv]|) \, d\mathcal{H}^1 \ < \ +\infty \ ,$$

the functional (6) achieves its minimum over the set

$$\left\{v \in SBH(\mathbf{R}^2) : v = w \text{ a.e. in } \mathbf{R}^2 \setminus \Omega\right\}.$$

Eventually the unilateral case is solved by using the sequential recession functional[3,4].

Theorem 8 - Let $U \subset \overline{\Omega}$ be a closed set. Then there is a constant $\eta(\Omega, U) > 0$, such that, if we assume (3) and

$$\langle L, 1 \rangle < 0$$

$$\langle L, x_j - c_j \rangle = 0 \qquad \text{where } c_j = \frac{\langle L, x_j \rangle}{\langle L, 1 \rangle}, \; j = 1, 2,$$

$$c = (c_1, c_2) \in \overset{\circ}{U},$$

$$|L|_T < \eta(\Omega, U), \tag{7}$$

then the functional (6) has a minimum over the set

$$\{v \in SBH(\Omega) : \; v \geq 0 \text{ on } U\}.$$

Remark 9 - The conditions $\langle L, 1 \rangle \leq 0$ and (in the strict inequality case) $c \in coU$ are necessary for the finiteness of the infimum of $E(v) - \langle L, v \rangle$ in the unilateral problem (here co denotes the closed convex hull).

Remark 10 - Existence statements analogous to Theorems 6, 7 and 8 hold true in dimension $m > 2$ too, provided $q > m/2$, $s > m - 1$ and $|L|_T$ is replaced in assumptions (5),(7) by

$$\|g\|_{L^q(\Omega)} + \|l\|_{L^s(\partial\Omega)}.$$

Remark 11 - Any local minimizer in $\mathbf{R}^m$ of the functional E, with finite energy and bounded gradient-discontinuity set, is an affine linear function, as it can be shown by a suitable Caccioppoli inequality.

PARTIAL REGULARITY AND THE STRONG FORMULATION

In order to simplify the notation, we set the Poisson coefficient $\nu = 0$ and consider only the Neumann problem in dimension 2. The hypotheses of Theorem 6 are assumed and in addition we consider $q > 2$.

The strong formulation of the free gradient-discontinuity problem takes the following form:

$$\text{minimize } G(v, K) =$$

$$\int_{\Omega \setminus K} |D^2 v|^2 dx + \int_{K \cap \Omega} (1 + |[Dv]|) \, d\mathcal{H}^1 - \int_\Omega g v dx - \int_{\partial\Omega} l v d\mathcal{H}^1 \tag{8}$$

over $K \subset \mathbf{R}^2$ closed and $v \in C^0(\overline{\Omega}) \cap C^2(\Omega \setminus K)$.

Theorem 12 - The problem (8) has a solution (u, C) : more precisely if ω solves the weak formulation

$$\min_{SBH} \{E(v) - \langle L, v \rangle\}$$

then the solution of (8) is obtained by taking

$$u = \omega \qquad \text{and} \qquad C = \overline{S_{D\omega}}$$

as a minimizing pair.

Remark 13 - The partial regularity of the weak solutions (and hence the existence of strong solutions) is proved by using a Poincare'-Wirtinger inequality[7] for functions of class SBV in a ball and showing a suitable monotonicity property for the scaled energy density of a minimizer

$$\rho^{-1} \left\{ \int_{B_\rho(x)} |\nabla^2 \omega|^2 dx + \int_{S_{D\omega} \cap B_\rho(x)} (1 + |[D\omega]|)\, d\mathcal{H}^1 \right\} .$$

REFERENCES

1. L.Ambrosio, Existence theory for a new class of variational problems, *Arch. Rational Mech. Anal.*, 111:291 (1990).
2. A.Blake and A.Zisserman, "Visual Reconstruction", The MIT Press, Cambridge, (1987).
3. C.Baiocchi, G.Buttazzo, F.Gastaldi, and F.Tomarelli, General existence theorems for unilateral problems in continuum mechanics, *Arch. Rat. Mech. Anal.*, 100:149 (1988).
4. G.Buttazzo and F.Tomarelli, Compatibility conditions for nonlinear Neumann problems, *Advances in Math.*, 89:127 (1991).
5. M.Carriero and A.Leaci, Existence theorem for a Dirichlet problem with free discontinuity set, *Nonlinear Analysis TMA*, 15:661 (1990).
6. E.De Giorgi, Free discontinuity problems in calculus of variations, *in:* "Analyse Mathématique et Applications" (Paris, 1988), Gauthier–Villars, Paris, (1988).
7. E.De Giorgi, M.Carriero and A.Leaci, Existence theorem for a minimum problem with free discontinuity set, *Arch. Rational Mech. Anal.*, 108:195 (1989).
8. F.Demengel, Fonctions a hessian borné, *Ann. Inst. Fourier*, 34:155 (1984).
9. F.Demengel, Compactness theorem for spaces of functions with bounded derivatives and applications to limit analysis problems in plasticity, *Arch. Rational Mech. Anal.*, 105:123 (1989).
10. D.Mumford and J.Shah, Optimal approximation by piecewise smooth functions and associated variational problems, *Comm. Pure Appl. Math.*, 42:577 (1989).

INDEX